Himalaya

(Geological Aspects)

Volume - 3

Himalaya

(Geological Aspects)
Volume - 3

Edited by :

Prof. P.S. Saklani
Department of Geology,
University of Delhi

SSPH

SATISH SERIAL PUBLISHING HOUSE

403, Express Tower, Commercial Complex, Azadpur, Delhi - 110033 (INDIA)
Phone : 011-27672469, 27672852 Fax : 91-11-27672046
E-mail : ssp@ndf.vsnl.net.in, hkjain@satishserial.com
Website : www.satishserial.com

Published by :

SATISH SERIAL PUBLISHING HOUSE
403, Express Tower, Commercial Complex, Azadpur, Delhi - 110033 (INDIA)
Phone : 011-27672469, 27672852 Fax : 91-11-27672046
E-mail : ssp@ndf.vsnl.net.in, hkjain@satishserial.com

First Published 2005

ISBN 81-89304-04-6

Printed at
SALASAR IMAGING SYSTEMS, DELHI-35

ACKNOWLEDGEMENTS

I express my sincere thanks to all contributers in this volume and to Prof. E. Appel of Tübingen University (Germany), Prof. M. Räsänen of Turku University (Finland), Prof. A. Steck of Lausanne University (Switzerland), Dr. G. Fuchs, Laimback (Austria) and very many well-wishers for their help in many ways.

P.S. SAKLANI

INTRODUCTION

The majestic Himalaya would always remain an attraction to the geoscientists, tourists, artists and a common man. I first ventured my publication activity about the Geology of the Himalaya in 1978. The combined wisdom of my fellow geoscientists is reflected in my edited volumes and due to overwhelming support and cooperation, I could publish many of them. Meanwhile, I got some academic/administrative assignments and after completing them, I noticed that still there is a need to continue with publication activity and the result is in your hands. In my opinion, Geology of the Himalaya is continuously being updated. Though the Himalaya would remain where it is, during my life time, yet interpretations would vary due to advancement of science and technology. This volume is being published under the aegis of the Geoscience Foundation, India.

In this volume, the formation of the Himalaya has been discussed by *M. Brunel et al.* They are of the view that the collision of Indian and Asian Plates took place around 53 m.y. ago (within sedimentary sequence-60 m.y.). The Tethys measured 6500 km in width. The MOHO is located at about 75 km beneath the Everest. Tomographic results indicate that Indian lithosphere went down for about 2000 km, associated with anatexis. The Precambrian Lesser Himalayan rocks occur in the form of an imbricate stack. The exhumation of the rocks to get exposed on the surface was for about 40 km. The ophiolites and associated flysch are also characterized by imbrications. The Main Central Thrust (movement for about 200 km) is a flat shear zone and the rocks were subjected to ductile deformation. Metamorphic minerals were formed around 36 m.y. ago but this formation activity was more intense between 25 to 10 m.y. and continued to be thermally active for about 2 m.y. Analogue modelling suggests that prior to collision the regime was marked by sedimentation. The neotethys was opened up in Late Permian and on the fore arc series of Tibetan margin, lavas and batholiths were emplaced.

A brief account of the history of Himalayan geology is given by *R. Sorkhabi.* The geographic explorations, geological maps and botanical catalogues of the Himalaya were made from 1175 to 1800. Colebrooke (1765-1837) was the first person to opine that the Himalaya is world's highest mountain than the Andes. W. Webb (1816-1820) determined the height of the Nada Devi (25669 feet, just 24 feet higher than today's measurement). Himalayan geology became an independent and systematic field of science between 1850-1900. The Geological Survey of India was established in Calcutta in 1851. T. Oldham (1816-1878) was the true architect of the Geological Survey of India. J.D. Hooker (1817-1911) explored Sikkim and Bengal during 1848-1852. A. Waugh (1810-1878) who succeeded G. Everest, discovered peak XV and named it as the Mount Everest (height 29002 feet which is less than 26 feet than that of the 1955 measurement). World I (1914-1918) and World War II hampered the investigations of the Himalaya. After the World War II. India became independent and since then many new advances were made in Himalayan geology. Presently native geologists of the Himalayan countries (India, Pakistan, Bangladesh, China, Nepal and Burma) and also from U.S.A., Canada, Europe, Australia, Japan etc. are actively engaged on the geological researches of the Himalayan and Tibetan regions.

Magnetotelluric techniques reveal significant information of the deep interior of earth and *G. Gupta et al.* has applied these in Mohand-Ramnagar and Phobrang regions of the

Northwestern Himalaya. Two layers of Siwalik sediments underlie the alluvium. The Upper Siwaliks are 700-1000 m in thickness while Middle and Lower Siwaliks measure 3000-4000m. The vertical conductive zone is delineated at shallow depth in Phobrang profile of the Ladakh, which is characterized by a deep-rooted fault at a depth of about 12 km.

Sedimentological studies of the Jammu region have been dealt with by *B.P. Singh et al.* The Palaeogene succession represents a regressive phase of a coastal regime. The Murrees are estuarine. The overlying sandstones belong to intertidal zone. The Palaeocene sediments were derived from the Himalayan arc and their derivation was subsequently followed from the Indus Suture Zone towards the foreland.

The biotite in granite gneiss, granites and leucogranites of Kinnaur, Himachal Pradesh have been studied by *S. Kumar and B. Singh.* The Rakcham granitoids intrude the gneiss of the Vaikrita group. The tourmaline granite is leucocratic and is a product of the Himalayan collision. The biotite from Rakcham ganitoids and granite gneiss is Fe rich while it, from the leucogranite, corresponds to siderophylite and lepidomelane field. Sheelite mineralisation is found in Rakcham granitoids.

S.C. Bhatt and R. Kumar have discussed about the mineralogical properties of the aggregates of gneiss, quartizites, schists etc. The strained quartz caused alkali-silica reaction in concrete. The rock aggregates of hard rounded quartz and feldspar produced high strength. The excess of strained quartz showed adverse effects on strength and chemical stability of aggregates of concrete.

A.K. Biyani has described the structure and tectonics of the Yamuna mylonite zone of Uttarkashi district. He is of the view that the Higher Himalayan granites were initially mylonitised in ductile conditions during the Middle Tertiary and were again later deformed under ductile-brittle conditions. Several microstructures in minerals suggest a southward displacement of the Main Central Thrust Sheet. The rocks of the hinterland experienced flattening type of strain where as in the toe region it was constrictional. The deformation continued after the thrusting and caused the formation of strike slip, reverse and normal faults. The river terraces, gorges etc. exhibit effects of Neotectonics.

A.C. Pandey et al. making use of satellite images discussed the geology and morphotectonics of a part of Tehri-Uttarkashi districts. The rocks of the area are represented by the Simla, Krol, Jaunsar and Crystalline groups delineated by thrusts. They have analysed the data of lineaments, which exhibit NE-SW, NW-SE, N-S, and E-W trends. The cross-valley and longitudinal profiles of the Bhagirathi river show breaks in the slopes, knick points etc. which were caused by dynamic rejuvenation. The river terraces were formed due to five phases of uplift.

The editor has authored the Tectonic Geology of the Main Central Thrust in Garhwal Himalayan region, which is characterised by the Vaikrita, Jutogh and Budhakedar units delineated at their base by the Main Central Thrust-I, II and III. Infact, these thrusts occur in the form of a large Duplex. The Bhagirathi Duplex was shortened for about 17 km while the shortening in the Yamunotri Duplex was about 8 km. The metamorphics, located at a depth of about 16-18 km, ramped for about 8 km during the thrusting.

The evolution of the Almora nappe along the Chhara-Someshwar Section has been contributed by *M. Joshi and A.N. Tiwari.* The structural history of this nappe is characterised by the structures of pre-, syn- and post-shear zones. The F_1, F_2 and F_3 folds are pre-shear zone while shear bands, asymmetric porphyroclasts, pressure shadows and quartz-c-axes were oriented in syn-shear zones. The post-shear zone structures are represented by steep brittle faults which were affected by Neotectonics.

Based on field and experimental studies *A.K. Dubey and R. Jayangondaperumal* have discussed the structural evolution of the Mussoorie syncline with reference to Satengal and Banali klippen. Their model explains the occurrence of older lesser Himalayan rocks over the younger ones due to combination of fault bifurcation, foreland propagation of thrust and superposed folding.

A.K. Shandilya has attempted on the sequence of thrusts (roof and floor) and imbrications. The compression caused by the northward movement of the Indian plate formed the Krol thrust. The Almora thrust was folded during the later phase of deformation. The Amri and Bijni thrusts are extensions of the North Almora Thrust.

Metamorphism and deformation of the Central Crystallines of the Alaknanda Valley have been described by *S.P. Singh et al.* The Barrovian type of metamorphism was M_1 which began at D_1 deformation and culminated at D_2 deformation. This was followed by granitisation. The M_2 metamorphism was mainly retrograde and was localized along the shear zones in the late stages of D_1 deformation. The inverted metamorphism noticed near the Main Central Thrust is the result of polyphase metamorphism.

V.K. Singh has dealt with the structural and metamorphic aspects of the Chails and the Main Central Thrusting. The structural and metamorphic studies reveal that the M_1-metamorphism took place at 600 °C to 700 °C (6-7kbar) during D_1- and D_2 deformation while the temperatures were at 350 °C-400 °C (4-5 kbar) during the M_2- metamorphism.

B.C. Joshi has contributed on the metamorphic history of Chamoli- Garhwal area which consists of repeatedly metamorphosed, folded and faulted rocks. The migmatites and gneisses were produced by load/regional metamorphism. The crystalline nappe of Chamoli-Garhwal constitutes a Duplex structure.

R. Bali has given an account of the neotectonic and morphotectonic evolution of the Dharchula - Tawaghat sector of the Kumaun Himalaya. Many microhydel schemes as well as the Dhauliganga hydroelectric project are adversely affected by the instability of hill-slopes. The area is characterized by unconsolidated and non-cohesive weak fans resting at about 45° (repose angle) and are instable. Neotectonics reactivated the debris cones.

A. Kumar and M. Sanoujam have carried out the Seismotectonic studies of Manipur region. The complex focal mechanism is due to strike-slip-movements. The NNW-SSE to NNE-SSW dragging of the Indian plate is confirmed by Harvard Centroid Moment Tensor Analysis. The focal depth in the strike slip mechanism varies from 62 to 140 km (Indian Plate) whereas this is located at 40-45 km towards the Burmese arc.

Structural patterns and geokinematics of the Mediterranean - Central Asian Orogenic belt has been contributed by *M. Kopp*, who attempted the structural analysis of the fold belt.

The Mediterranean - Central Asian orogenic belt was formed due to indentation of Gondwana micro-continents into Eurasia. He has established that block rotational movements played a significant role in the formation of Caucasus, Alps, Pamirs etc. Geokinematically the Alpine-Himalayan belt is characterised by compressional structures, strike slip faults, shear zones, rotational movements and lateral squeezing out of the deformed rocks.

It is hoped that the volume would be useful to researchers, academicians and geoscientists in governmental and non-governmental organisations of the country and abroad.

P.S. SAKLANI

CONTENTS

THE FORMATION OF THE HIMALAYA

Maurice Brunel *. Stephane Dominguez*. Georges Mascle**

Uniersite Montpellier II
UFR Science Destem & UMr. 5573 Geoph. Tecto. Sedimento.
Cc. 058 Pl.E. Bataillon. 34094- Motpellier Cedex 05
** Universite Joseph Fourier Grenoble 1 LGIT, Maison Des Geoscience
1381, rue de la Piscine BP 53, 3841 Grenoble Cedex 9

ABSTRACT

The geology of the Himalaya has been described keeping in view the relief, seismic tomography, structural units, the rocks at depth and their exhumation. Attempt has also been made to suggest a possible model for the formation of the Himalaya.

INTRODUCTION

The Himalaya is considred a case study mountain belt in understanding the mechanisms of orogenesis. Switch over from the outcrop to the mountain belt and from the mountain belt to the continent is easy. The strong and variable relief allows the three dimensional understanding of its structure.

The overall kinematics can be described by the location of India-Europe rotation pole at 19^0 43'N, 38^0 28'E (situated in the Red sea, about 200 km SSE of La Mecue), the direction of the transform faults (Owen-Murray fault, 90^0 E fault) oriented N20^0 E, and earthquake focal mechanism. The actual convergence was thus directed between N0^0 and N20^0 E , roughly perpendicular to the belt (Patria and Achache, 1984)

The Himalaya was formed by a relatively simple assemblage: to the north, the Asian margin, built during the subduction of the large Tethys ocean, similar to the Andes being formed by the subduction of the Pacific ocean underneath the South America. To the south, the Indian margin, and old (oceanic) passive margin is characterized by a nearly continuous sedimentation during 500 Million Years ranging, from Cambrian to Eocene times (Gansser, 1964). Inherited structures are rare along the whole Indian Margin and, the tectono-metamorpic structure is essentially of the Himalayan type.

The Himalaya: The highest relief on earth is due to the thickening of the Northern edge of Indian plate during its Collision with Asia.

The Himalaya is a type example of mountain belt formed by two colliding continental plates: Indian and Asian (Le fort, 1975; Valdiya, 1984, 1988). Results form palaeontological studies, absolute dating of tectonic and metamorphic events within Himalayan rocks as well as the palaeomagneic data obtained from onshore and offshore of the Indian ocean indicate the collision of the two continents during the early Tertiary, (around-53 Million years). By this time, the large Tethys ocean, 6500 km wide existed during the Mesozoic, Between Asian and India Still sticking to Africa and Australia, (Patriat and Achache 1984). The ocean was consumed by the subduction underneath the Tibet leading to the formation of large granitic coastal masifs (of Andean type): the Lhasa or Ladakh granites. Remnants of this wide ocean are preserved only in a few locations, along the suture that was more or less followed by the large Indus and Tsang Po/Brahmaputra rivers.

The Himalaya in itself is a mountain belt resulting from the collision between the two continents; it is situated south of the disappearance line of the Tethys ocean (the Indus/ Tsang Po suture line). The highest mountain belt of the world thus resulted by the deformation on the edge of the Indian continental plate squeezed against Asia. Its relief allows the largest differences in level on the planet: 8500m from the Terai lowlands in Nepal (with the banana trees) to the top of Makalu (figure1)

Along its 2500km length the mountainous Himalayan arc is located between the Naga Parbat in the west and the Namche Bawa in the east and comprises of ten peaks higher than 8000m. The belt extends south of the Indus-Tsang Po suture to the Ganges plain with an average width of about 250 to 300 km. From the experiments of George Everest, we know that the gravitational anomaly associated with the bulk of the mountain, can only be explained by the existence of a root underneath the belt forming a bulge of light material balancing the overweighing relief(figure 2). Gravity modelling , and also seismic experiments are helpful to calculate accurately the depth of root and to locate the MOHO (boundary between the crust and the mantle) at abouth 75 km deep under the Everest. This measurement indicates that the Indian crust was deformed by doubling or even sometimes may be tripling its original thickness. This estimate is fundamental and gives us a reasonable limit to the amount of displacement of India towards the north for the formation of the Himalaya. On this basis it is generally admitted that the actual relief of the Himalaya cannot explain more than 500 km of convergence between India and Asia since 50 Ma . This means that the Himalayan relief was outlined by an average convergence speed of the order of some cm/yr.

(Frame1) Seismic tomography : new data to be taken into account.

Recently, the new techniques of seismic tomography have allowed to image the lithosphere at great depth (Van der Voo et al. 1999). The images obtained bring a revolutionary result to the understanding of the globe and, in particular, of the subduction zones. The deep tomographic section running through central Nepal presented here (figure3) shows that the Indian lithosphere went down into the mantle to more than 1500km or possibly 2000 km . Moreover, on the basis of reconstruction from oceanic magnetic anomalies maps, we can consider that the Indian continent, before the collision , was large for about a thousand kilometers(Matte et al ., 1997). In this case the Himalaya would represent the result of a convergence larger than expected between India and Asia and a large amount of continental lithosphere disappeared into the mantle . Following the hypothesis of the subduction of a large amount of Indian continental lithosphere, an important part of the crust should have disappeared either by erosion or due to the subduction.

Figure. 1: Large differences in levels on the planet: 8500m from the Terai lowlands in Nepal (having banana trees) towards the top of the Makalu.

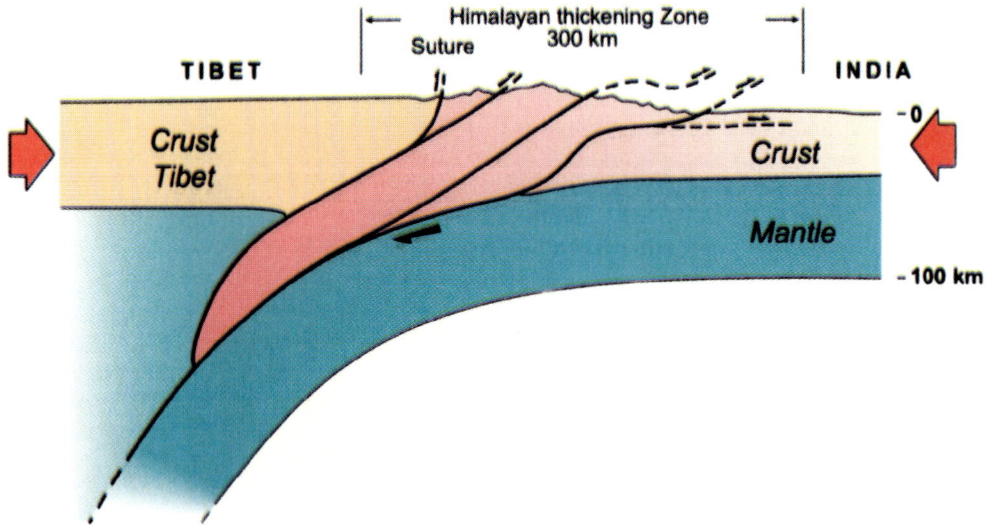

Figure. 2: Sketch section of the Himalayan accretionary thickened zone.

The Himalayan Structure: The Structural Units of the Everest Traverse (Figure 4)

Since about fifty years international teams have been working within the Himalayan belt to understand its functional processes and its construction. These works are useful to understand better the complexity of the palaeogeographic domains from which the structure and major mechanisms of crustal thickening have been reported. Though we have chosen the Everest section yet , in fact, all along the transverse sections, the structural units remain relatively constant throughout the belt. These units resulted due to the superimposition of portions of the Indian continental crust, one on top of the other along the north verging faults or thrusts , from south to north:

The Siwaliks, formed by detrital material of the molasse type were deposited on the old Indian shield and corresponded to the destruction of the Himalayan relief from Neogene to Quaternary times .These deposits are structured in a group of sheets thrusted southward along the Main Frontal Thrust or MFT on to the Quaternary sediments in the Ganga basin.

The Lesser Himalaya is thrusted on top of the Siwaliks along the Main Boundary Thrust or MBT. It is composed of series of mainly Precambrian rocks forming a nappe stack. This group of monotonous rocks is characterised by a slate like foliation or schistosity. On a regional scale,the schistosity planes tend to become parallel to the large faults and thus display a gentle northward dip. This foliation of the rocks usually displays a sub-meridian mineral lineation (Figure 6) indicating the direction of flow and transport of the rock masses towards the south , parallel to the direction of movement of India (Brunel, 1986, Pecher, 1989) . The Lesser Himalaya is buried under the High Himalaya along the Main Central Thrust or the MCT.

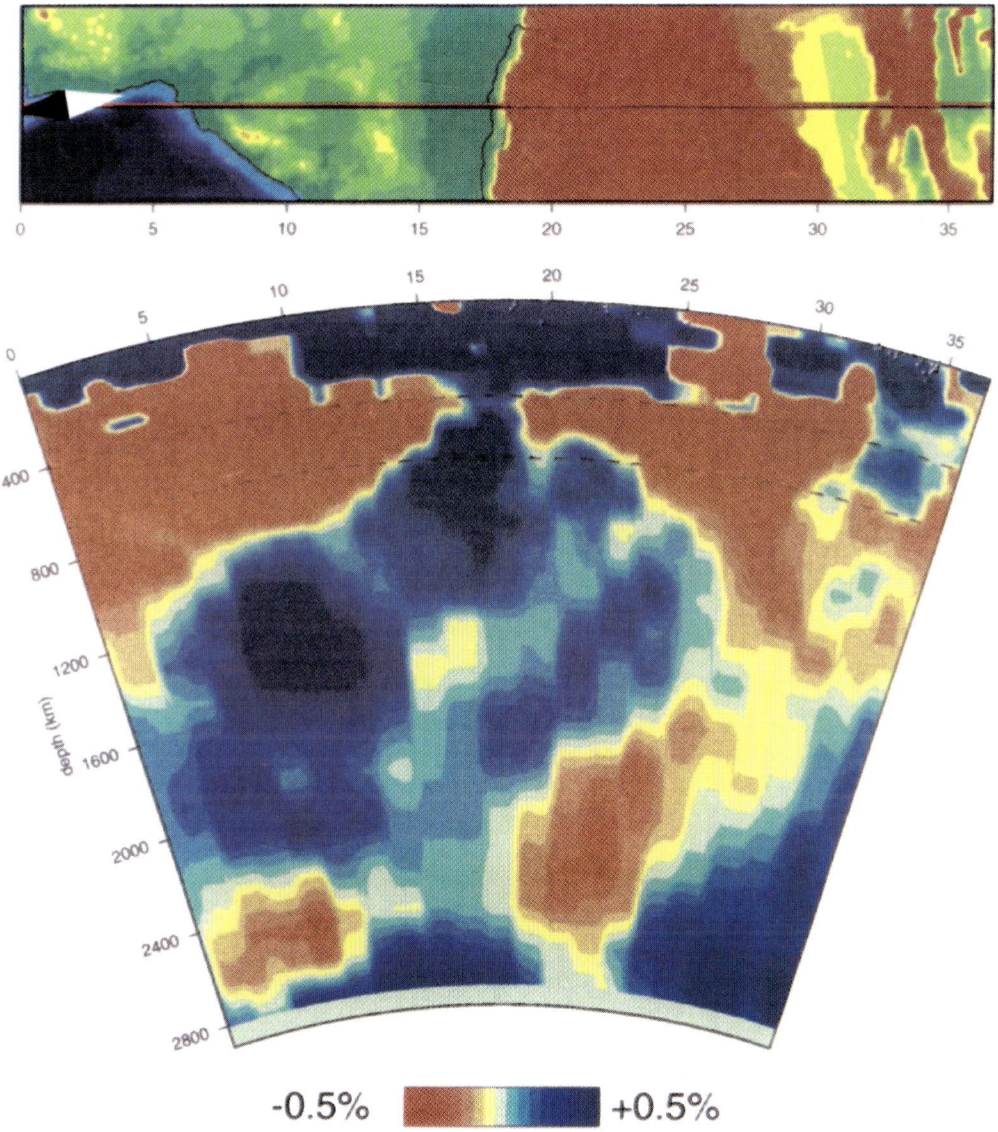

Figure. 3: Tomography across Central Nepal showing the Indian slab dipping in the mantle (after Vander Voo et al.)

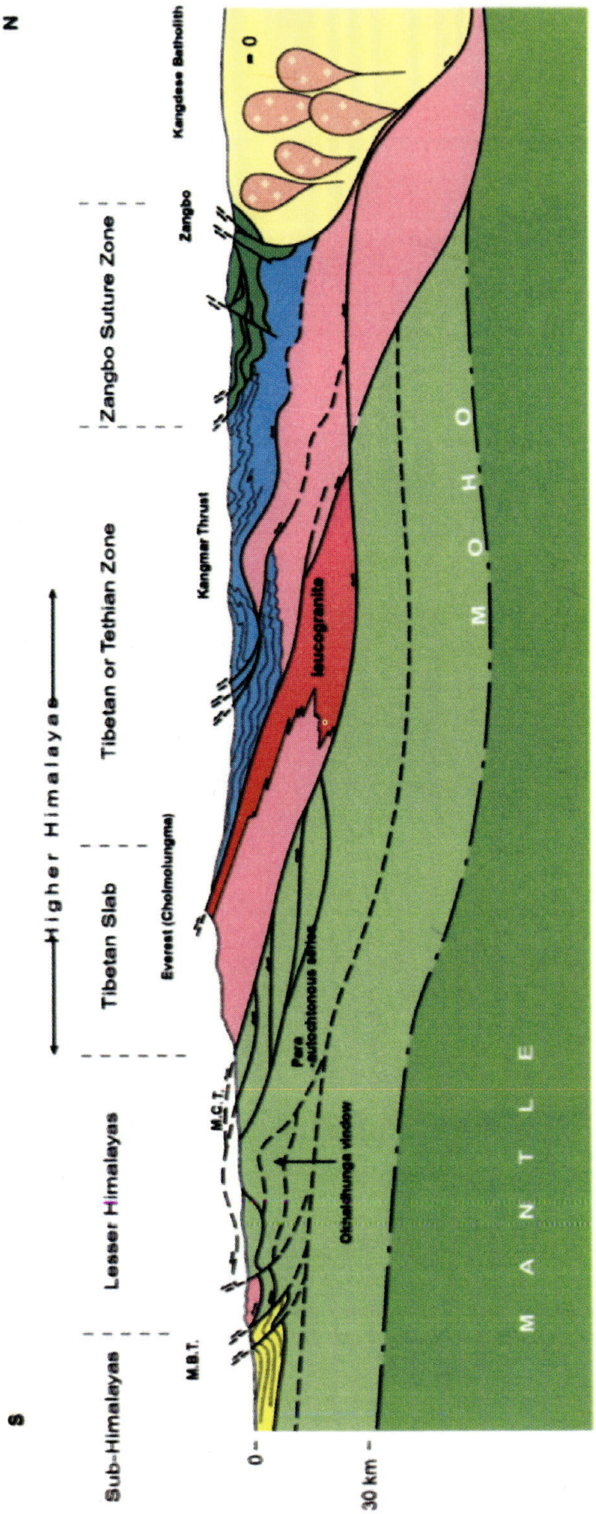

Figure. 4: Sketch section of the Everest traverse.

(Frame 2) Buying of the rocks at depth and their exhumation toward the surface.

To understand the evolution of the architecture of the belt through time and the kinematics of the continental deformation, it was necessary to describe and measure deformation at all possible scale, so that it can be linked with the physics of rock deformation from crystalline to the mountain belt scale, and finally can characterise the evolution of the physical variables such as pressure and temperature.

We know that during tectonic deformation, and evolution of the belt, the rocks at the surface are carried towards the depth where they are subjected to metamorphic modifications and even to anatectite melting to generate granitic magmas (burying process). Due to the effects of erosion and to the classical process of isostatic rebound (progressive uprising of the crustal root induced during erosion of the relief) or to more complex tectonic processes, the rocks which were at great depth, i.e., 20-30 or even 40km could be brought back towards the surface (exhumation process). The analysis of mineralogical assemblages contained within these metamorphic rocks define the pressure and temperature conditions experienced by rocks during their history. Determination of their age using absolute dating methods helps to find the path of the rock throughout the continental crust. We will thus be able to draw the PTt path (Pressure / Temperature / time path) followed by a particular rock or several rocks from the belt during its history (figure 5). This, in turn controls the burying and exhumation tectonic models.

(Frame 3) Figure 6 : picture of mineral transport lineation, Figure 7 : picture of fault zone.

At depth, the increase of temperature and pressure causes the rock to melt forming magmas that can migrate upward and infiltrate into the fault ones forming dykes (figure 7) (Pecher, 1988; Hodges et al., 1988, Guillot et al., 1994). Movement along those faults is shown by crystallization of fibrous minerals here, sillimanite fibers (figure 6).

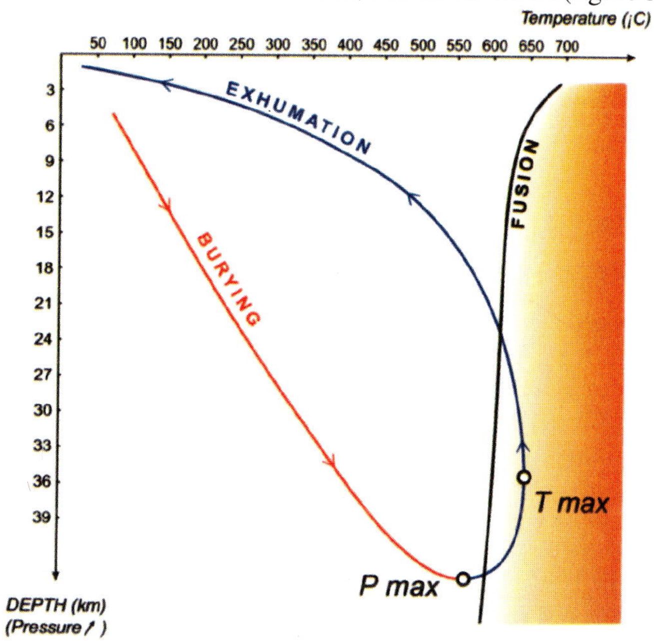

Figure.5: Models of burying and exhumation .

Within the MCT fault zone , the rocks were strongly deformed under pressure of 9 / 10 Mpa and temperature from 45°to 650° C . On the crustal scale it is possible to get the physical conditions of the deformation associated with shearing within the fault zone . Every mineral displays the characteristics which determine the direction of shearing or the direction of transportation, for example , the rotation of the so called " snow ball " garnets (figure 8) that can roll (Brunel, 1986) .

The High Himalaya overrides the Lesser Himalaya along the Main Central Thrust (MCT) (Le fort, 1975) . It is composed of a Precambrian crystalline basement, 5 to more than 10 km thick, of the Higher Himalayan Crystalline or HHC . This basement made by gneisses slides slightly dipping towards the north like the schistose structure which underneath is characterised by well - marked mineral lineation parallel to the direction of transport ("Tibet Slide") . Above the HHC a sedimentary unit (the Tibetan Series or SST) about 10 km thick, ranging in age from Cambrian to Eocene, with a subsiding continental shelf facies is exposed. The Himalayan metamorphism (45 to Ma) is observed within the gneisses (upto 700° C,1200 Mpa) which rapidly decreases in the sedimentary cover.

(Cross section of the Anapurna, Figure 9)

Within the Anapurna cross-section, the sedimentary cover is structured by slides and south-dipping folds (Bordet et al., 1972, 1975), which were reworked by the north dipping folds. The nappes were mainly formed by the sedimentary cover of the various parts of the Indian Tethys margin; two types were distinguished; some from the upper part of the margin (the shelf), some from the deeper part (slope) and, towards the West. Their basement was formed by the Inner Crystalline Massif (Tso Morari).

The Indus- Tsangpo suture zone corresponds to the old Tethyan oceanic area which disappeared due to subduction towards the north and then was deformed by the collision. It displays nappes and slides of flysch , ophiolites and radiolarites, and of blue schists covered by deformed continental conglomerates (Gansser,1964).

The Tectonic History.

The main stages are (Figure 10) :

1. *The disappearance of the oceanic domain*: the subduction of the Tethyan ocean borught the Indian block closer to north Tethyan Asian margin where the TransHimalayan belt was formed (from Middle Cretaceous times until the collision in Lower Paleocene times). The TransHimalayan magmatism (Ladakh and Kangdese belts) was typical of an active margin. To the West, the Dras-Kohistan intra-oceanic arc was docked by the Tibetan margin during the Cretaceous.

2. *Collision of the two continental margins*: collision between the active Tibetan margin to the North and the passive Indian margin to the South occurred during the Eocene (53 Ma). It was characterised by: a sharp decrease of the expanding rate of the Indian ocean (from10 to 5 cm/yr); end of the marine hiatus with

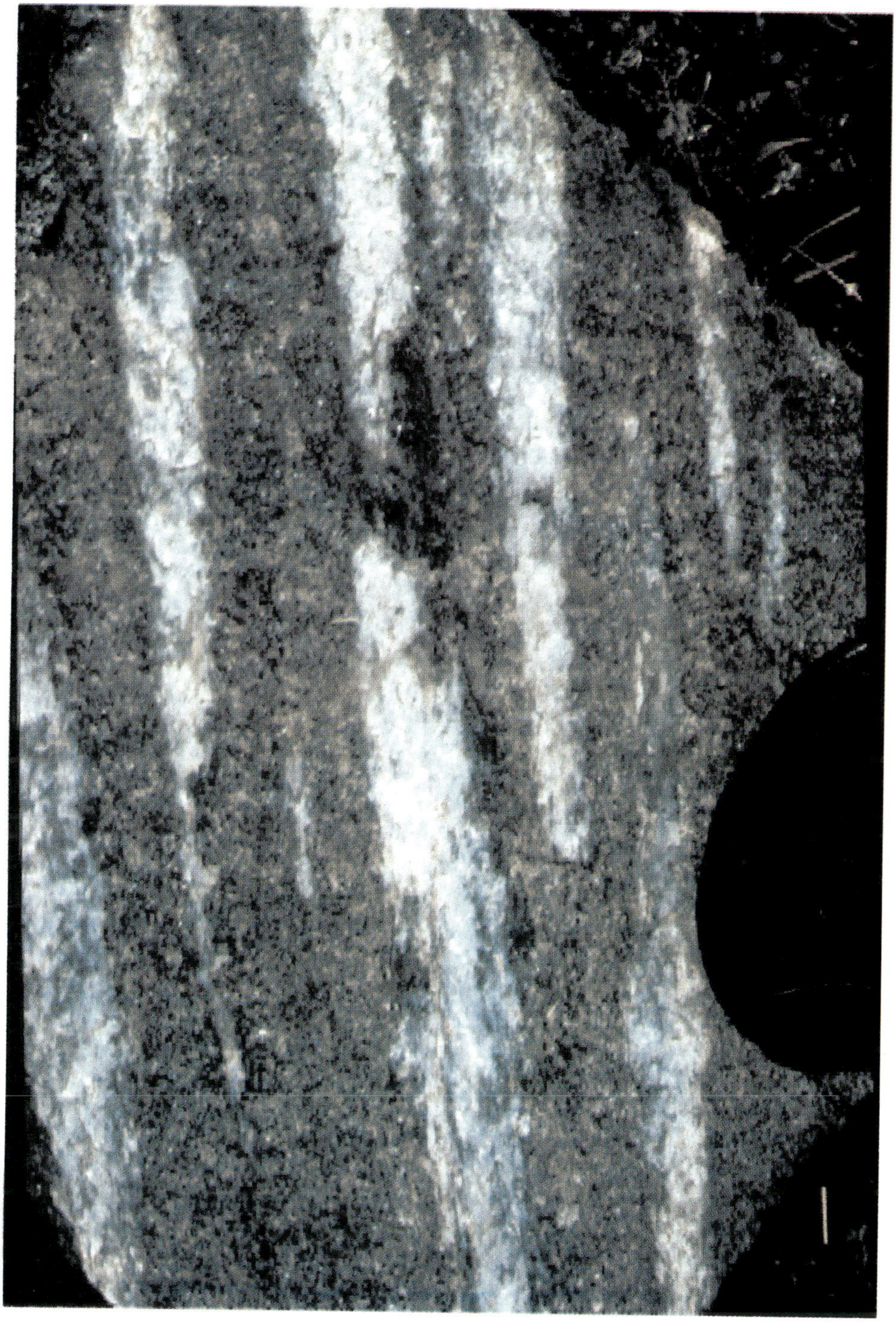

Figure 6: Sillimanite fibers in the Tibetan slab indicative of the stretching towards the transport direction.

Figure 7: General view of the Everest's normal fault detachment at Pum Qu (South Tibet). The main shearing foliation is indentified by numerous leucogranitic sills.

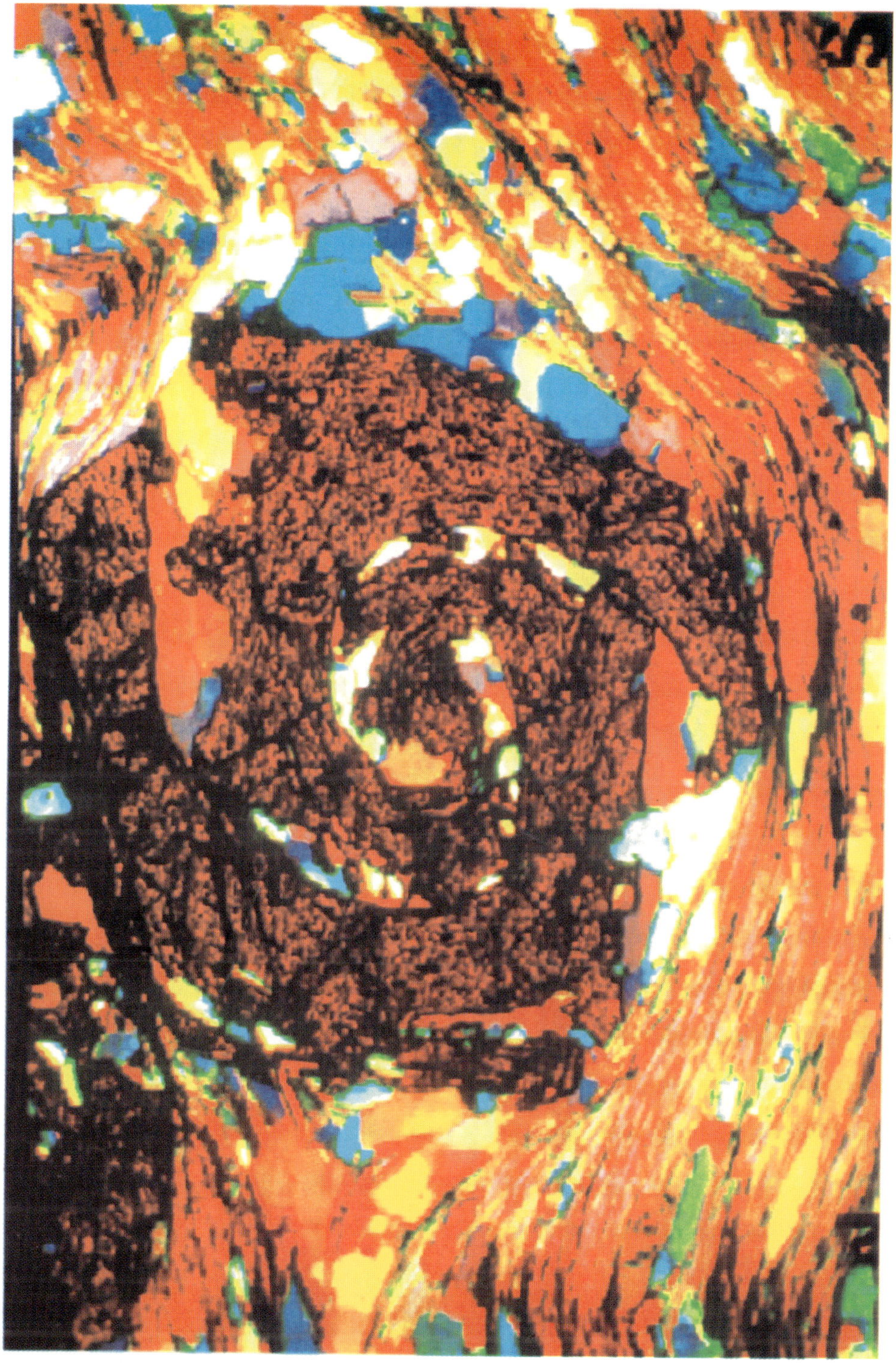

Figure 8: Typical Snow-ball garnet of the Tibetan slab showing shear sense southwards.

Figure 9: Geological cross-section of the Anapurna through Nilgiri: Back fold and NHF (Normal Himalayan Fault).

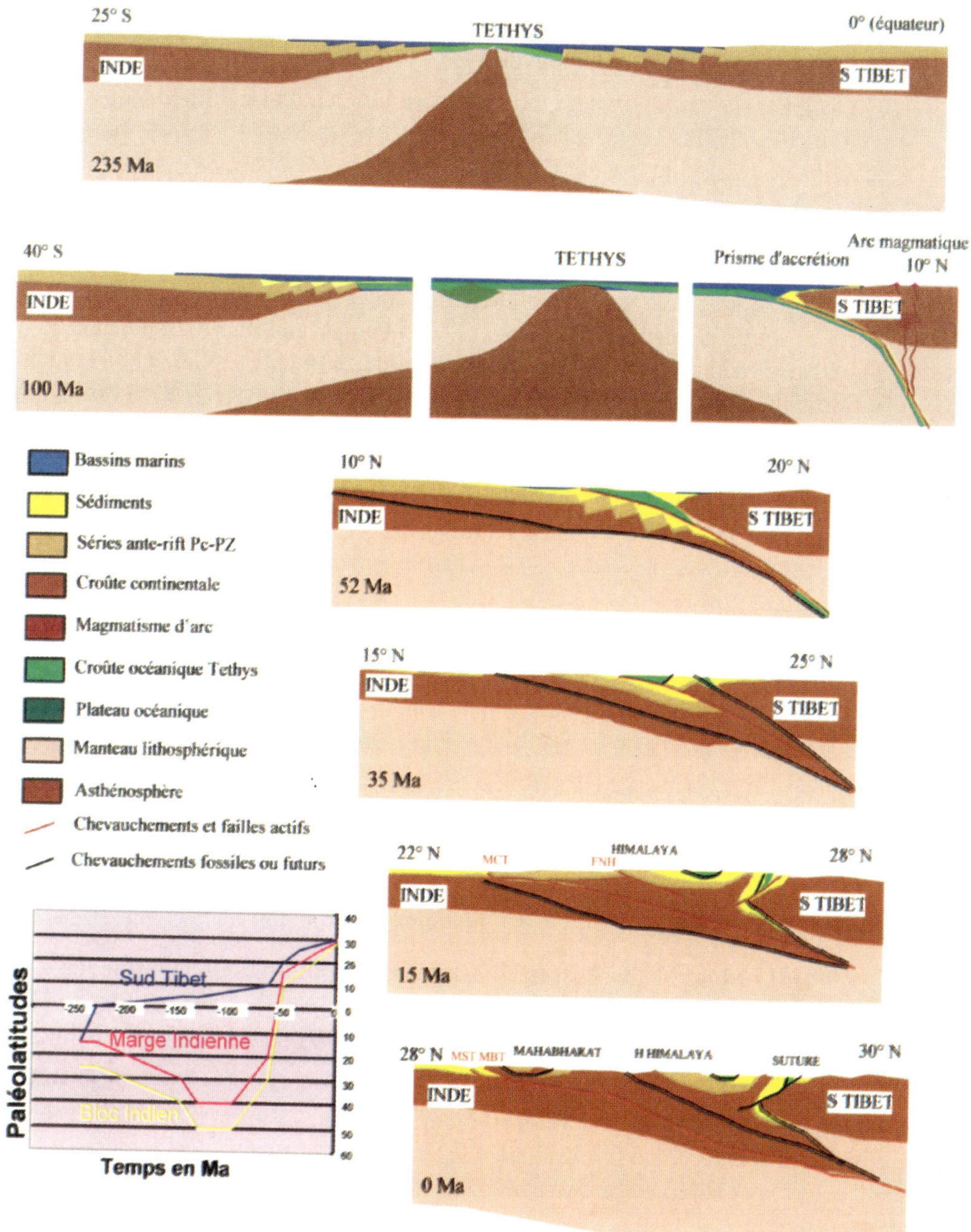

Figure 10 : Retrotectonic sketch model for the formation of the Himalaya, since 235Ma to present time.

homogenisation of the large mammalian fauna on both sides of the suture; a second tectonic thickening phase of the Indian margin through stacking of oceanic units (ophiolites) or flysch of the Tibetan margin with the emplacement of nappes and structures from the Indian margin.

3. *Cleavage of the Indian plate*: Once the India-Eurasia block formed, convergence continued with the expansion of the Indian ocean. Part of the shortening was accommodated by deformation of the South-Asia crust towards far north of the suture. A second part of the shortening was accommodated towards south of the suture by fracturing of the Indian plate and thickening through stacking which resulted into large crustal thrust zones slightly north-dipping and formed the Himalaya. Cartography and absolute dating show that these thrusts appeared in a north - south sequence; they had different ages and their associated relief propagated with time from the north towards the south . The main feature , towards the south, at the foot of the high range , was the MCT. It controlled the whole tectonic and thermal structure of the Higher Himalaya: The MCT is a "flat" shear zone, several km thick , with strong ductile deformation shown by numerous expressions of "rotational" deformation at every scale (Valdiya, 1984; Brunel, 1986). Due to this, a thrust outcrop measuring for about 25 km in thckness can be seen.

The thrust carried hot rocks from the middle or lower crust onto a non metamorphic foreland with induction of inverted Himalayan metamorphism (Hodges et al., 1988, Pecher, 1989), and partial melting displayed in numerous pegmatite dykes, with small tourmaline bearing leucogranite bodies (Manaslu) which were emplaced at the edge of the High Himalayan crystalline and the Tibeatan sedimentary series (Le Fort, 1975).

The amount of movement on the MCT was determined to be more than 200 km. Its age is indirectly deduced from radiochronological data of metamorphic minerals and those of leucogranites (Hubbard, 1988; Harrison, et al., 1977); it was initiated around 36Ma but was mainly active between 25 and 10 Ma and remained thermally active (K/Ar ages up to 2Ma) near the thrust fault, and along numerous thermal springs; the "Tatopani".

Underthrusting of the Lesser Himalaya underneath the Higher Himalaya was blocked around 10 Ma with the formation of cleavages in external zone: mainly the MBT, north of Siwaliks (with a displacement of 100 km), and also in the making of the MFT on the front of the actual Himalaya.

These various cleavages developed more easily due to the weight of the growing range of the lithosphere in the foreland area, forming the Miocene Siwalik molasse-basin and the Quaternary Ganga valley basin (200 to 300 km wide and upto 7 km thick at the foot of the range).

Furthermore , the crustal over - thickening resulted due piling up of slides of the northern sedimentary cover which had metamorphic basement along a large fault (North Himalayan Fault (NHF) or Everest fault; Burg et al., 1984).

(Frame 3, Figure11) Analogue modelling of the formation of the Himalaya: Formation of the domains

Prior to the collision the sediments were deposited:

i. On the wide NE margin of the Gondwana , the south polar palaeocontinent, on the southern edge of the Tethys ocean , at mean latitudes (55-41-28-S). After the opening of the Neo-Tethys in late Permian (285 MA) , a sequence of moderate thickness (>2 km) of syn- to post - rift sediments were deposited from Lower Triassic to Late Valanginian when the proto Indian ocean came into excitence through the breaking of the Gondwana with the splitting of India and Australia.

ii. On the Tibetan margin of the fore-arc series, the batholith and the lavas were emplaced in Ladakh, showing the active margins.

iii. On oceanic floors it was possible to distinguish between the sediments resting on the true oceanic crust and sediments of tectonic accretionary prism and of an old oceanic plateau.

The collision.

The collision is recorded within the sedimentary sequence in late Paleocene (60 Ma). It mainly involved shortening of the domain , induced the thickening of the crust and the deformation of the older series led to the migration and disappearance of the sedimentary basins. The first stage was distinguished during the Paleocene/ Lower Eocene (50 Ma), when a narrow and narrowing marine basin existed (it could be compared to the actual Arafura Sea, north-west of Australia).In Mid-Miocene times, the crust was thick enough and the basin was not marine anymore; the suture zone was then an intracontinental basin. The mountain belt behaved like a prism thickened by tectonic shortening which increased in its altitude and erosion . An unstable balance was thus established. The load induced by the range bent the lithosphere and with the formation of flexural sedimentary basins, specially to the south, marine environment was temporarily installed.

Cleavage of the Indian crust.

The initiation of the MCT system towards the south disturbed the previous balance and was responsible for the renewed deformation to the north (Indus back thrusts)in Oligocene times. The blockade, during the Miocene times (10 Ma), initiated the MBT system and displacement towards the south of the flexural system (Siwaliks). Then , onwards the active structural system was established.

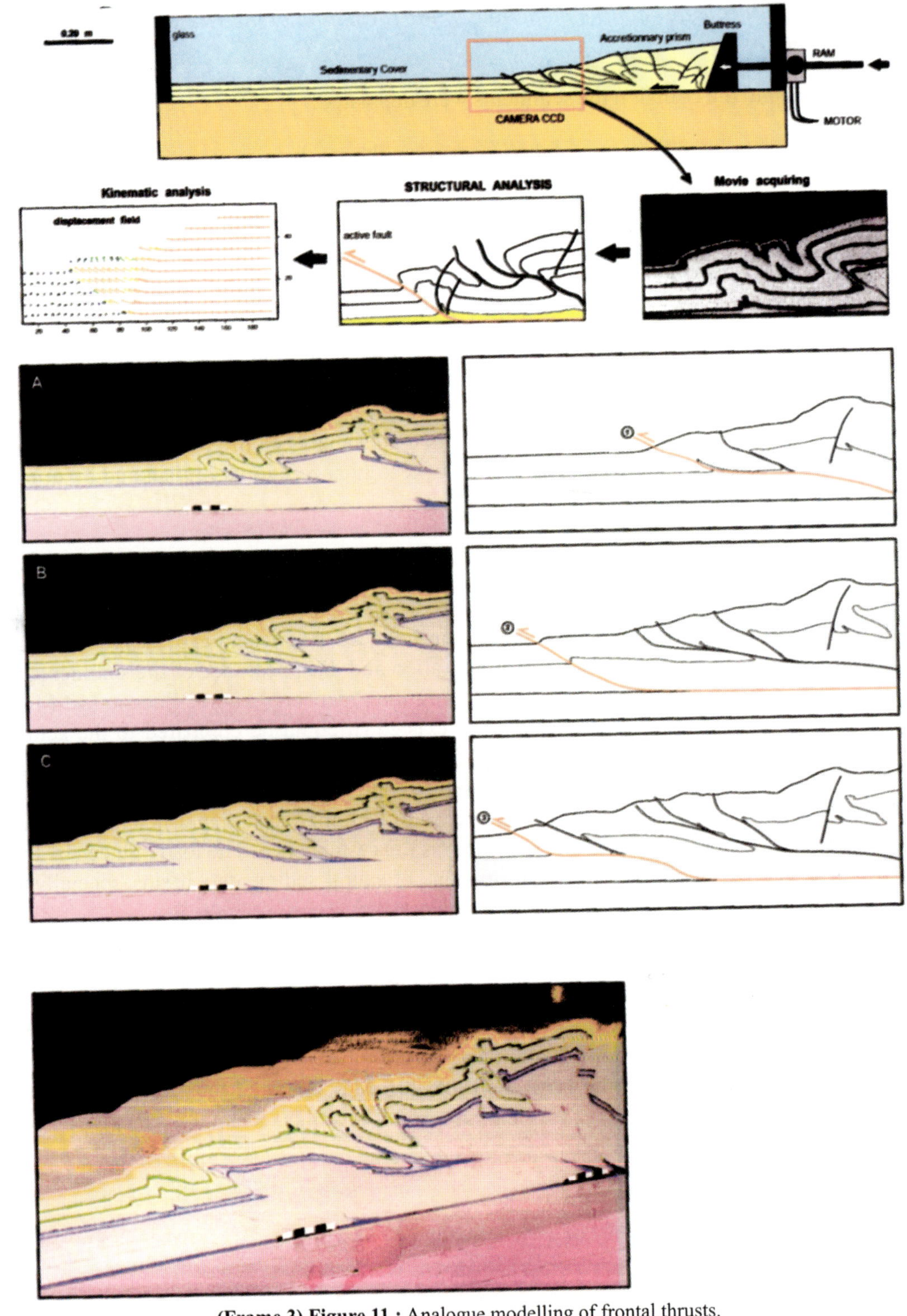

(Frame 3) Figure 11 : Analogue modelling of frontal thrusts.

REFERENCES

Bordet, P; Colchen, M; Le Fort, P. [(Monograph) Recherches geologiques dans I ' Himalaya du Nepal du Nyi-Shang. Geological studies in the Nyi-Shang region of the Nepalese Himalayas 138pp., 1975, Edts. CNRS Paris.

Bordet, Pierre; Colchen, Michel; Le Fort, Patrick. Some feature of the geology of the Annapurna Range, Nepal Himalaya. Himalayan Geology, vol.2, pp. 537-563, 1972.

Brunel, Maurice. Ductile thrusting in the Himalayas; shear sense criteria and stretching lineations. Tectonics, vol.5, no. 2 pp. 247-265, 1986.

Gansser, August. (Monograph) The Geology of the Himalayas, 289pp., 1964, Wiley & Sons.

Guillot, Stephane; Hodges, K.V., Le Fort, Patrick; Pecher, A,. New constraints on the age of the Manaslu Leucogranite; evidence for episodic tectonic denudation in the central Himalayas. Geology, vol. 22, no pp. 559-562, Jun 1994.

Guillot, Stephane; Cosca, Michael; Allemand, Pascal; Le Fort, Patrick.. Contrasting metamorphic and geochronologic evolution along the Himalayan Belt, Special Paper -Geological Society of America, vol. 328, pp. 117-128, 1999.

Harrison, T., Mark; Ryerson, F J; Le Fort, P; Yin, An; Lovera, Oscar M; Catlos, E J. A late Miocene - Pliocene origin for the central Himalayan inverted metamorphism. Earth and Planetary Science Letters, vol. 146, no.1-2, pp. E1-E7, Jan 1997.

Hodges, K V; Hubbard, Mary S; Silverberg, D. Metamorphic constraints on the thermal evolution of the central Himalayan Orogen. Philosophical Transactions of the Royal Society of London, Series A: Mathematical and Physical Science, vol. 326, no.1589, pp. 257-280, 01 Sep 1988.

Hubbard, Mary. (Monograph) Thermobarometry, (super 40) Ar/ (super) Ar geochronology, and structure of the Main Central Thrust zone and Tibetan slab, eastern Nepal Himalaya. 169pp., 1988.

Matte, P; Mattaure, M; Olivet, J M; G riot, D A. Continental subductions beneath Tibet and the Himalayan orogeny; a review. Terra Nova, vol. 9 no. 264-270, Dec 1997.

Patriat, Phillipe; Achache, Jose. India-Eurasia collision chronology has implications for crustal shortening and driving mechanism of plates., Nature (London), vol. 311, no. 5987, pp. 615-621, 18 Oct 1984.

Pecher, A. The metamorphism in the central Himalaya. Journal of Metamorphic Geology, vol .7, no. 1 pp. 31-41, Jan 1989.

Valdiya, K S. Evolution of the Himalaya. Tectonophysics, vol. 105, no. 1-4 pp. 229-248, 01 Jun 1984.

Valdiya, K S. Tectonics and evolution of the central sector of the Himalaya. Philosophical Transactions of the Royal Society of London ,Series A: Mathematical and Physical Science, vol. no. 1589, pp. 151-175, 01 Sep 1988.

Van der Voo, Rob; Spakman, Wim; Bijwaard, Harmen. Tethyan subducted slabs under India. Earth and Planetary Science Letters, vol. 171, no. 1, pp. 7-20, 15 Aug. 1999.

HIMALAYAN GEOLOGY: A HISTORICAL PERSPECTIVE

Rasoul B. Sorkhabi

Department of Geology, Arizona State University, Tempe, AZ 85287-1404, USA
Present Address: Japan National Oil Corporation, Technology Research Center,
2-2-1 Hamada, Mihamaku, Chiba 261-0025, JAPAN

ABSTRACT

The history of Himalayan geology as a scientific activity is followed through five stages, each spanning half a century: 1750-1800 (preparations); 1800-1850 (the heroic age); 1850-1900 (the golden age); 1900-1950 (ascends, depressions and transition); 1950-2000 (indigenisation and globalisation). In each stage, the major institutions, personalities, events and publications shaping Himalayan geology are briefly discussed.

"In geology, no less than in other sciences, it is desirable to be able to trace the stages of knowledge."

Henry B. Medlicott (1882)

INTRODUCTION

Geologic literature on various aspects of the Himalaya accumulated over the last two centuries is so abundant and growing so rapidly that an individual can no longer maintain in his (or her) command all the information, bibliography and developments in even a single field of Himalayan geology. Furthermore, geologic research in the Himalaya in the recent decades has been internationalised and diversified.Oddly, however, little effort has been made to systematically document and analyze the historical development of Himalayan geology as a scientific activity. This article is an effort towards this end. The long history and vast spectrum of Himalayan geology can easily fill a volume. Therefore, this article is not meant to be comprehensive, but rather to provide a chronological framework of significant events, personalities, institutions and publications that have contributed to the evolution of Himalayan geology. This article also offers a base bibliography for further research in this fascinating field.

PREPARATIONS : 1750-1800

Several trend-setting events took place in the late 18th century, which prepared the grounds for geographic and geologic studies in the Himalaya. To understand the roots of Himalayan geology these historical processes should be considered at the outset.

The first trend in the late 18th century was a major change in European attitude towards mountains. To borrow the terms used by John Ruskin in the fourth volume of his *Modern Painters* (1856), "Mountain Gloom" gave way to "Mountain Glory." Nicolson (1959) has shown that during the first seventeen centuries, Europeans viewed mountains mainly as "warts, wens, blisters, imposthumes" upon the otherwise fair face of the Earth (in a sharp contrast to the Asian views of sacred mountains). Even the poets like Virgil, Horace, Dante, Shakespeare or Milton did not praise high mountain landscape. The rapid change in European perception of mountains can be seen in arts (painting and poetry, especially in the romanticism of Wordsworth, Shelley and Byron) as well as in sciences (natural history and geology) of the late 18th century. Mountains then came to be regarded as suppliers of fresh water and mineral resources; they fascinated human spirit and evoked awe and wonder; and they held the keys ("written in the stone") to unravel the mysteries of the Globe. A leading personality in Mountain Glory was the Swiss mountaineer and geologist, Horace Benedict de Saussure (1740-1799), whose *Voyages dans les Alpes* (1779-1796) influenced many intellectuals, including geologists (James Hutton) and artists (John Ruskin). De Saussure's most famous phrase, "It is above all the study of mountains which will accelerate progress in the theory of the earth", served as the epigraph of many textbooks in the 19th century (Greene, 1982).

The second trend was the consolidation of geology as a science independent from natural philosophy or natural history. Indeed the very word "Geology" was first used in a modern sense in 1778 by Jean Andre Deluc, a Swiss geologist who lived in England as advisor to Queen Charlotte (It was also mentioned by de Saussure in his book). During this period, several founders of geology lived: James Hutton and William Smith in Britain, Jean de Lamarck, Georges-Louis Buffon and Georges Cuvier in France, and Abraham Werner in Germany.

The third historical process was the political and military consolidation of the British East India Company in India (see e.g., Edwardes, 1968; Keay, 1991), following the formal surrender of sovereign rights ("Diwan") of Bengal by the Indian Moghul Emperor, Shah Alam, to Captain Robert Clive in 1765. This was a major turning point in the British encroachment in the subcontinent and the beginning of the "Great Game" among Britain, France, Russia and China to dominate the strategic lands of High Asia (e.g., Edwardes, 1975; Hopkirk, 1990; Myer and Brysac, 1999). It also initiated the geographic surveys necessary for the Great Game (see histories, e.g., Mason, 1955; Cameron, 1984). Indeed in 1765 itself, Clive asked Major James Rennell to prepare a map of Bengal. Two years later, Rennell was appointed as the first Surveyor General of Bengal. Although Rennell returned to England in 1777 (after having suffered from malaria and impairing climate of Bengal), his leading role in Indian geography continued for several decades. He published *A Bengal Atlas* in 1777 and the *Map of Hindoostan* in 1782 (In 1794, Rennell retried and Robert Colebrooke succeeded him as Surveyor-General of Bengal). Clive also sent Captain Kinloch to Nepal in 1767 (at the request of the Raja of Nepal), who brought back the first sketches of the Nepalese Himalaya to the Company.

Clive's system of "dual government" (with the native rulers in action, but dependent on the British in the background) gave way to direct British rule in India in 1772, when

Warren Hastings was appointed as the first Governor-General of Bengal. In 1774 (the same year Clive committed suicide), Hastings despatched George Bogle to Bhutan and Tibet. After a year, Bogle returned to Calcutta. His journey was followed in 1783 by that of an army officer, Samuel Turner (Hastings' cousin), whose *Account of an Embassy to the Court of the Teshoo Lama in Tibet* (1800), remained for nearly a century the standard European book on Tibet because that country was soon invaded by the Chinese and its borders were closed to foreigners. In 1792, Hastings sent William Kirkpatrick to Nepal (for help with a minor war with Tibet) and John Gerard, who was in charge of the escort to Nepal, prepared the route map to Kathmandu. Colonel Kirkpatrick, the first Englishman to visit Nepal, compiled all the geographic information about Nepal available to him, although his *Account of the Kingdom of Nepaul* was published in 1811.

The fourth trend in the late 18th century was more ideological than political. It was the rise of Utilitarianism and Orientalism. Hastings and other Utilitarians were sympathetic to the Indian culture and social institutions, and saw no reason to Anglise India. (In contrast, the Evangelicals, such as Hastings' successors, Charles Cornwallis and Richard C. Wellesley, had little respect for the Indian norms and cultural heritage, and favoured a radical transformation of India into the British etiquette, education and Christianity.) The Utilitarians encouraged Oriental studies in order to understand India; and indeed some of them were eminent Orientalists, such James Mill (author of *History of India*, 1817) and Sir John Malcolm (author of *History of Persia*, 1815). (Stokes, 1959, presents a scholarly analysis of the English Utilitarians in India.) The word "Orientalism" appeared in the English language in 1769 (Said, 1979). In the late 18th century, Orientalists felt the need and opportunity for systematic, scholastic studies of India rather than relying on travel tales of a few adventurers. This led to the establishment of the Asiatic Society of Bengal by Sir William Jones (1746-1794) in Calcutta in 1784. The Society had a strong support from Hastings himself, although he declined to be its president (Kejariwal, 1988). "The intended objects of the Asiatic Society were Man and Nature — whatever is performed by the one, or produced by the other — within the geographic limits of Asia, with Hindustan as a centre" (Jones, 1788).

While the first two were philosophical and scientific trends taking place in Europe, the last two were socio-political events that occurred in India. However, it was through the socio-political changes in India that the effects of European Enlightenment, Romanticism, Industrial Revolution and developments in natural sciences were soon felt in the Himalaya in the form of systematic geographic explorations, geological mapping, and botanical cataloguing.

THE HEROIC AGE: 1800-1850

During the first half of the 19th century, geographic survey and geological work in the Himalaya relied largely upon the enthusiasm and endeavour of the individuals (with or without official support); the geological landscape was virgin and the mapping was romantic as well as risky. It was the period of heroes and pathfinders.

The combination of "Man and Nature", which was the ideal of Jones in launching the Asiatic Society of Bengal, was indeed reflected in many leading personalities of the

19th-century Indologists. Henry T. Colebrooke (1765-1837), who succeeded Jones as a champion of Sanskrit literature and the Society's president, also carried out many topographic surveys in the Himalaya. Indeed he was the first person to document in 1816 the Himalaya as the world's highest mountain range, surpassing the Andes which were then believed to be the highest (Colebrooke, 1816), although the British geographers at home did not accept Coolbrooke's findings. James Prinsep (1799-1840), the Society's secretary, first began working on the topography of the Bengal area, and then became an authority on Indian inscriptions. Early geological papers on India and the Himalaya were published in the Asiatic Society's periodicals: the *Asiatic Researches* (1798-1839) and *Journal of the Asiatic Society of Bengal* (1832-). The latter was a continuation of *Gleanings in Science*, launched by James D. Herbert in 1829. The Royal Geographical Society (RGS), founded in London in 1830, was also a major institution supporting exploration in the Himalaya. The Society's gold medals were given to many explorers of High Asia (see Cameron, 1980, for details).

Keay (1981) has noted that "no subject people, no conquered land, was ever as exhaustively studied as was India during the period of British rule." The Survey of India illustrates this point remarkably: It was one of the most prolonged and meticulous scientific projects ever conducted in human history ("big science" projects like NASA's space missions and the Genome project are 20th century phenomena).

Following the defeat of Tippu Sultan of Mysore by the East India Company forces led by Arthur Wellesley in 1799, two survey works were initiated in southern India. The first was a route and topographic survey in Mysore conducted by Francis Buchanan and Colin Mackenzie (1753?-1821, who became Surveyor-General of India in 1815 and assembled the Topographical and Revenue Survey under his authority in 1817). The second (in 1800) was a triangulation survey by William Lambton (1756-1823), initially around Bangalore, and then beginning with 1802 from coast-to-coast equipped with a huge theodolite. Lambton's vision was to cover the entire country, and his Trigonometric Survey of Indian Peninsula was expanded into the officially-called Great Trigonometric Survey (GTS) of India in 1818, when George Everest (1790-1866) succeeded Colonel Lambton. Everest was promoted to the Superintendent of the GTS in 1823 and to the Surveyor-General of India in 1830. By 1841, the Great Indian Arc of Meridian from Cape Comorin (Kanya Kumari) to the Himalayan foothills in Dehra Dun was completed, and Everest retired in 1843. This Great Indian Arc (about 2500-km long) served as a baseline for surveying the Himalayan peaks in the following decades. Primary sources of information on the heroic activities and results of the Survey of India include Markham (1878), Phillimore (1945-1962) (consisting of Five volumes: I. 18th century; II. 1800-1815; III. 1815-1830; IV. 1830-1843; and V. 1844-1861), the *Records of the Survey of India* and *Survey of India Professional Papers*. Popular accounts of the history of the Survey of India are given in Styles (1970), Edney (1997) and Keay (2001).

Simultaneously with the survey work in southern India, geographic expeditions continued in the Himalayan region. During 1801-1804, Charles Crawford mapped parts of Nepal, and during 1802-1803, Francis Buchanan Hamilton visited Nepal with the mission of Captain Knox. Although like Kirkpatrick before him, Hamilton stayed only in the Kathmandu valley, he derived much information from the native people, Hamilton published *An Account of the Kingdom of Nepal and of the Territories Annexed to This Domain by the House of Gorkha* in 1819, containing his research as well as the mapping by Colonel Crawford.

In 1808, Robert Colebrooke planned to explore the source of the Ganges in Garhwal, although due to illness he could not partake in the expedition. (Robert Colebrooke died in 1808 and was succeeded by John Garstin as Surveyor-General of Bengal.) William Webb, Captain Felix Raper, and Hyder Hearsey journeyed Garhwal. They produced the first map of the Bhagirathi valley as far north as the Bhatwari village, but could not of Gangotri as was intended by Colebrooke. Nevertheless, they sent Hearsey's Hindu munshi (assistant) to Gangotri and used his observations in their report (in *Asiatic Researches*, volume 11, 1810), which showed that the source of the Ganges lay within the Himalaya not in the Manasarowar lake in southern Tibet, as then was widely believed.

In 1811, Thomas Manning became the first Englishman to enter Lhasa. (His travelogue as well as that of Bogle were not published until Sir Clements Markham, a geographer at the India Office in London, put them together and sent them to print in 1876 under the title, *Narratives of the Mission of George Bogle to Tibet and the Journey of Thomas Manning to Lhasa*. It became so popular that a second edition was brought out in 1879. In 1811-12, William Moorcroft and Hyder Hearsey made the first trip to Manasarovar (*Asiatic Researches*, volume 12, 1816). The knowledge of the Himalaya until this time was compiled by Walter Hamilton in *The East India Gazetteer containing particular description of Hindostan and the adjacent countries, India beyond the Ganges* (1815).

After the triumphant war with the Gurkhas of Nepal in 1814-16, the western boundary of Nepal was fixed at Kali, and the Company thus annexed the Garhwal and Kumaun regions of the Himalaya. (These regions were included in what was initially called the Ceded and Conquered Provinces, later the Northwestern Provinces in 1836, the United Provinces of Agra and Oudh in 1902 and finally Uttar Pradesh in 1950.) The English painter, John B. Fraser, was the first European to reach Gangotri in 1815; and his *Journal* and paintings of the Sutlej and Ganga valleys were published in 1820. From 1816-1820, William Webb spent five lonely years surveying Kumaun (he determined the height of Peak XIV or Nanda Devi to be 25,669 feet, just 24 feet higher than today's measurement), while John Hodgson and James Herbert mapped Garhwal. In 1821, John Hodgson (several years junior to Webb) succeeded Colin Mackenzie as Surveyor-General of India, and in the following year, Webb resigned from the Survey and left for England. Meanwhile, the three brothers, James, Alexander and Patrick Gerard (nephews of John Gerard), made their specialty in exploration of the Sutlej Valley and Simla Hills (Himachal).

In 1817, Ranjit Singh of the Punjab conquered Kashmir, and the British through their friendly relationship with Ranjit Singh could then explore Kashmir, Ladkah and Baltistan. Grey and Garrett (1929) and Keay (1977, 1979) gave accounts of explorers of the western Himalaya and the Karakoram throughout the 19th century. Moorcroft (1765-1825) was probably the greatest Himalayan explorer of the 19th century. During 1819-1825, he and George Trebeck (a geologist and surveyor from Calcutta) journeyed Ladakh, Kashmir, Baltistan and Afghanistan. Although both of them died mysteriously on their way back from Afghanistan, their painstaking report was preserved and published in two volumes (*Travels in the Himalayan Provinces of Hindoostan and the Punjab*) in 1841. The French botanist, Victor Jacquemont, in 1831-32 and the British spy, Godfrey Vigne, in 1834-39 explored Kashmir and Baltistan, and returned with documented observations, both of which were published in 1840 (*Voyages dans l'Inde 1828-32* by Jacquemont, Paris; and *A Personal Narrative of a Visit to Ghazni, Kabul and Afghanistan* by Vigne, London). During 1831-33, Alexander Burnes, a political agent-cum-explorer travelled across the Punjab and Afghanistan as far north as Bokhara. One of his companions and a sailor with Indian Navy, John Wood, reached the source of the Oxus River in 1838. Burnes in 1834 (*Travels into Bokhara*) and Wood in 1841 (*A Journey to the Source of the River Oxus*) published their travelogues. Burnes in 1835 and Wood in 1841 were awarded the RGS' gold medals.

Following the first Anglo-Burmese war of 1824, the East India Company paid more attention to the northeastern parts of the Himalaya. Richard Wilcox and James Burlton were the pioneers of exploring Assam and the Brhamaputra valley during 1825-1828. Wilcox's Memoir was published in *Asiatic Researches* (volume 17). Elwin (1959) has described the exploration history of India's North East borders during the 19th century.

From 1831-1854, Captain Proby Cautley (1802-1871) engineered the construction of the Ganga Canal in order to get water out of the Ganga into the Doab ("two water"), the land between the Yamuna and the Ganga. The Ganga Canal project put an end to the serious and frequent famines that hit the region. (Cautley described this project in a book, *The Ganges Canal*, published in 1860.) As a result of the surveys and excavations made for the Ganga Canal, the Siwalik mammalian fossils were unearthed in the Siwalik Hills (those between Ganga and Yamuna rivers near Hardwar, named after Lord Shiva). Cautley and the palaeontologist Hugh Falconer (1808-1865), who studied the Siwalik fossils for two decades, gave the name of "Siwaliks" to the sedimentary rocks of the hills in which the fossils were found. The Siwalik fossils captured the attention of many British geologists, including Sir Roderick I. Murchison (1792-1871), who obtained a collection of these fossils for the British Museum.

In 1846, Henry Strachey, and two years later his younger brother Richard Strachey, visited Lake Manasarovar via Kumaun. Like Moorcroft and Hearsey, Henry Strachey disguised himself as a Hindu pilgrim; his full report (*On The Physical Geography of Western Tibet*) was published in 1853 (Henry Strachey was awarded the RGS' gold medal in 1852). Henry Strachey emphasised that the watershed of the Indian rivers does not correspond with the Great Himalayan peaks as seen from India, but lies beyond the Himalaya (Strachey, 1853). Richard Strachey mapped the geology of Kumaun and southern Tibet, and determined the position of Mount Kailas and calculated its height to be 22,000 feet (only 28 feet lower than later measurements) (Strachey, 1851a,b; 1900).

Following the 1845-1847 wars with the Sikhs, the British annexed the Punjab to their rule. In 1847-48, Thomas Thomson made a journey to western Himalaya and Tibet, and scaled the Karakoram pass. His account (*Western Himalaya and Tibet*) was published in 1853 (for which the RGS awarded him a gold medal in 1866). From 1848 to 1852, Andrew Fleming conducted a geological survey of the Punjab and worked out the Palaeozoic stratigraphy of the Salt Range.

The two German "founders of modern geography", Alexander von Humboldt (1769-1859) and Carl Ritter (1779-1859) also contributed to Himalayan science in the context of Asian geography. Although Humboldt did not visit the Himalaya, he journeyed the Urlas and Siberia in 1829 at the invitation of the Russian Tsar, Nicholas I. Humboldt's report, *Asie centrale* (3 volumes, 1843) discussed the impact of orography of High Asia (especially the Himalaya) on the variations of regional climate. Later in his *Kosmos* (volume 1, 1845), Humboldt wrote about the perpetual snow line in the Himalaya. Ritter's contribution was *Die Erdkunde von Asien* (1832). (Incidentally, in 1959, on the one-hundredth anniversary of the death of Humboldt and Ritter, another German geographer, Carl Troll, who himself worked in the Himalaya, presented a comparative analysis of the thoughts and works of Humboldt and Ritter, which is an interesting paper to read; Troll, 1960.)

The Himalayan geology in the first half of the 19th century was subordinate to other fields of British activities in India, such as geographic exploration, archaeology, botany, or engineering. Therefore, an understanding of Himalayan geology during this period cannot be separated from these fields. Nevertheless, several important events directly concerned with the geology also took place during this period. Geology and mineralogy were taught in the Company's colleges in India (Larwood, 1958). The Company founded an Indian Museum in London in 1801, and the rock and fossil specimens from the Himalaya and the subcontinent were shipped to Britain by amateur and official collectors were shared between the Indian Museum, the British Museum and the Geological Society (Moore, 1982). Another Indian Museum was founded in Calcutta itself in 1814. The Asiatic Society of Bengal expanded its mineral and fossil collections into a full-scale Museum of Economic Geology in 1841. Of the many surveyors in India, four were officially regarded as geologists: Laidlaw (mineralogist to the Survey of Kumaun, 1817-21); Dr. Voysey (geologist-cum-physician to the GTS, 1818-23); Dangerfield (Geological Surveyor of the Himalayan Mountains, 1820-23) and Herbert (Geological Surveyor of the Himalayan Mountains, 1823-29) (Fermor, 1951).

The first attempt to present a coherent geological picture of Himalayan physiography and geology was made by James D. Herbert. In 1818, Herbert surveyed the areas between Sutlej and Kali. Herbert's paper was published in 1842, nine years after his death (his map was published in 1844). Herbert divided the Himalayan rocks into two "primary" formations (central gneiss and schists) and a "secondary" formation (sandstones). He noted that the formations dip towards the central gneiss, and that gneiss lies on top of the schist, and schist on top of the sandstones — an arrangement which was not true stratification, but produced by "concretionary action" (Herbert, 1842).

Brian Hodgson (1800-1894) also published his ideas on the physiography of the Himalaya. According to Hodgson (1849), numerous streams flow from the snowy peaks at right angle to the strike of the Himalaya, but converge to form huge rivers close to the plains. He divided the Himalaya into spurs on the basis of transverse river gorges, and mistakenly argued against the continuity of the Himalaya as a mountain range.

THE GOLDEN AGE: 1850-1900

Himalayan geology as an independent, systematic field of science emerged in the 1850s. The second half of the 19th century was the "Golden Age" of Himalayan geology (Gansser, 1991). This was largely due to the establishment of the Geological Survey of India (GSI) in Calcutta in 1851. British geological work in India hardly lagged behind that practiced in the home islands (Stafford, 1984; 1989).

The original purpose of the GSI was limited to exploring India's coal reserves. In 1836, the Government of Bengal formed a Coal Committee to study the coal supplies for use in the steamers on the Ganges. In the early 1840s, the Coal Committee's secretary, John McCelland, contacted the leading British geologists, Sir Charles Lyell and Roderick Murchison (then President of the Geological Society in London), as well as the Government of Bengal and persuaded them to hire trained geologists and setup a proper geological survey in India. Both Lyell and Murchison were supportive of a survey in India in order to explore the mineral reserves as well as to advance the knowledge of the earth's history, and both of them appreciated the importance of geological discoveries in the Himalayan region (Stafford, 1989). In 1843, the Government of Bengal requested Murchison to select a geologist for the survey in India. Murchison turned this selection over to Sir Henry De la Beche, then the head of the British Geological Survey. (Prior to that, in 1841, De la Beche had supported the foundation of Museum of Economic Geology in the Asiatic Society of Bengal). At De la Beche's recommendation, David H. Williams was appointed in 1846 as Geological Surveyor to the Government of Bengal. Williams worked nearly three years in India, exploring coal reserves. He died of fever during fieldwork in 1848. John McCelland took his place until 1850, and finally in 1851, Thomas Oldham was appointed in Calcutta as a successor to Williams. There has been a debate about GSI's beginning date. Fox (1948) regards 1848 as the beginning year and David Williams as the first superintendent, while Fermor (1951) supports the year 1851. The GSI itself has supported the latter position (Geological Survey of India, 1976, 2001).

Thomas Oldham (1816-1878), a former professor of geology at Dublin and a local director of the Geological Survey of Ireland, was the real architect of the GSI, and aptly gave himself the title of the Superintendent of the Geological Survey. During his 25 years of service, Oldham expanded the GSI's activities from merely finding coal reserves to fuel the great Indian railway system, to working out the stratigraphy of India and the Himalaya. The GSI became one of the finest geological research centers; and its publications became the main reference on the geology of the Indian subcontinent and Asian highlands for eminent authors, such as Eduard Suess in his masterpiece, *Das Antlitz der Erde* (1885-1909; *The Face of the Earth*, 1904-1909).

Oldham hired several active geologists, including the two brothers John G. Medlicott and Henry B. Medlicott (1829-1905, served 1854-1887), the two brothers William T. Blanford (1832-1905; served 1855-1882) and Henry F. Blanford (1834-1893; served 1855-1862), William Theobald (served 1848-1881), R. Bruce Foote (1834-1912; served 1858-1891), Arthur B. Wynne (1835-1906, served 182-1883), Ferdinand Stoliczka (1838-1874; served 1862-1874), Frederick R. Mallet (1841-1921; served 1859-1889), Ottokar Feistmantel (1848-1891; served 1875-1885), Wilhelm Waagen (served 1870-1875) and Richard Lydekker (1849-1915; served

1874-1883). The well-known publications of the GSI, *Memoirs of the GSI* (1856~), *Palaeontological Indica* (1865~), and *Records of the GSI* (1868~) were also launched by Oldham.

The pioneer Himalayan geologists at the GSI were very prolific, although the conditions to do field work then were much more difficult than these days. H.B. Medlicott, who coined the terms "Sub-Himalaya" and "Main Boundary Fault", worked out single-handedly on the structure of the Sub-Himalaya between the Ravi and Ganga rivers in a Memoir (1864) containing 212 pages! (Medlicott succeeded Oldham as Director of the GSI in 1878.) Stoliczka's Memoir (1865) on Himachal was of 154 pages. Mallet's 50-page Memoir (1875) still serves as the foundation of geological knowledge on Darjeeling. Wynne's Memoir (1878) on the Salt Range was of 314 pages. Lydekker's Memoir (1883) on Kashmir and Chamba was of 344 pages. Griesbach's Memoir (1891) on the Central Himalaya was of 232 pages. Middlemiss (who served at the GSI from 1883-1917) produced a 302-page Memoir (1896) only on the Hazara area, and his Memoir (1910) on the 1905 Kangra earthquake ran 409 pages (probably still a record for a single-author report on a single earthquake!)

Of the several Stracheys who served in the British India, Richard Strachey (1817-1908) was the most talented in geoscience (although he was originally trained as an engineer in the Bombay Army). The year the GSI was established, he published two influential papers on the Himalaya: one on the physical geography and the other on the geological structure and stratigraphy of the Garhwal-Kumaun regions (Strachey, 1851a,b). After Herbert (1842), this was the next major attempt to interpret the geological setup of the Himalaya. Strachey (1851a) correlated the marine formations he observed in southern Tibet with those of Great Britain that had been described by Murchison in *The Silurian System* (1839). Strachey's fossil collections from the Himalaya were described by J.W. Salter and H.F. Blanford in Calcutta, and were transmitted to the Indian Museum in London. Later in 1862, when Eduard Suess visited London, he observed these fossils and noted the stratigraphic correlations of Triassic fauna between the Himalaya and Europe. Suess, who eventually proposed the idea of Tethys Sea in 1893, was very impressed with the results of the geological studies in the Tibetan Himalaya (later called the Tethys Himalaya by John Auden) such as those made by Ferdinand Stoliczka in 1864-66. Indeed Suess supported a joint expedition in 1892 to the Himalaya by the Imperial Academy of Vienna and the GSI. Led by Professor Carl Diener (of Austria) and the GSI geologists, Carl L. Griesbach (a former student of Suess at Vienna who had joined the GSI in 1878) and Charles S. Middlemiss, this expedition contributed greatly to the Tethyan stratigraphy of the western Himalaya and to the discovery of the Tethys Ocean. For the history of the Tethys concept, see Jenkyns (1980), Sengör (1984, 1985), and Sorkhabi (1995, 1997). (Griesbach became Director of the GSI in 1894.)

Sir Joseph Dalton Hooker (1817-1911) explored Sikkim and Bengal during 1848-52 and published a fascinating account, *Himalayan Journals,* in 1854. In the same year, the first geological map of India was published by George B. Greenough (1778-1855; founder of the Geological Society of London), based on the data provided by the Englishmen working in India. Also in 1854, Alexander Cunningham (1814-1893) published his famous book, *Ladak: Physical, Statistical and Historical*, in which he introduced the term "trans-Himalaya" for a range immediately beyond the Himalaya (later Cunningham became the first Archaeological Surveyor of India).

During the 1850s, the three Prussian brothers, Adolf, Herman and Robert Schlagintweit, were sent by Alexander von Humboldt and King of Prussia to survey India and the Himalaya. They were not well received by the Survey folks in India (because of professional jealousy or political cautiousness) although their work was officially supported by the East India Company. The Schlagintweits made expeditions to Lake Manasarovar, Ladakh and the Karakoram, and documented a wide range of important observations. However, their work drew little attention in India, partly because of the political climate of that time: When they had finished their work, it was the beginning of the Indian Mutiny in 1857. A year later, the power to rule India was transferred from the Company directly to the British Crown and Parliament, and the Governor-General became Queen's Viceroy. (The Schlagintweits published a four-volume report in 1861-1866, and this monumental work awaits an analysis by a historian of science.)

After it was known that Chimborazo of the Andes was not the highest summit, this honour went for some time to Dhaulagiri and then to Kangchenjunga. During the late 1840s, the Great Trigonometric Survey under Andrew Waugh (1810-1878) (who had succeeded George Everest as Surveyor-General) pushed into the foothills of the Himalaya and triangulated major peaks of the range. As a result, it was discovered in 1852 that Peak XV, then estimated to be 29,002 feet tall (26 feet less than the 1955 measurement), was the world's highest mountain. At Waugh's suggestion (and despite George Everest's objection), the new peak was named Mount Everest (for Nepal being a closed country, the native name of Chomolungma was unknown to the surveyors at that time).

A very important finding of the Survey of India under the leadership of George Everest was the gravitational deficiency beneath the Himalaya to deflect the plumb lines. This led to the discovery of isostasy. Although the negative mass anomaly beneath a mountain range was first observed in the Andes in the mid-18th century by the French geographers (when Pierre Bouguer was measuring a meridian arc in Peru), it was on the basis of the results of Everest's survey in the Indian subcontinent (the Kaliana-Kalianpur traverse) that two different (but complementary) concepts of isostasy were formulated in 1855 by George Airy, Astronomer Royal of England, and John Pratt, Archdeacon of Calcutta (both published in the *Philosophical Transactions of the Royal Society of London*). Isostasy (a term coined by Charles Dalton in 1893) has proved to be one of the most fundamental concepts to explain the distribution of the Earth's topography, including the high elevation of the Himalaya, the Andes and other orogenic belts (on the account of crustal roots or Airy's isostasy) and the low elevation of oceanic floor (due to density variation or Pratt's isostasy).

During 1855-1864, the Great Trigonometric Survey began a major work in Kashmir. In charge of the survey was Thomas G. Montgomerie (1830-1878), who was aided by two energetic colleagues, William H. Johnson (?-1892) and Henry H. Godwin-Austen (1834-1923). This team surveyed Kashmir as well as the Karakoram, and calculated the height of the world's second highest peak, K2, to be 28,287 feet (overestimated by 37 feet). Godwin-Austen, a geologist-mountaineer, pioneered the mapping of the Karakoram (he was awarded a gold medal by the RGS in 1910).

Thomas Montgomerie, who received the RGS' gold medal in 1865, extended his work to Garhwal and Kumaun in 1867. One of his great achievements was the training and despatch of the soealled "Pundits" to explore and fix correctly the southern part of Tibet on map. The Pundits (e.g., Nain Singh, Kishen Singh and Kinthup), who journeyed in disguise as traders or pilgrims, were native Himalayan men. The Pundits, sometime referred to as British spies, actually advanced the geographic knowledge about the courses of great Himalayan rivers, the Trans-Himalayan Range, and southern Tibet. They covered a vast mountainous region from the River Kabul on the east to the Yangtze on the west, and from Darjeeling as far north as Kashgar. In 1877, the RGS awarded a gold medal to Nain Singh "for his great journeys and surveys in Tibet and along the Upper Brahmaputra, during which he determined the position of Lhasa and added largely to our knowledge of the map of Asia." Accounts of the Pundits' lives and geographic works are some of the most interesting stories of Himalayan exploration during the 1860s-1890s. Rawat (1973) (Colonel Rajit Singh Rawat is grandson of Nain Singh), and Waller (1990) gave the details of the travels and contributions made by the Pundits.

In 1868, the Great Trigonometric Survey began its work in Assam, following the course of the Brahmaputra. The Assam triangulation was completed in 1878. In the same year, James T. Walker (1826-1896) was appointed as Surveyor-General of India. He combined the Great Trigonometric Survey, and the Topographical and Revenue Surveys as the Survey of India, headquartered in Dehra Dun.

During 1879-1887, *A Manual of the Geology of India* was published by the GSI, bringing together a vast geologic knowledge of India and the Himalaya. It consisted of four parts: Part I. Peninsular India (by H.B. Medlicott and W.T. Blanford, 1879), Part II. Extra-Peninsular Area (H.B. Medlicott and W.T. Blanford, 1879-1881); Part III. Economic Geology (V. Ball, 1881); and Part IV. Mineralogy (F.R. Mallet, 1887). The second edition of the *Manual* was published in 1893, revised by Richard D. Oldham (1858-1936). Other important encyclopedic publications included: *The Imperial Gazetteer of India* (10 volumes, 1881-1882; and second edition in 14 volumes, 1885-1887), all edited by William W. Hunter (1840-1900) (*The Imperial Gazetteer* was revised and published in 1907-1909 in 26 volumes); and *The Himalayan Districts of the North Western Provinces of India* (3 volumes, 1882-1886), edited by Edwin F.T. Atkinson (1840-1890).

THE ASCENDS, DEPRESSIONS, AND TRANISTION: 1900-1950

As far as the GSI geologists were still active players, the period 1900-1950 was the continuation of the Golden Age described above. However, this period also witnessed new developments (notably mountaineering) and depressions (world wars), and was, in a sense, a transition for the following period (described below).

The beginning of this period was marked by three important events. The first was the British invasion of Tibet during 1903-4 by Colonel (and later Sir) Francis Younghusband (1863-1942), which opened the "forbidden city" of Lhasa and the Tibetan highland to the western world after centuries of isolation. Younghusband himself was a noted explorer of High Asia (he had got the RGS' gold medal in 1890) and was the author of popular books such as *The Heart of a Continent* (1896), *India and Tibet* (1910), *Wonders of the Himalaya* (1924) and *Everest: The Challenge* (1936). For the history of the exploration of Tibet, refer to Holdich (1906), Bell (1924), MacGregor (1970), Allen (1982, 1999), Hopkirk (1982), and Bishop (1989).

The second event directly related to Himalayan geology was the publication of *A Sketch of the Geography and Geology of the Himalaya Mountains and Tibet* in 1907. Written by Sidney G. Burrard and Henry H. Hayden, this book summarised the geological and geographical knowledge of the Himalaya gathered during the 19th century and set the stage for new surveys and researches. (The book was revised by S.G. Burrard and Alexander M. Herron and published in 1934.)

The third event was what may be called the "Great Dying" of one generation and the "Rise of other Stars" in Himalayan geology. Within half-a-decade several outstanding figures passed away, including C.A. McMahon in 1904, W.T. Blanford and H.B. Medlicott in 1905, A.B. Wynne in 1906, C.L. Griesbach in 1907 and R. Strachey in 1908. A new generation of the GSI geologists in the early 20th century continued the studies of the Himalaya. Thomas H. Holland (1868-1947) succeeded C.L. Griesbach as Director of the GSI (1903-1910). H.H. Hayden published works on Spiti (1904), Tibet (1907), northern Afghanistan (1911) and Gilgit (1915). (Hayden, the GSI Director from 1910 to 1921, was killed by a rock fall in the Alps in August 1923. He was, in turn, succeeded by Sir Edwin Pascoe.) Guy E. Pilgrim (1875-1943) reconstructed the stratigraphic classification of the Siwalik Group (1910, 1913), which is used to this day. In a series of publications in *Palaeontological Indica*, Carl Diener and Frederick R.C. Reed made detailed studies of Tethyan fauna in the Kashmir, Spiti and Kumaun basins. During the 1930s, D. N. Wadia in Kashmir, Ladakh, and the Punjab, William D. West in Himachal and John B. Auden in Garhwal produced some of the finest works in Himalayan geology.

Sir Sidney Burrard was Surveyor-General of India from 1910-1919. He ordered a re-survey of Kashmir (1909-13, led by J. Hunter and H.G. Bell) and Assam (1911-13, led by C.P. Gunter, E.B. Cardew, F.M. Bailey, H.T. Morshed, and O.H.B. Trenchard). Much later, during 1927-30, Kohistan (now in Pakistan) was surveyed by C.G. Lewis, and during 1935-38, Kumaun was re-surveyed by Gordon Osmaston.

The greatest explorer of High Asia in the early decades of this century was undoubtedly Sven Hedin (1865-1952), from Sweden, whose accounts of expeditions and observations ran into numerous volumes, including *Scientific Results of a Journey in Central Asia* (6 volumes of text and 2 volumes of maps, 1899-1902), *Southern Tibet* (9 volumes of text and 3 volumes of maps, 1906-1908) and *Trans-Himalaya* (3 volumes, 1909-1912). Few explorers in history have been so prolific as Hedin; however, he has been highly criticised for his support of Kaiser Wilhelm II in the First World War and Adolf Hitler in the Second. (Hedin was awarded two medals by the RGS in 1898 and 1903; later, however, the Society cancelled his membership.)

World War I (1914-1918) and World War II (1939-1945) hampered all geological research, including those in the Himalaya. However between the two wars, several significant contributions were made. One major development was the British expeditions to Mount

Everest during the 1920s and 1930s. Although unsuccessful in reaching the summit at that time, it initiated a new wave of expeditions to the Higher Himalaya, with geological studies carried out by A.M. Heron (in 1921), Noel E. Odell (in 1924 and 1938) and Lawrence R. Wager (in 1933) in the Tibetan side of Mt. Everest. Mountaineering expeditions and publications, such as those contributed by the Günther Dyhrenfurth (a Swiss geologist/mountaineer), Herbert Tichy (an Austrian geologist/explorer), Francis S. Smythe, Eric Shipton and Harold W. Tilman (all British climbers), promoted the knowledge about the Himalaya and attracted public attention. (Unswroth, 2000, gives the most comprehensive history of the mountaineering in Everest.)

In 1928, the Himalayan Club (on the model of the Alpine Club) was founded in Bombay by a group of British officers and mountaineers (notably Geoffrey Corbett, Kenneth Mason and William Birdwood) in order "to encourage and assist Himalayan travel and exploration, and to extend knowledge of the Himalaya and adjoining mountain ranges through science, art, literature and sport." *Himalayan Journal*, published by the Club, has enriched the Himalayan literature to this day (58 volumes as of 2002).

Some of the Continental Europe geologists working in the Himalaya-Karakoram-Tibetan region in the early decades of this century deserve special mention here. Fillipo de Filippi, Giotto Dainelli and Adito Desio (all Italians) published numerous volumes on their expeditions to the Karakoram, and the Himalaya; Helmut de Terra (a German-American) investigated the Pliocene-Quaternary foreland deposits in Kashmir; Peter Misch (another German-American) studied the granites of Nanga Parbat; Erik Norin (a Swedish) explored Tibet; Arnold Heim and Augusto Gansser (both Swiss) published, *Central Himalaya* and *The Throne of the Gods* in 1939, which represent some of the highly original and fascinating publications in Himalayan geology. The mapping and concept of the Main Central Thrust was an outcome of Heim and Gansser's work in Garhwal (see Sorkhabi, 1996, for the Himalayan research of Gansser).

Some general publications of this period (treating the Himalaya in the overall context of India, China or Asia) include D.N. Wadia's *Geology of India* (1919, fourth edition in 1975), L. Dudley Stamp's *Asia* (1929, eleventh edition in 1962), John W. Gregory's *The Structure of Asia* (1929), J.S. Lee's *Geology of China* (1932), H.L. Chibber's *Geology of Burma* (1934), D. Mushketov's *Modern Conception of the Tectonics of Asia* (1936), K. Lauchs' *Geologie von Asien* (1939), G.B. Cressey's *Asia's Lands and Peoples* (1944), and O.H.K. Spate's *India and Pakistan* (1954).

Two important geological concepts applied to the Himalaya during this period were the Nappe Theory and Continental Collision. The Alpine nappe theory won international recognition at the International Geological Congress in Vienna in 1903 — thanks to the geological research of Alpine geologists Eduard Suess, Marcel Bertrand, Hans Schardt, Maurice Lugeon, Emile Argand, and Pierre Tremier. Pioneering mapping by Lajos V. Lóczy (in Sikkim), D.N. Wadia (in Kashmir and Zanskar), J.B. Auden, A. Heim and A. Gansser (in Garhwal and Kumaun) revealed the existence of nappe structures in the Himalaya. Following Alfred Wegener's continental drift hypothesis (1915), the Swiss geologist Emile Argand in his pioneering work, *Tectonique de l'Asie* (1924), argued that the whole Alpine-Himalayan belt was originated due to the drift of Indo-Africa, closure of Tethys, subsequent continental collision with Asia, and large-scale underthrusting and plastic deformation of the crust. Among a few geologists in the world of that time who accepted Wegener's and Argand's concepts was the Indian palaeobotanist, Birbal Sahni (1891-1949), who also published several important geological papers on the Himalaya in the Indian journal, *Current Science*, in 1936, and founded a reputed Institute of Palaeobotany in 1946, which is now called in his honour.

INDIGENIZATION AND GOLOBALIZATION:- 1950-2000

After World War II, India attained independence from Britain in 1947, and simultaneously Pakistan separated from India as a new state in the subcontinent. China witnessed Mao Tse-tung's Communist Revolution in 1949. These changes brought about the "indigenisation" of many activities, including geological work, in the Himalayan region.

The last British Director of the GSI was W. D. West, who served from 1945-47 (Dr. West later joined Saugar University in 1955, and developed a geology department, serving there until his death in 1994). The first Indian Director of the GSI was M.S. Krishnan (1898-1970), author of a well-known textbook, *Geology of India and Burma* (1943, fifth edition in 1968). His tenure began in the same year as a new journal, *Indian Minerals*, joined the family of the GSI periodicals. (The GSI also began publishing the *Miscellaneous Publications* on its conferences and the *Bulletin series on Economic and Engineering Geology* in 1950.) Another significant event in the GSI during this period was the publication of third edition of *Manual of Geology of India and Burma* (3 volumes) revised by Sir Edwin Pascoe, Director of the GSI from 1921-32. Pascoe had worked on the new edition during his retirement but the publication was delayed until 1950 due to World War II (Pascoe, 1950-1964).

In 1958, the Geological Society of India was founded. Based in Bangalore, the Society has been publishing a *Journal* (initially yearly and later monthly) as well as *Memoirs* (47 as of 2001), and state geology *Textbooks* (14 as of 2002). Some of these publications are on the Himalaya, e.g., *Geology of Himachal Pradesh* (by S.V. Srikanita and O.N. Bhargawa, 1988) and *Geology of Assam* (by A.B. Das Gupta and A.K. Biswas, 2000). The Society organises its annual meetings at various places in India.

After the partition, Pakistan did not inherit any geological institution of British India. Therefore, the Geological Survey of Pakistan was founded in 1947 and its first director was H. Crookshank (who was soon to retire from the Indian Survey). The Geological Survey of Pakistan is now headquartered in Quetta with five regional offices across the country. The Institute of Geology (established in 1951-52) at Punjab University in Lahore publishes *Geological Bulletin of the Punjab University* since 1961 (Pakistan's oldest journal); National Centre of Excellence in Geology (1974) at University of Peshawar publishes *Geological Bulletin of the Peshawar University* (1966~); Institute of Geology (1980) at Muzaffarabad publishes *Kashmir Journal of Geology* (1983~); National Geological Society of Pakistan in Islamabad is a young society (established only in 1989), although Punjab Geological Society in Lahore has been in existence since 1974. In 1990, the Geoscientific Laboratory (a research arm of the Geological Survey of Pakistan) was established in Islamabad as a grant-in-aid project by the Japan International Cooperation Agency. *Stratigraphy of Pakistan* (by S.M.I. Shah, 1977), *Geology of Kohistan, Karakorum Himalayas, Northern Pakistan* (by R.A.K. Tahirkheli and M.Q. Jan, 1979), *Geodynamics of Pakistan* (edited by A. Farah and K.A. DeJong, 1979), *Granites of Himalaya, Karakoram and Hindukush* (edited by F.A. Shams, 1983), and *Geology of Pakistan* (edited by F. Bender and H.A. Raza, 1995), are some noted efforts by Pakistani geologists.

In 1949, Nepal opened its borders to outsiders, and the following year, the Swiss geologist Toni Hagen went to Nepal for geological mapping. He worked there for more than a decade and published a pictorial book, *Nepal* (1967), and *Reports on the Geology of Nepal* (2 volumes, 1969-70). (Hagen, 1994, describes his memories of life and work in Nepal.) In 1954, a team of French geologists, led by Pierre Bordet, visited Nepal. Initially working in the Makalu area, the French then went on to study other parts of Nepal, especially the Manaslu area (the Manaslu is said to be the most studied granite on earth). Le Fort (1994) gave a comprehensive account of four decades of the French studies in the Himalaya. During the 1960s-1970s, Japanese geologists from the University of Hokkaido carried out extensive work in Nepal, the results of which were documented in *Geology of the Nepal Himalayas* (edited by S. Hashimoto, Y. Ohta and C. Akiba, 1973). Japanese contribution to Nepal geology has continued to this day (see special issues of the *Journal of Nepal Geological Society*, volumes 2-1982, and 25-2001, 4-1982, and *Bulletin of Department of Geology, Tribhuvan University*, volume 2, number 1, 1992). In Nepal, the Department of Mines and Geology is the main geological survey organization. Nepal Geological Society, established in Kathmandu in 1980, publishes a *Journal* annually. The late Chandra K. Sharma was a champion of disseminating the geological knowledge of Nepal Himalaya by authoring several books, including *Geology of Nepal Himalaya and Adjacent Countries* (1990).

The Himalaya's association with mountain lovers and mountaineering challenges has been fruitful for both sport and science. G.O. Dyhernfurth's *Zum Dritten Pol* (1952; *To the Third Pole: A History of the High Himalaya*, 1955) summarised the geographic and geologic knowledge of the high peaks of the Himalaya prior to Edmund Hillary's and Tenzin Norgay's conquest of Mount Everest (Chomolungma). Following this successful climbing of Everest from the Nepali side in 1953, an unprecedented attention was drawn for mountaineering and trekking in the Higher Himalaya. In 1954, K2 was summitted by an Italian group led by Ardito Desio. The mountaineering has produced a genre of folk literature about Himalayan landscape and natural history (e.g., the works of James R. Ulman, Wilfrid Noyce, William O. Douglas, Edmund Hillary and Reinheld Messner), thus promoting a public support for Himalayan studies. Baume (1979) has compiled the history of mountaineering in the "Sivalaya" (8,000-m peaks) of the Himalaya. The tourist and trekking industry has led to the construction of roads and bridges, which have made geologists' job less arduous; it has also generated many guidebooks to almost any region of the Himalaya.

While India took a socialist democratic path after Independence, Chinese communists followed an autocratic fervour, ultimately leading to the notorious Cultural Revolution (1966-1969). In 1959, two seemingly unrelated events took place: Tibet was violently annexed to the People's Republic of China, and the Chinese Academy of Geological Sciences was founded in Beijing. Over the past three decades, the Academy has carried out extensive studies of the Tibetan Plateau and its surrounding mountains. However, with a few exceptions, the bulk of these studies have been published in Chinese. The major Chinese writings on the Himalayan-Tibetan geology available in the English language include *The Evolution of the Tethys in China and Adjacent Regions* by Huang Jiqing and Chen Bingwei (1987); *Geology and Tectonics of Qinghai-Xizang Plateau* by Chang Chengfa (1996); *Crust-Mantle Thermal Structure and Tectonothermal Evolution of the Tibetan Plateau* by Shen Xianjie (1996).

During the 1960s, the theory of plate tectonics emerged from several important discoveries in the dynamics and structure of the ocean floor, notably sea-floor spreading along mid-oceanic ridges, subduction along the trenches, and magnetic anomalies on the ocean floor. Drifted continents could also be restored back to their palaeo-positions based on palaeomagnetic data and geological evidence; convectional currents in the mantle were considered as a mechanism for plate motions; seismicity, volcanism, rifting and orogenesis were associated with active plate boundaries. The plate tectonic theory revolutionized all branches of earth sciences; Himalayan geology was no exception. Gansser (1966), Dewey and Bird (1970), Dietz and Holden (1970), Powell and Conaghan (1973), and Molnar and Tapponnier (1975) offered influential interpretations of the Himalayan orogeny in terms of plate tectonics, although their fundamental concepts had been formulated by Argand (1924). (Argand's work was first translated into English by A.V. Carozzi in 1977.) All these authors regarded the Himalaya as a type-example of continental collision orogenesis, and this image has been enhanced since then.

In 1964, the 22nd International Geological Congress was held in New Delhi, which was an important achievement for the Indian geological community. In the same year, Augusto Gansser published *Geology of the Himalayas*, which synthesised the geological knowledge of the Himalaya gathered from the 1850s to the early 1960s (Gansser, 1964).

New Delhi also hosted the 21st International Geographical Congress in 1968, presided by the late Professor S. P. Chatterjee (Law, 1968). In the same year, the Institute of Himalayan Geology Society was established through the efforts of D.N. Wadia in the Geology Department of University of Delhi. Later it was re-named the Wadia Institute of Himalayan Geology (WIHG) and headquartered in Dehra Dun in 1976. WIHG has conducted numerous conferences since 1970, and their proceedings have been published in *Himalayan Geology* (22 volumes as of 2001). From 1990-1995, WIHG also brought out the semi annual *Journal of Himalayan Geology*, which later merged with *Himalayan Geology*. WIHG's workers have been prolific. Its current president, K. S. Valdiya, has authored several volumes (*Geology of Kumaun Lesser Himalaya*, 1980; *Aspects of Tectonics, Focus on South-Central Asia*, 1984; *Environmental Geology, Indian Context*, 1987; *Kumaun Land and People*, 1988; *Geology and Natural Environment of Nainital Hills, Kumaun Himalaya*, 1988; *Dynamic Himalaya*, 1998). V.C. Thakur, a former director of WIHG, has synthesized the *Geology of Western Himalaya* (1992) and has co-edited (with K.K. Sharma) the *Geology of Indus Suture Zone of Ladakh* (1983). K.K. Sharma has edited two volumes on *Geodynamic Evolution of the Himalayan Collision Zone* (1991), and A.K. Sinha, a former geologist at WIHG and currently director of the Birbal Sahni Institute, has authored *Geology of the Higher Central Himalaya* (1989) and has edited the volumes, *Contemporary Geoscientific Researches in the HImalaya (2 Volumes, 1983) Himalayan Orogen and Global Tectonics* (1992) and *Geodynamic Domains in the Alpine-Himalayan Tethys* (1997). Other Indian efforts in Himalayan geology are reflected in the volumes of *Recent Researches in Geology* (1975-1995; 16 volumes), *Contributions to Himalayan Geology* (1979-1989; 4 volumes), and *Current Trends in Geology* (1978-1991; 13 volumes, most of which have been edited by P. S. Saklani), all published in Delhi.

An important factor in the development of Himalayan geology during the 1970s was the opening of the restricted areas of the Himalayan region. In 1974, India opened Ladakh, and Pakistan opened the Karakoram to foreign visitors. This facilitated many studies of the Himalayan Suture Zones in Ladakh and Kohistan (e.g., Gansser, 1980). Plate tectonic models for the origin of the Himalaya thus became geologically testable and popular.

In 1979, the construction of the Karakoram Highway was completed after twenty years of a Chinese-Pakistani collaborative work, and its opening to foreigners in 1986 resulted in modern geological studies of the Karakoram and western Tibet (see Searle, 1991, for a synthesis of the recent research). In 1980, the International Karakoram Project was organised by the Royal Geographical Society of London in the Hunza valley of Pakistan; its results have been published in the two volumes of *International Karakoram Project* (edited by Keith J. Miller, 1984) and in a popular account, *Continents in Collision*, by K.J. Miller (1982). (In 1991, Professor K.J. Miller was awarded the RGS' gold medal — the latest geographer working on the Himalayan region to obtain this medal.)

In 1979, the Chinese opened Tibet to outsiders; the following year, Academia Sinica organised an international meeting of Himalaya-Tibetan geologists, which resulted in the two volumes of *Geological and Ecological Studies of Qinghai-Xizang Plateau* (1981). Since then, collaborative research between the Chinese geologists and those from France, Britain and United States has produced several important reports on the Tibetan Plateau in the pages of *Nature* (3 January 1981, 5 January 1984, and 10 October 1985 issues documenting the 1980-1983 French-Chinese research results; 9 October 1986 issue on the British-Chinese work) and *Science* (6 October 1996 issue reporting on the results of the American-Chinese INDEPTH project). In addition, .. the French-Chinese collaboration has resulted in the publication of ... the *Geology of the Himalayas* (2 volumes, edited by Li Guangcen et al., Beijing, 1991), and the British-Chinese 1985 project has resulted in a proceedings volume, *Tectonic Evolution of the Himalayas and Tibet* (edited by R.M. Shackleton, J.F. Dewey and B.F. Windley; Royal Society, London, 1998). Chinese involvement in international geology was culminated in hosting the 30th International Geological Congress in Beijing in 1996, which drew further attention to the geology of Tibet.

In 1983, the International Centre for Integrated Mountain Development (ICIMOD) was established in Kathmandu. Funded by several European and Asian governments, ICIMOD is a multidisciplinary research and information centre to promote the conservation of mountain ecosystems and sustainable economic development in the Himalaya-Karakoram-Hindu Kush region. ICIMOD has published numerous papers, reports and conference proceedings, although they are largely geographical, ecological and agronomic research. (An integration of various branches of geosciences towards solving socioeconomic problems seems to be a long way to go, but ICIMOD's experience is a positive sign)....the American geologists have published *Notes on the Geology of tibet and Adjacent areas* (edited by A.W. Bally and other, USGS open file report, 1980).

Since 1985, a series of annual international meetings, the Himalaya-Karakoram-Tibet Workshops, have been held: Leicester, UK (1985); Nancy, France (1986); London, U.K. (1987); Lausanne, Switzerland (1988); Milan, Italy (1990); Grenoble, France (1991); Oxford, U.K. (1992); Vienna, Austria (1993); Kathmandu, Nepal (1994); Ascona, Switzerland (1995); Flagstaff, U.S.A. (1996); Rome, Italy (1997); Peshawar, Pakistan (1998); Kloster Ettal, Germany (1999); Chengdu, China (2000), Schloss Seggau, Austria (2001) and Sikkim, India (2002). These meetings and their proceedings volumes have contributed significantly to Himalayan research and geologic literature: A.C. Barnicoat and P.J. Treloar, eds., "Himalayan Metamorphism" (*Journal of Metamorphic Geology*, volume 7, number 1, 1989); A. Baud, ed., "Colloque Himalaya-Karakoram-Tibet, Lausanne, 1988" (in *Eclogae Geologicae. Helv*., volume 82, part 2, 1989); P.J. Treloar and M.P. Searle, eds., *Himalayan Tectonics* (Geological Society of London Special Publication 74, 1993); B.N. Upreti and M.R. Dhital, eds., "Proceedings of the 9th Himalaya-Karakoram-Tibet Workshop" (*Nepal Geological Society Journal*, volume 11, 1995); J.P. Burg, ed., "Uplift and Exhumation of Metamorphic Rocks, the Himalayan Tibetan Region" (*Tectonophysics*, volume 260, 1996) and Special Issue of *Geodinamica Acta* (volume 9, numbers 2-3, 1996); A. Macfarlane, R.B. Sorkhabi, and J. Quade, eds., *Himalaya and Tibet: Mountain Roots to Mountain Tops* (Geological Society of America Special Paper 328, 1999); M. Gaetani et al., eds., Special Issue of *Journal of Asian Earth Sciences* (volume 16, number 2-3, 1998); M. Khan, P.J. Treloar, M. P. Searle, and M.Q. Jan, eds., *Tectonics of the Nanga Parbat Syntaxis and the Western Himalaya* (Geological Society of London Special Publication 170, 2000); C. Wang, C.J.L. Wilson and M.P. Searle, eds., Special Issue of *Journal of Asian Earth Sciences* (volume 20, number 3, 2002).

Some notable publications of the recent decades arising from international meetings or individual works on the Himalaya and Tibet (and not cited above) include: *Himalaya: Science de la Terre* (CNRS, Paris, 1976); *Albert F. de Lapparent volume* (Geological Society of France Memoir 8, 1977); *Orogenic Evolution of Southern Asia from Turkey to Indonesia: Pierre Bordet volume* (edited by P. Le Fort, M. Colchen and C. Montenat, Sciences de le Terre Memoir 47, Paris, 1986); *Zagros-Hindu Kush-Himalaya: Geodynamic Evolution* (edited by H.K. Gupta and F.M. Delany, American Geophysical Union Geodynamic Series 3, 1981); *Geology of the Bhutan Himalayas* (by A. Gansser, 1983); *Tectonics of Asia* (Volume 5 of the 27th International Geological Congress, Moscow, 1984); *The Cimmeride Orogenic System and the Tectonics of Eurasia* (by A.M.C. Sengör, Geological Society of America Special Paper 195, 1984); *Tectonics of Western Himalaya* (edited by L.L. Malinconico and R.J. Lillie, Geological Society of America Special Paper 232, 1989); *Geodynamics of the Indian Peninsula and the Indian Plate Margin* (by R.K. Verma (1991); *Geomorphological Record of the Quaternary Orogeny in the Himalaya and the Karakoram* (by J. Kalvoda, 1992); *Geology of the Lower Himalaya – Garhwal* (by P.S. Saklani, 1993); *Himalaya to the Sea: Geology, Geomorphology and the Quaternary* (edited by J.F. Shroder, 1993); *The Tectonic Evolution of Asia* (edited by A. Yin and M. Harrison, 1996); *Geodynamics of the NW Himalaya* (edited by A.K. Jain and R.M. Manickavasagam, 1999); ...two volumes edited by P.K. Verma (*Metamorphic Rock Sequences of the Eastern Himalaya*) 1980; *Geological Studies in the Eastern Himalaya*, 1999) and *Foreland Sedimentation in Himalayan Tectonic Regime* (by V. Raiverman, 2002)

In 1992, the First South Asian Geological Congress (GEOSAS) was held in Islamabad (Pakistan). It was followed by GEOSAS-II in Colombo (Sri Lanka) in 1995, GEOSAS-III in Lahore (Pakistan) in 2000, and GEOSAS-IV in New Delhi (India) in 2002. The proceedings of these meetings have been published as *Geology in South Asia-I* (edited by R. Ahmad and A.M. Sheikh, Hydrocarbon Development Institute of Pakistan, Islamabad, 1994); *Geology in South Asia-II* (edited by N.P. Wijayananda, P.G. Cooray and P. Mosley, Geological Survey and Mines Bureau, Sri Lanka, 1997) and *Siwaliks of South Asia: Proceedings of the Third GEOSAS Workshop* (edited by M.I. Ghaznavi, S.M. Raza, and M.I. Hasan, Records of Geological Survey of Pakistan, Islamabad, 1998). The GEOSAS provides a platform for exchange of geoscientific information between nine countries in South Asia (from Turkey to Sri Lanka).

The most significant characteristic of Himalayan geology over the past two decades has been the applications of analytical techniques, including geochemical, geochronological, geophysical, and quantitative structural techniques, although none of these should replace the traditional field work ("seeing for oneself") and masterly syntheses ("armchair geology"). Himalayan geology has become a subject of regular sessions in international meetings such as those of the International Geological Congress, American Geophysical Union, Geological Society of America, and European Geoscience Union. Not only native geologists from the Himalayan countries (India, Pakistan, Bangladesh, China, Nepal, and Burma) but also those from various countries in Europe, USA, Canada, Australia and Japan are now involved in geoscientific studies in the Himalaya-Tibetan region. A major obstacle to geologic research in the Himalaya in the past decade has been the political conflicts and the consequent insecurity for fieldwork. Scientific research itself can be a factor to promote peace in the Himalayan region. We are in the Global Age of Himalayan geology, which perhaps would not surprise the great poet Kalidasa, where he would like to return and listen to our modern observations and ideas of his beloved "Nagadhiraj" (King of Mountains).

ACKNOWLEDGEMENTS

This article stems from the author's studies of the historical and geologic literature on Himalayan exploration and geology for more than a decade, and especially during the editing of the Himalayan Notes newsletter (1993-1996). For lack of space, it is not possible to reference all the original sources on Himalayan geology and exploration; nevertheless, the references cited below provide a portal of entry to further research and studies in this field. I appreciate Dr. P.S. Saklani's efforts in bringing out this volume on Himalayan Mountain Geology, for his comments on the manuscripts, and the opportunity given to me for writing this article.

BIBLIOGRAPHIC REFERENCES

Allen, C. (1982) A Mountain in Tibet: The Search for Mount Kailas and the Sources of the Great Rivers of India. André Deutsch, London, 254 p.

Allen, C. (1999) The Search for Shangri-La: A Journey into Tibetan History. Little, Brown & Co., London, 305p.

Argand, E. (1924) La tectonique de l'Asie. C.R. 13 Congr. Géol. Intl., Belgique, 1922, Vaillant-Carmanne, Liège, 171-372.

Auden, J.B. (1933) On the age of certain Himalayan granites. Rec. Geol. Surv. India, 66(4), 461-471.

Auden, J.B. (1934) The geology of the Krol belt. Rec. Geol. Surv. India, 67(4), 357-454.

Auden, J.B. (1935) Traverses in the Himalaya. Rec. Geol. Surv. India, 69(2), 123-167.

Auden, J.B. (1937) The structure of the Himalayas in Garhwal. Rec. Geol. Surv. India, 71, 407-433.

Baker, J.N.L. (1967) A History of Geographic Discovery and Exploration. Cooper Square Publishers, New York, 552 p.

Baume, L.C. (1979) Sivalaya, the 8,000-Metre Peaks of the Himalaya: A Chronicle and Bibliography of Exploration. The Mountaineers, Seattle, 346 p.

Bell, C. (1924) Tibet, Past and Present. Oxford University Press, Oxford, 326 p.

Bishop, P. (1989) The Myth of Shangri-La: Tibet, Travel Writing and the Western Creation of Sacred Landscape. Athlone Press, London, 308 p.

Bordet, P. (1961) Recherches géologiques dans l'Himalaya du Népal, région du Makalu. Centr. Nation. Rech. Sci., Paris, 275 p.

Bordet, P., Colchen, M., Krummenacher, D., LeFort, P., Mouterde, R. and Rémy, J.M. (1971) Recherches géologiques dans l'Himalaya du Népal, région du Thakkhola. Centr. Nation. Rech. Sci., Paris, 279 p.

Bordet, P. and LeFort, P. (1975) Recherches géologiques dans l'Himalaya du Népal, région du Nyi-shang. Centr. Nation. Rech. Sci., Paris, 138 p.

Burrard, S.G. and Hayden, H.H. (1934) A Sketch of the Geography and Geology of the Himalaya Mountains and Tibet (revised by S.G. Burrard and A.M. Heron). Government of India Press, Calcutta, 359 p.

Cameron, I. (1980) To the Ends of the Earth: 150 Years of World Exploration by the Royal Geographical Society. Donald Payne, London, 288 p.

Cameron, I. (1984) Mountains of the Gods: The Himalaya and the Mountains of Central Asia. Hammond Inns, London, 248 p.

Carozzi, A.V. (1977) Emile Argand: Tectonics of Asia. Hafner, New York, 218 p.

Colebrooke, H.T. (1816) On the height of the Himalaya mountains. Asiatic Res., 12, 251-285.

Crone, G.R. (1972) The Discovery of the East. Hamilton, London, 178 p.

De Terra, H. (1935) Geological studies in the north-west Himalaya between the Kashmir and Indus valleys (Yale north India expedition). Mem. Connecticut Acad. Arts Sci., 8, 18-76.

De Terra, H. (1936) Himalayan and Alpine orogenesis. Proc. Intl. Geol. Congr. 16th Session, Section 2, 859-871.

De Terra, H., and Paterson, T.T. (1939) Studies on the Ice Age in India and Associated Human Cultures. Carnegie Institution, Washington D.C., 354 p.

Dewey, J. F. and Bird, J.M. (1970) Mountain belts and the new global tectonics. Jour. Geophys. Res., 75, 2625-2647.

Diener, C. (1912) The Trias of the Himalaya. Mem. Geol. Surv. India, 36(3), 1-159 p.

Dietz, R.S. and Holden, J.C. (1970) Reconstruction of Pangaea: breakup and dispersion of continents, Permian to Present. Jour. Geophys. Res. 75B, 4939-4955.

Dyhrenfurth, G.O. (1955) To the Third Pole: The History of the High Himalaya. Werner Laurie, London, 133 p.

Edney, M. (1997) Mapping an Empire: The Geographic Construction of British India 1765-1843. University of Chicago Press, 424 p.

Edwardes, M. (1968) British India: 1772-1947. Tapliner Publishing Co., New York, 396 p.

Edwardes, M. (1975) Playing the Great Game: A Victorian Cold War. Hamilton, London, 176 p.

Elwin, V. (1959) India's North-East Frontier in the Nineteenth Century. Oxford University Press, Oxford, 473 p.

Fermor, L. (1951) Geological Survey of India. Nature, 167, 10-12.

Fox, C.S. (1947) The Geological Survey of India, 1846 to 1947. Nature, 160, 889-891.

Fuchs, G. (1967) Zum Bau des Himalaya. Denkschr. österr. Akad. Wiss, math.-naturw. Kl, Wien, 113, 1-211.

Fucsh, G. (1975) Contributions to the geology of the north-western Himalayas. Abh. Geol. Bundesanst., Wien, 32, 1-59.

Fucsh, G. (1982) Explanations of the geologic-tectonic map of the Himalaya. Jahrb. Geol. Bundesanst., Wien, 125, 1-50.

Fuchs, G. and Frank, W. (1970) The geology of west Nepal between the rivers Kali Gandaki and Thulo Bheri. Jahrb. Geol. Bundesanst., Wien, 18, 1-103.

Gansser, A. (1964) Geology of the Himalayas. Interscience Publishers, London, 289 p.

Gansser, A. (1966) The Indian Ocean and the Himalaya: a geological interpretation. Eclog. Geol. Helv., 59, 831-848.

Gansser, A. (1980) The significance of the Himalayan suture zone. Tectonophysics, 62, 37-52.

Gansser, A. (1983) The Geology of the Bhutan Himalayas. Birkhauser Verlag, Basel, 181 p.

Gansser, A. (1991) Facts and theories on the Himalayas. Eclog. Geol. Helv., 84, 33-59.

Geological Survey of India (1976) One Hundred Twenty Five Years of the Geological Survey of India (1851-1976). Diti Printing and Binding Works, Calcutta, 58 p.

Geological Survey of India (2001) The Story of GSI: Comparative Volume on the Occasion of 150 Years of the GSI. Calcutta.

Gill, W.D. (1951a) The stratigraphy of the Siwalik series in the northern Potwar region, Punjab, Pakistan. Quart. Jour. Geol. Soc. London, Ser. 4, 107, 375-394.

Gill, W.D. (1951b) The tectonics of the Sub-Himalayan fault zone in the northern Potwar region and in the Kangara district of the Punjab. Quart. Jour. Geol. Soc. London, Ser. 4, 107, 395-421.

Gole, S. (1976) Early Maps of India. Humanities Press, New York, 126 p.

Grey, M., and Garrett, H.L.O (1929) European Adventurers of Northern India 1785-1849. Government Printing, Lahore, 361 p.

Greene, M.T. (1982) Geology in the Nineteenth Century. Cornell University Press, Ithaca, 324 p.

Griesbach, C.L. (1880) Palaeontological notes on the Lower Trias of the Himalayas. Rec. Geol. Surv. India, 13, 94-113 and 14, 154-155.

Griesbach, C.L. (1891) Geology of the Central Himalaya. Mem. Geol. Surv. India, 23, 1-232.

Hagen, T. (1971) Nepal: The Kingdom in the Himalayas, 3rd ed. Kummerly and Frey, Bern, 180 p.

Hagen, T. (1969) Report on the geological survey of Nepal: vol. 1. Preliminary reconnaissance. Denkschr. Schweiz. Naturf. Ges., 86, 1-185.

Hagen, T. (1970) Report on the geological survey of Nepal: vol. 2. Geology of Thakkhola. Denkschr. Schweiz. Naturf. Ges., 86, 1-159.

Hagen, T. (1994) Building Bridges to the Third World: Memories of Nepal 1950-1992. (Translated by Philip Pierce). Delhi, 386 p.

Hall, J.S. (Ed.) (1981) The Himalaya: Aspects of Change. Oxford University Press, Delhi, 482 p.

Hashimoto, S., Ohta, Y., and Akiba, C. (Eds.) (1973) Geology of the Nepal Himalayas. Himalayan Committee of Hokkaido University, Sapporo, Japan, 286 p.

Hayden, H.H. (1904) The geology of Spiti with parts of Bushair and Rupshu. Mem. Geol. Surv. India, 36(1), 1-129.

Hayden, H.H. (1907) The geology of the provinces Tsang and U in Tibet. Mem. Geol. Surv. India, 36(2), 122-201.

Hayden, H.H. (1911) The geology of northern Afghanistan. Mem. Geol. Surv. India, 39(1), 1-97.

Hayden, H.H. (1913) Notes on the relationship of the Himalaya to the Indo-Gangetic plain and the Indian peninsula. Rec. Geol. Surv. India, 43(2), 138-157.

Hayden, H.H. (1915) Notes on the geology of Chitral, Gilgit and the Pamirs. Rec. Geol. Soc. India, 45(4), 271-335.

Heim, A., and Gansser, A. (1939a) Central Himalaya: Geological Observations of the Swiss Expedition 1936. Denkschr. Schweiz. Naturf. Ges., Mem. Soc. Helv. Sci. Nat., 73(1), 1-245.

Heim, A., and Gansser, A. (1939b) Throne of the Gods. Macmillan, London, 233 p.

Herbert, J.D. (1842) Report of the mineralogical survey of the region lying between the rivers Satlaj and Kali, with a geological map. Jour. Asiatic Soc. Bengal, Vol. 11(I), Pt. X (map in Vol. 13, p. 734).

Heron, A.M. (1922) Geological results of the Mount Everest reconnaissance expedition. Rec. Geol. Surv. India, 54(2), 215-234.

Hodgson, B. (1849). On the physical geography of the Himalayas. Jour. Asiatic Soc. Bengal, 18(II), 761-788.

Holdich, T. (1906) Tibet, the Mysterious. Alston Rivers, London, 356 p.

Holland, T.H. (1897) Report on the geological structure and stability of hillslopes around Nanital. Misc. Publ. Geol. Surv. India, 1-85.

Hopkirk, P. (1982) Trespassers on the Roof of the World: The Race for Lhasa. Oxford University Press, Oxford, 274 p.

Hopkirk, P. (1990) The Great Game: On Secret Service on High Asia. John Murray, London, 562 p.

Hopkirk, P. (1996) Quest for Kim: In Search of Kipling's Great Game. John Murray, London, 274 p.

Ives, J.D. and Messerli, B. (1989) The Himalayan Dilemma. Routlege & Kegan Paul, London, 295 p.

Jenkyns, H.C. (1980) Tethys: past and present. Proc. Geol. Assoc., 91, 107-118.

Jones, W. (1788) A discourse on the institution of a society, for inquiring into the history, civil and natural, the antiquities, arts, sciences, and literature, of Asia. Asiatic Res., 1, 10.

Kapadia, H. (Ed.) (2001) A Passage to Himalaya. Oxford University Press, Delhi, 352 p.

Keay, J. (1977) When Men and Mountains Meet: The Explorers of the Western Himalayas 1820-1875. John Murray, London, 278 p.

Keay, J. (1979) The Gilgit Game: The Explorers of the Western Himalayas 1865-1895. John Murray, London, 278 p.

Keay, J. (1981) India Discovered. Collins, London, 222 p.

Keay, J. (1991) The Honourable Company: A History of the English East India Company. Simon & Schuster, New York, 475 p.

Keay, J. (2001) The Great Arc: The Dramatic Tale of How India was Mapped and Everest Named. Harper Collins, New York, 182 p.

Kejariwal, O.P. (1988) The Asiatic Society of Bengal and the Discovery of India's Past 1784-1838. Oxford University Press, Delhi, 293 p.

Kraft, A. von (1902) Notes on the "Exotic blocks" of Malla Johar in the Bhot Mahals of Kumaon. Mem. Geol. Surv. India, 32(3), 127-183.

Krishnan, M.S. (1953) The structural and tectonic history of India. Mem. Geol. Surv. India, 81, 1-109.

Krishnan, M.S. (1968) Geology of India and Burma, 5th ed. Higginbothams, Madras, 536 p.

Larwood, H.J.C. (1958) Science in India before 1850. British Jour. Ed. Stud., 7, 36-49.

Law, B.C. (ed.) (1968) Mountains and Rivers of India. 21st Intl. Geogr. Congr., Calcutta.

LeFort, P. (1975) Himalayas - the collided range: present knowledge of the continental arc. American Jour. Sci., 275A, 1-44.

LeFort, P. (Ed.) (1994) French Earth Sciences Research in Himalayan Regions. Alliance Française, Kathmandu, 174 p.

Lewis, C.G. (1937) A new Siwalik correlation. American Jour. Sci., Ser. 5, 33, 191-204.

Loczy, L. von (1907) Beobachtungen im östlichen Himalaya (vom 8 bis 28, Febr. 1878). Földr. Közlem, 35 (9), 1-24.

Lydekker, R. (1876) Notes on the geology of the Pir Panjal and neighbouring districts. Rec. Geol. Surv. India, 9(4), 155-162.

Lydekker, R. (1879) Notes on the geology of Kashmir, Kishtwar and Pangi. Rec. Geol. Surv. India, 11(1), 30-63.

Lydekker, R. (1880) Geology of Ladak and neighbouring districts. Rec. Geol. Surv. India, 13(1), 26-58.

Lydekker, R. (1883) The geology of the Kashmir and Chamba territories and the British district of Khagan. Mem. Geol. Surv. India, 22, 1-344 p.

MacGregor, J. (1970) Tibet: A Chronicle of Exploration. Routledge & Kegan Paul, London, 373 p.

Mallet, F.R. (1875) On the geology and mineral resources of the Darjiling district and the western Duars. Mem. Geol. Surv. India, 11, 1-50.

Markham, C.R. (1878) A Memoir of the Indian Surveys, 2nd ed. Allen, London, 303 p.

Mason, K. (1955) Abode of Snow: A History of Himalayan Exploration and Mountaineering. Rupert Hart-Davis, London, 372 p.

McMahon, C.A. (1875) The Blaini group and the "Central Gneiss" in the Simla Himalayas. Rec. Geol. Surv. India, 10(4), 204-223.

McMahon, C.A. (1882) The geology of Dalhousie. Rec. Geol. Surv. India, 15, 34-51.

McMahon, C.A. (1895) The geological history of the Himalaya. Proc. Geol. Assoc., 14, 80-96.

Medlicott, H.B. (1864) On the geological structure and relations of the southern portion of the Himalayan ranges between the rivers Ganges and the Ravee. Mem. Geol. Surv. India, 3(2), 1-212.

Medlicott, H.B. (1866) The Alps and the Himalaya: a geological comparison. Quart. Jour. Geol. Soc. London, 24, 34-52.

Medlicott, H.B. (1882) Geology. In: The Himalayan Districts of the North Western Provinces of India (E.T. Atkinson, Ed.), 1(I), 111-168.

Medlicott, H.B. (1886) Boulder beds of Salt Range. Rec. Geol. Surv. India, 19(2), 97-142.

Medlicott, H.B., and Blanford, W.T. (1879-1887) Manual of the Geology of India. Part I. Peninsular India, 1-444; Part II. Extra-Peninsular Area, 445-817. Government of India Press, Calcutta.

Middlemiss, H.B. (1885) A fossiliferous series in the lower Himalayas. Rec. Geol. Surv. India, 18(2), 73-77.

Middlemiss, H.B. (1887-1888) Crystalline and metamorphic rocks of the lower Himalaya, Garhwal and Kumaon (sections I,II, and III). Rec. Geol. Surv. India, 20(3), 134-143; 20(4), 161-167; and 21(1), 11-28.

Middlemiss, H.B. (1890a) Physical geology of the Sub-Himalaya of Garhwal and Kumaon. Mem. Geol. Surv. India, 24(2), 59-200.

Middlemiss, H.B. (1890b) Geological sketch of Nani Tal, with some remarks on the natural conditions governing mountain slopes. Rec. Geol. Surv. India, 23(4), 213-234.

Middlemiss, H.B. (1896) The geology of Hazara and Black mountain. Mem. Geol. Surv. India, 26, 1-302.

Middlemiss, H.B. (1910a) The Kangra earthquake of 4th April 1905. Mem. Geol. Surv. India, 38, 1-409.

Middlemiss, H.B. (1910b) A revision of the Silurian-Trias sequence in Kashmir. Rec. Geol. Surv. India, 40(3), 206-260.

Misch, P. (1949) Metasomatic granitization of batholithic dimensions. American Jour. Sci., 247, 209-245, 372-405 and 673-705.

Molnar, P., and Tapponnier, P. (1975) Cenozoic tectonics of Asia: effects of a continental collision: Science, v. 189, p. 419-426.

Moore, D.T. (1982) Geological collectors and collections of the India Museum, London, 1801-79. Archiv. Natur. Hist., 10, 399-427.

Myer, K.E., and Brysac, S.B. (1999) Tournament of Shadows: The Great Game and the Race for Empire in Central Asia. Counterpoint, Washington, D.C., 672 p.

Nicolson, M.H. (1959) Mountain Gloom and Mountain Glory. Cornell University Press, Ithaca, 403 p.

Norin, E. (1946) Geological explorations in western Tibet. Report Sino-Swedish Expedition, 29, Aktiebolaget Thule, Stockholm, 229 p.

Odell, N.E. (1925) Observations on the rocks and glaciers of Mount Everest. Geogr. Jour., 66, 299-315.

Odell, N.E. (1948) Geological and some other observations in the Mount Everest region. In Mount Everest 1938 (H.W. Tilman, Ed.), Cambridge University Press, Cambridge, 143-154.

Oldham, R.D. (1893) A Manual of the Geology of India, 2nd ed. Government of India Press, Calcutta, 554 p.

Oldham T. (1883) A catalogue of Indian earthquakes from the earliest times to 1869. Mem. Geol. Surv. India, 19(3), 1-53.

Pascoe, E. (1919) The early history of the Indus, Brahmaputra and Ganges. Quart. Jour. Geol. Soc. London, 75, 138-155.

Pascoe, E. (1956-1964) A Manual of the Geology of India and Burma, 3rd ed. (3 vols.). Geological Survey of India, Calcutta.

Phillimore, R.H. (1945-1962) Historical Records of the Survey of India (5 vols.). Survey of India, Dehra Dun.

Pilgrim, G.E., (1910) Preliminary note on a revised classification of the Tertiary freshwater deposits of India. Rec. Geol. Surv. India, 40(3), 185-205.

Pilgrim, G.E. (1913) The correlation of the Siwaliks with mammal horizons of Europe. Rec. Geol. Surv. India, 43(4), 264-326.

Pilgrim, G.E. and West, W.D. (1928) The structure and correlation of the Simla rocks. Mem. Geol. Surv. India, 53, 1-140.

Powell, C.Mc.A. and Conaghan, P. J. (1973) Plate tectonics and the Himalayas. Earth Planet. Sci. Lett., 20, 1-12.

Rawat, I.S. (1973) Indian Explorers of the 19th Century: Account of Exploration of the Himalayas, Tibet, Mongolia, and Central Asia. Ministry of Information and Broadcasting, New Delhi, 228 p.

Reed, F.R.C. (1910) Cambrian fossils of Spiti. Palaeontol. Indica, Ser. 15, 7(1), 1-70.

Reed, F.R.C. (1912) Ordovician and Silurian fossils of the Central Himalayas. Palaeontol. Indica, Ser. 15, 7(2), 1-77.

Sahni, B. (1936a) The Karewas of Kashmir. Curr. Sci., 5(1), 10-16.

Sahni, B. (1936b) The Himalayan uplift since the advent of man: its cultural significance. Curr. Sci., 5(2), 57-61.

Said, E. (1979) Orientalism. Vintage Press, New York, 368 p.

Saklani, P.S. (Ed.) (1978) Tectonic Geology of the Himalaya (Current Trends in Geology 1). Today and Tomorrow's Printers and Publishers, Delhi, 340 p.

Saklani, P.S. (Ed.) (1979) Structural Geology of the Himalaya (Current Trends in Geology 2). Today and Tomorrow's Printers and Publishers, Delhi, 394 p.

Saklani, P.S. (Ed.) Current Trends in Geology 3: Shortening Structures in Eastern and North-Western Himalayan Rocks (by W. Schwan). Today and Tomorrow's Printers and Publishers, Delhi, 62 p.

Saklani, P.S. (Ed.) (1981) Metamorphic Tectonites of the Himalaya (Current Trends in Geology 4). Today and Tomorrow's Printers and Publishers, Delhi, 370 p.

Saklani, P.S. (Ed.) (1983) Himalayan Shears. Himalayan Books, New Delhi, 113 p.

Saklani, P.S. (Ed.) (1986) Himalayan Thrusts and Associated Rocks (Current Trends in Geology 9). Today and Tomorrow's Printers and Publishers, Delhi, 290 p.

Saklani, P.S. (Ed.) (1988) Seminar on Himalayan Crystallines, Metamorphics and Structures. Jour. Geol. Soc. India, 31(2), 185-251.

Saklani, P.S. (Ed.) (1989) Himalayan Mountain Building (Current Trends in Geology 11). Today and Tomorrow's Printers and Publishers, Delhi, 198 p.

Saklani, P.S. (Ed.) (1990) Metamorphism, Ophiolites and Orogenic Belts. (Current Trends in Geology 12). Today and Tomorrow's Printers and Publishers, Delhi, 416 p.

Saklani, P.S. (Ed.) (1991) Current Trends in Geology 13: Tectonic and Metamorphic Investigations of Kumaun-Garhwal-Himachal Lesser Himalaya (by I.C. Pande). Today and Tomorrow's Printers and Publishers, Delhi, 216 p.

Saklani, P.S. (1993) Geology of the Lower Himalaya (Garhwal). International Books and Periodicals, New Delhi, 254 p.

Saklani, P.S., and Verma, V.K. (Eds.) (1982) Himalaya: Landforms and Processes. Today and Tomorrow's Printers and Publishers, Delhi, 116 p.

Searle, M.P. (1991) Geology and Tectonics of the Karakoram Mountains. John Wiley, Chichester, 358 p.

Sengör, A.M.C. (1984) The Cimmeride Orogenic System and the Tectonics of Asia. Geol. Soc. America Sp. Paper 195, pp. 1-82.

Sengör, A.M.C. (1985) The story of Tethys: How many wives did Okeanos have? Episodes, 8, 3-12.

Severin, T. (1976) The Oriental Adventure: Explorers of the East. Little, Brown & Co., Boston, 240 p.

Sharma, C.K. (1990) Geology of Nepal Himalaya and Adjacent Countries. Kathmandu, 482 p.

Sinha, A.K. (1989) Geology of the Higher Central Himalaya. John Wiley, Chichester, 236 p.

Sorkhabi, R.B. (1995) From Eduard Suess to Emile Argand: Reflections on Tethys and India-Asia tectonics. Curr. Sci., 68, 853-858.

Sorkhabi, R.B. (1996) The Himalaya of Augusto Gansser. Jour. Geol. Soc. India, 47, 1-10.

Sorkhabi, R.B. (1997) Emile Argand and Tethyan tectonics of Asia. Himalayan Geol., 18, 17-32.

Sorkhabi, R.B., and Macfarlane, A. (1997) Himalaya and Tibet: mountain roots to mountain tops. Geol. Soc. America Sp. Paper 328, 1-7.

Sorkhabi, R.B., Valdiya, K.S., and Arita, K. (1999) Cenozoic uplift of the Himalayan orogen: chronologic and kinematic patterns. In: Geodynamics of the NW Himalaya (A.K. Jain and R.M. Manickavasagam, Eds.), Gondwana Research Group Mem. 6, 189-206.

Srivastava, R.A.K. (Ed.) (1984) Sedimentary Geology of the Himalaya (Current Trends in Geology 5). Today and Tomorrow's Printers and Publishers, Delhi, 246 p.

Staford, R.A. (1984) Geological surveys, mineral discoveries, and British expansion, 1835-71. Jour. Imperial Commonwealth Hist., 12, 5-32.

Staford, R.A. (1989) Scientist of Empire: Sir Roderick Murchison, Scientific Exploration and Victorian Imperialism. Cambridge University Press, Cambridge, 293 p.

Stocklin, J. (1980) Geology of Nepal and its regional frame. Jour. Geol. Soc. London, 137, 1-34.

Stokes, E. (1959) The English Utilitarians and India. Clarendon Press, Oxford, 350 p.

Stoliczka, F. (1886a) Geological sections across the Himalayan mountains from Wangtu bridge on the river Sutlej to Sungdo on the Indus, with an account of the formations in Spiti accompanied by a revision of all known fossils from that district. Mem. Geol. Surv. India, 5(1), 1-154.

Stoliczka, F. (1866b) Summary of the geological observations during a visit to the provinces Rupshu, Karnag, south Ladak, Zanskar, Suruoo and Dras of western Tibet. Mem. Geol. Surv. India, 5(3), 337-354.

Strachey, H. (1853) On the physical geography of western Tibet. Jour. Royal Geogr. Soc., 23, 1-69.

Strachey, R. (1851a) On the geology of part of the Himalaya mountains and Tibet. Quart. Jour. Geol. Soc. London, 7, 292-310.

Strachey, R. (1851b) On the physical geography of the provinces Kumaun and Garhwal in the Himalaya mountains and the adjoining part of Tibet. Jour. Royal Geogr. Soc., 21, 57-85.

Strachey, R. (1900) Narrative of a journey to the lakes Rakas-Tal and Manasarowar in western Tibet undertaken in September 1848. Geogr. Jour., 15, 150-170, 243-264 and 394-415.

Styles, S. (1970) The Forbidden Frontiers: The Survey of India from 1765 to 1949. Hamilton, London, 160 p.

Suess, E. (1904-1918) The Face of the Earth (4 vols., translated by H.B.C. Sollas). Clarendon Press, Oxford.

Sykes, P.M. (1939). Explorers All, Famous Journeys in Asia. G. Newnes, London, 426 p.

Thakur, V.C. (1992) Geology of Western Himalaya. Pergamon, Oxford, 426 p.

Theobald, W. (1880) The Kumaon lakes. Rec. Geol. Surv. India, 13(3), 161-175.

Troll, C. (1960) The work of Alexander von Humboldt and Carl Ritter: A centenary address. Advancement of Sci., 64, 441-452.

Unsworth, W. (2000) Everest: A Mountaineering History, 3rd ed. Mountaineer Books, Seattle, 736 p.

Valdiya, K.S. (1980) Geology of Kumaun Lesser Himalaya. Wadia Institute of Himalayan Geology, Dehra Dun, 292 p.

Valdiya, K.S. (1984) Aspects of Tectonics: Focus on South-Central Asia. Tata McGraw-Hill, New Delhi, 320 p.

Valdiya, K.S. (1998) Dynamic Himalaya. Universities Press (India), Hyderabad, 178 p.

Valdiya, K.S. and Bhatia, S.B. (Eds.) (1980) Stratigraphy and Correlation of Lesser Himalayan Formations. Hindustan Publishing Corporation, New Delhi, 340 p.

Waagen, W. (1878) On the geological distribution of the fossil organisms in India. Rec. Geol. Surv. India, 11(4), 267-311.

Wadia, D.N. (1928) The geology of Poonch state (Kashmir) and adjacent portions of the Punjab. Mem. Geol. Surv. India, 51(2), 185-370.

Waida, D.N. (1931) The syntaxis of the northwest Himalaya: its rocks, tectonics and orogeny. Rec. Geol. Surv. India, 65(2), 189-220.

Wadia, D.N. (1932) The Tertiary geosyncline of north-west Punjab. Quart. Jour. Geol. Min. Soc. India, 4(3), 69-96.

Waida, D.N. (1933) Note on the geology of Nanga Parbat (Mt. Diamir) and adjoining portions of Chilas, Gilgit district, Kashmir. Rec. Geol. Surv. India, 66(2), 212-234.

Wadia, D.N. (1934) The Cambrian-Trias sequence of north west Kashmir (parts of Muzaffarabad and Barumula districts). Rec. Geol. Surv. India, 68(2), 121-176.

Wadia, D.N. (1937) The Cretaceous volcanic series of Astor-Deosai, Kashmir, and its intrusions. Rec. Geol. Surv. India, 72(2), 151-161.

Wadia, D.N. (1938) The structure of the Himalayas and the north Indian foreland. Proc. 25th Indian Sci. Congr. (Presidential address to the geology section). Pt. II, 91-118.

Wadia, D.N. (1942) The making of India. Proc. 28th Indian Sci. Congr. (Presidential address to the geology section). Pt. II, 3-23.

Wadia, D.N. (1974) Geology of India, 4th ed. Tata McGraw-Hill, Delhi, 508 p.

Wager, L.R. (1934) A review of geology and some new observations. In Everest 1933 (H. Ruttledge, Ed.), Hodder and Stoughton, London, 312-337.

Wager, L.R. (1937) The Arun river drainage pattern and the rise of the Himalaya. Geogr. Jour., 89, 239-250.

Wager, L.R. (1939) The Lachi series of N. Sikkim and the age of the rock forming Mount Everest. Rec. Geol. Surv. India, 74(2), 171-188.

Waller, D.J. (1990) The Pundits: British Exploration of Tibet and Central Asia. University Press of Kentucky, 327 p.

West, W.D. (1934) Recent advances in Indian geology. 3. The geology of the Himalayas. Curr. Sci., 3(6), 286-291.

West, W.D. (1939) Structure of the Shali "window" near Simla. Rec. Geol. Surv. India, 74(1), 133-163.

West, W.D., Wadia, D.N. and Auden, J.B. (1931) Hazara, Simla hills correlation. Rec. Geol. Surv. India, 65, 125-132.

Wynne, A.B. (1874) Notes on the geology of the neighbourhood of Mari hill station in the Punjab. Rec. Geol. Surv. India, 7, 64-74.

Wynne, A.B. (1878) On the geology of the Salt range in the Punjab. Mem. Geol. Surv. India, 14, 1-314.

MAGNETOTELLURIC STUDIES IN NORTHWESTERN HIMALAYA

Gautam Gupta, S.G. Gokarn, C.K. Rao and C. Selvaraj

Indian Institute of Geomagnetism, Colaba, Mumbai- 400 005

ABSTRACT

This paper describes the theoretical principles underlying the magnetotelluric technique with two case studies from the seismically active regions in the northwestern Himalaya.

Magnetotelluric studies in the Siwalik sediments have delineated about 5 km thick sedimentary sequences in the Mohand-Ramnagar region in the foothills of the Himalayas. Two layers of Siwalik sediments were observed underlying the alluvium. The Upper Siwaliks were resistive (30-200 Ù-m) and had a thickness of 700-1000 m whereas; the Middle and Lower Siwaliks were 3000-4000 m thick and had a resistivity of 9-15 Ù-m. These studies could not differentiate between the Middle and Lower Siwaliks, presumably because both these sequences have similar resistivities. A deep crustal conductive intrusive was delineated below the central part of the profile between Gendikhata and Saneh near Kotdwar. Magnetotelluric studies over the Pang-Phobrang profile indicated a high resistive crust. A vertical conductive feature was delineated on the south of Tso Morari crystallines. A vertical conductive feature at shallow depths separates the Tso Moraris from the Ladakh batholith on the NE. A fault zone was delineated below the stations Lukung and Phobrang.

INTRODUCTION

The highest mountain chain in the world- the Himalaya has attracted attention for its uniqueness and the subject of its origin is an active area of research. This is a seismically active region and the crust underneath is undergoing an extensive folding and deformation from times immemorial. Our concepts of the Himalaya have undergone unprecedented advances during the last few decades. However, a consistent picture of the substructure of the Himalaya has been slow in coming.

The knowledge of the deep crustal structure is of paramount importance in understanding the crustal evolutionary processes involved in the formation of the present day crust. Although the geological studies do provide a good insight into these processes, many ambiguous situations arise where the need for the deeper crustal information is desired. The various geophysical techniques used for the exploration of the deep crust include the gravity, seismic and the electromagnetic induction studies. The electromagnetic (EM) studies probe the diffusion of the electromagnetic waves into the earth's interior and yield the lateral and vertical electrical resistivity distribution in the earth's crust. There are two major variants of EM techniques used in the geoexploration: the magnetovariational (MV) and magnetotelluric (MT) studies.

In the magnetovariational technique, an array of magnetometers are operated, each of which records all the three components of the magnetic field variations along the magnetic N-S, E-W and the vertical (denoted as H_x, H_y and H_z). The dependency of the vertical component (H_z) on the two horizontal components (H_x and H_y) is studied through the transfer functions and the corresponding induction arrows are plotted. The magnitude and direction of these induction arrows are then used to determine the lateral resistivity contrasts that are then interpreted to delineate the conductive zones in the earth's interior.

Magnetotelluric technique is a method for determining the electrical conductivity distribution in the earth's interior from the surface measurements of the naturally occurring electric and magnetic field variations (Vozoff, 1972). The theoretical concepts underlying the MT technique are derived for a plane horizontally stratified earth on which a plane unpolarised EM wave front is normally incident. Here the two horizontal components of the electric field (along the magnetic N-S and E-W, denoted as E_x and E_y, respectively) and three orthogonal components of the magnetic field (along magnetic N-S, E-W and vertical, denoted as H_x, H_y and H_z, respectively) are measured synchronously in the form of time series. With this information, the impedance tensor (Z) is computed. The impedance tensor is related to the electric and magnetic field components through the vector relation,

$E = Z\ H$, or

$$\begin{pmatrix} E_x \\ E_y \end{pmatrix} = \begin{pmatrix} Z_{xx} & Z_{xy} \\ Z_{yx} & Z_{yy} \end{pmatrix} \begin{pmatrix} H_x \\ H_y \end{pmatrix}$$

The impedance tensor is the central parameter in the MT studies and contains all the available geoelectric information. Thus, the impedance tensor can be rotated to determine the regional geoelectric strike direction. It is also used to determine the apparent resistivity of the subsurface that can then be used to obtain the geoelectric structure. The penetration of the EM wave in the interior of the earth is controlled among other parameters by the frequency of the EM wave under consideration. Thus, the high frequencies penetrate only the shallow strata and as the frequency is decreased, the depth of penetration increases. In regions of conductive sedimentary covers, longer periods are required for significant depths, whereas in the older exposed basement rocks, which normally have very high resistivities, large depths can be penetrated using not so low frequencies. Usually with a frequency band of 100 – 0.0001 Hz, the crust and part of the upper mantle can be generally studied in most of the regions on the surface of the earth. The lateral inhomogeneities of interest in the earth's interior are normally located in the lithosphere and thus this frequency range suffices for most of the MT studies. In recent times, need is felt for lower frequency studies, especially in the Himalayan region and the Tibetan plateau, where the crustal thickness may be in excess of 70 km and thus the conventional frequency band mentioned above may not be adequate to probe the crustal depths.

As mentioned earlier, the earth is assumed as horizontally layered, which is seldom the case. There are many faults criss-crossing the continental regions and due to the presence of mineralized water in the fractures associated with these faults, conductive structures appear in many parts of the continent. Some of the lineaments may also have markedly different resistivities on either side of them. In all these cases, the assumption of horizontally layered earth are violated and thus the geoelectric structure becomes two or three-dimensional. In MT studies, this problem is addressed to by the rotation of the impedance tensor in the spatial coordinates. When the geoelectric structure is one-dimensional (1-D), i.e. no lateral resistive contrasts exist; the diagonal elements of the impedance tensor Z_{xx} and Z_{yy} are zero and the off-diagonal elements, Z_{xy} and Z_{yx}, are equal in magnitude and are opposite in sign for all the rotations of the impedance tensor. In the two-dimensional (2-D) case, the impedance tensor elements in general are non-zero. However, if the axes of the two-dimensionality coincide with the axes of measurement, the diagonal elements of the impedance tensor (Z_{xx} and Z_{yy}) can be reduced to zero with a proper rotation. The off-diagonal elements (Z_{xy} and Z_{yx}) in this case are however not equal. When the lateral resistivity variation is more complex and exists in more than one horizontal direction (three-dimensional earth), it may not be possible to reduce Z_{xx} and Z_{yy} to zero by any rotation in spatial coordinates.

Once the dimensionality information is obtained, the apparent resistivity and phase curves are rotated parallel and perpendicular to the geoelectric strike. The apparent resistivities and phases parallel to the geoelectric strike are known as the E-polarization values and those perpendicular to the strike, the H-polarization values. The apparent resistivity and phase curves are then visually inspected to identify the possible number of layers as well as the possible locations of the conductive/resistive lateral contrasts. Generally one-dimensional inversion scheme can be used to quickly obtain the preliminary geoelectric structure, which can then be refined using two/three-dimensional model, or inversion schemes to obtain the final geoelectric cross-section. The geoelectrical model thus obtained is then integrated with the available geological and geophysical data from other techniques, normally the surface geology and tectonics, gravity, magnetic, DSS and heat flow studies. Thus a combined interpretation of the results of different techniques helps to reduce the degree of nonuniqueness and evolving a geophysical model that is closer to the true structure.

Over the past decade, the experimental as well as some of the theoretical model studies on the MT response functions have shown that electrical inhomogeneities which are commonly encountered on the surface of the earth have strong influence on the impedance tensor to the extent of corrupting the geoelectric models even at deeper strata. It was also observed that the presence of these distortions do not in any way obliterate the information on the deeper structures and the data can still be interpreted after the effects of the distortions caused by the near-surface inhomogeneities. Thus the importance of identifying and correcting the effects of these near-surface distortions was realized. The theoretical treatment of the tensor decomposition procedure was developed by Groom and Bailey (1989).

Lateral variations in resistivity at or near the surface have an effect on MT data. The MT static shift effect may be manifested by a parallel shift of the two apparent resistivity curves on log-log scale without any change in the phase curves. There are no theoretical prescriptions available to correct the static shift effect. However, there are many empirical methods employed to correct for the static shift such as, the key layer approach, spatial filtering, global curve matching, central-loop transient EM sounding, continuous profiling technique etc. In the key layer approach (Jones, 1988), the depth to one of the resistive interphases is assumed to be known from some other technique (e.g., well logging) at one of the points on the survey profile. The model parameters are constrained using this information and the regional structure is obtained. Normally the depth to the bottom of resistive layers is a good constraining parameter, because it is a well-resolved parameter in MT studies. This approach is possible only when the well-logged data are available in the vicinity of the survey profile. In the spatial filtering approach, the surface inhomogeneities are assumed to be randomly distributed and the static shifts are also expected to be random. Thus, the apparent resistivities in the high frequency end of the available spectrum at all the stations are spatially filtered to remove the short wavelength spatial features. The correction factors obtained at high frequencies are then applied to the entire frequency band and the corrected data are interpreted. The global curve approach is based on the fact that when all the MT sounding curves in the horizontally layered (isotropic) crustal regions are extended to very long periods (> 10000 s), the apparent resistivities at any place on the globe have similar values (e.g., at 10000 s, ñ$_a$= 100±10 Ù-m). This is because the asthenosphere and core are devoid of any major lateral resistive inhomogeneities. The apparent resistivity curves are thus shifted such that the low frequency asymptote passes through this point. One or more of these techniques are used for correcting the static shift, depending upon the nature of the geoelectric sub-structure and availability of information.

Case Studies

The MT method came into prominence following the formulation of the theoretical basis by Tikhonov (1950) and Cagniard (1953). With the rapid strides witnessed in the electronics and high-speed computers, the MT technique has become more viable as a survey tool. Availability of high sensitivity magnetometers and data recorders made a remarkable contribution towards the development of reliable field worthy magnetotelluric systems. This technique is being actively pursued nowadays world over.

The seismic activity is a result of relative displacement of crustal blocks across the fault zones or it may be the result of rejuvenation of dormant faults or creation of new faults. In all the cases, the frictional forces over the large contact surfaces may lead to heat generation. The fractured regions in the presence of mineralized water increase their conductivity in relation to the surroundings. Also, the conductivity normally increases with the increase in temperature. Thus it is felt that there is a close link between the electrical conductivity and seismic activity. The MT studies thus provide good support for enhancing our understanding of the crustal evolutionary processes, particularly in geologically and structurally complex regions.

The Himalaya is a result of the collision between the Indian plate and the Eurasian plate. The tectonic activities in the Himalayan range of mountains and its foothills have resulted in a large number of faults and folds in this region. This orogenic belt is the highest mountain range with an average height of about 5000 m and extends about 2500 km in length. The Himalayan tectonic framework consists of four lithotectonic subprovinces (Valdiya, 1964). All these subprovinces extend from east to west and are sharply delimited by many boundary thrusts such as the Indus Tsangpo Suture Zone (ITSZ), the Trans Himadri Fault, the Main Central Thrust (MCT), the Main Boundary Thrust (MBT) and the Himalayan

Frontal Fault. Subsequent to the intercontinental collision, the subduction of the Indian plate is known to have occurred at the ITSZ. However, the seismicity studies indicate that the ITSZ is essentially an aseismic contact at present (Ni and Barazangi, 1984). Most of the subsequent subduction seems to be occurring along the MCT and MBT (Gansser, 1964). In this section, two case studies are presented from the Himalayan region.

Mohand-Ramnagar Profile

The Himalayan foreland is the product of the tectonic processes that have been taking place in the orogenic belt since the Cenozoic era. Deformation and sedimentation are the active processes in the Himalayan foreland. The Siwalik basin, a part of the Himalayan foreland, was formed as a foredeep in front of the newly risen Himalaya during the Middle Miocene orogeny and was the site of deposition of the Siwalik sediments. The Siwalik system is 5-6 km thick and is composed mainly of sandstones, grits, conglomerates, clays and silts, which were deposited in a brackish and freshwater environment (Gupta et al., 1994).

Tectonic activity leading to the collision between the Indian and the Eurasian plates have resulted in a large number of fractures, thrust and folds. Magnetovariational studies (Arora, 1990) conducted over a rectangular array of 24 stations in this region indicated the presence of a strong elongated conductive anomaly (Trans Himalayan conductor, THC) that was interpreted as the extension of the Aravallis in north-north easterly direction (Fig. 1). This conductive structure coincides with the break in negative isostatic gravity contours, which otherwise follow the strike of the Himalaya (Qureshy, 1969). The spatial distribution of earthquake epicenters (Khattri and Tyagi, 1983) superposes the area of the anomaly in a broad band. The conductive region is also associated with a transverse zone of high seismicity (Kaila and Narain, 1976) indicating its linkage with neotectonic activity or to an ancient structure that may be reactivating now.

Based on the existing geologic and geophysical data (Vozoff, 1984), MT studies were conducted over a 150 km long linear profile in the Himalayan foothills between Mohand and Ramnagar across the THC with an intention to understand the electrical properties of the substructure that may have an important role in characterizing the tectonic activity in this region. Data were collected at 13 stations with an interstation spacing of 8-12 km in the frequency range 0.01-100 Hz using the SPAM (Short Period Automatic Magnetotelluric) system. The MT station location map is shown in Fig. 2.

The apparent resistivities and phase indicated predominantly one-dimensional electrical behaviour at all stations. However, at frequencies below 0.1 Hz, stations Kotdwar (KOT), Kangri (KAN), Aurangabad (AUG) and Bandarjudh (BAN) showed some deviation from one dimensionality. To determine the possible geoelectric strike direction, the impedance tensors were rotated at each frequency for all the four stations mentioned above. The major and minor axes were predominantly aligned along N 80° W and N 10° E, respectively, which is approximately perpendicular and parallel to the conductive anomaly delineated using MV studies. In view of this fact, the apparent resistivities parallel to the conductive anomaly were assumed to be the E-polarization values and those perpendicular, the H-polarization values. The E-polarization values were used for 1-D inversion because of the fact that for a

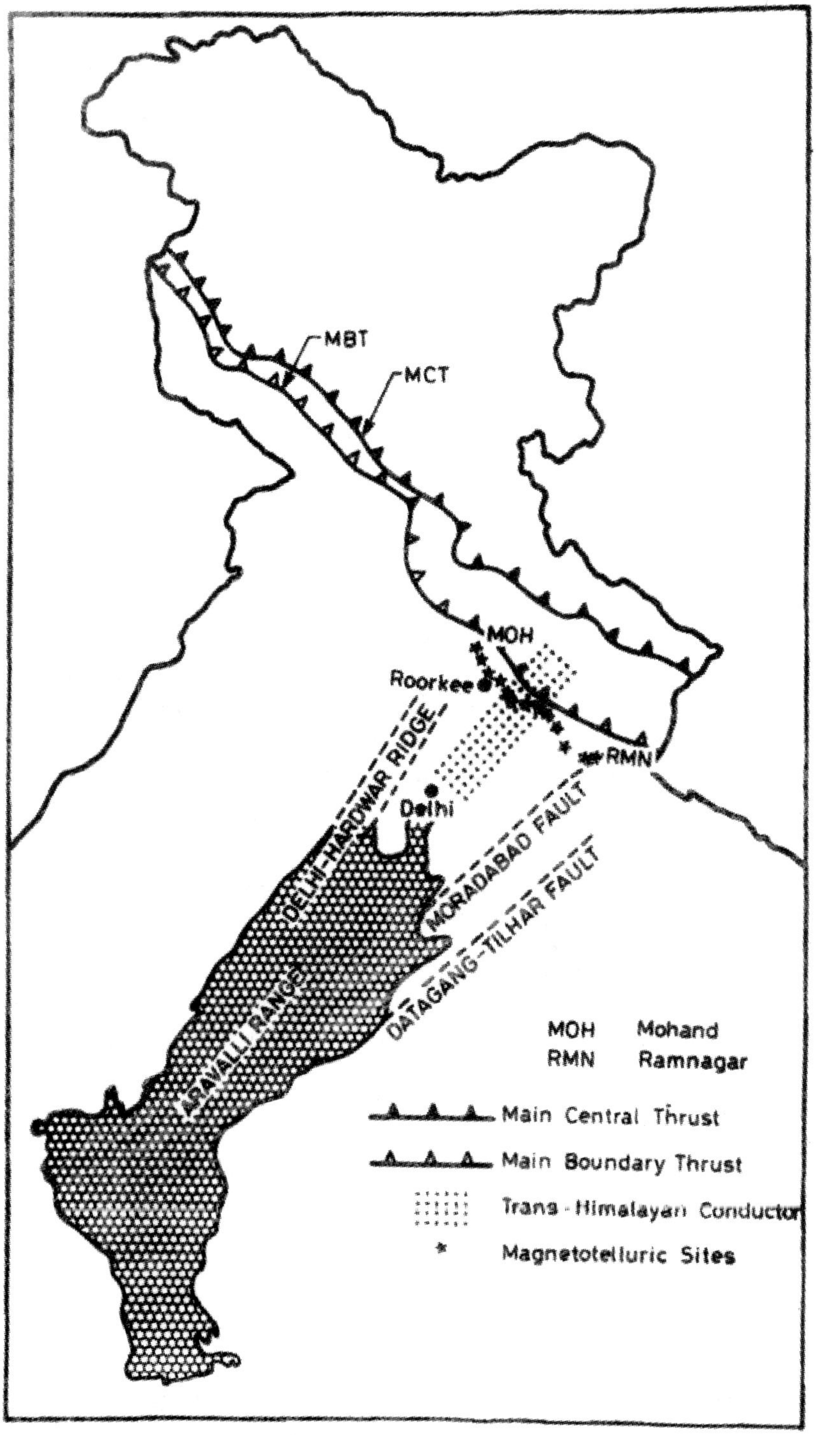

Figure 1: Map showing the major tectonic elements in and around the survey region.

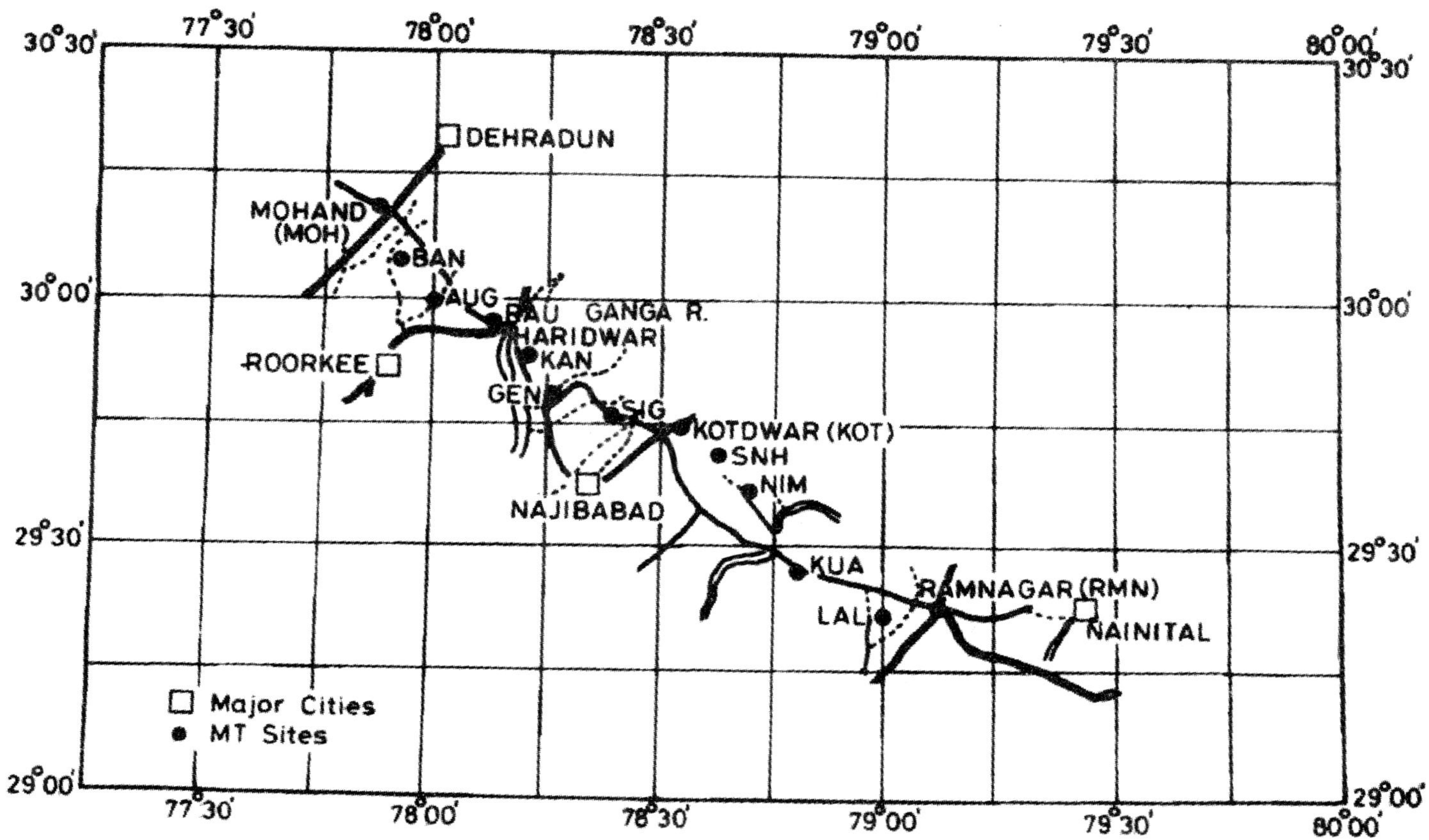

Figure 2: Location map showing the magnetotelluric stations over the Mohand-Ramnagar profile.

strictly two-dimensional situation, parallel resistivity and phase estimate are the most representative one to invert.

The depth-resistivity profiles were computed using the Marquardt inversion method (Marquardt, 1963) as described by Cerv and Pek (1981). The period dependence of apparent resistivity indicated that the geoelectric substructure is four layered at all the stations except between Gendikhata (GEN) and Saneh (SNH), where only three layers were observed. The depth-resistivity estimates for different stations were then plotted together and layers with similar resistivities were interpolated to obtain the geoelectric cross-section (Fig. 3). The solid lines below each station are the resistivity discontinuities obtained from the inversion scheme and the dashed lines are the interpolations between the adjacent stations.

The top layer in the survey region was conducting with a resistivity of 6-20 Ù-m and thickness of about 50 m overlying a more resistive (30-200 Ù-m) second layer having a thickness of 700-1000 m. The third layer had a resistivity of 9-15 Ù-m. At all stations except those between GEN and SNH, this layer was about 3-4 km thick, overlying a more resistive basement with a resistivity of 30-250 Ù-m. The resistive layer below stations GEN and SNH was absent. The resistive basement was observed at a depth of 5-8 km from the surface. The top conducting layer with a thickness of about 50 m was understood to be due to the alluvial and post Siwalik deposits, overlying about 1 km thick Upper Siwalik sediments. The conductive layer below the second layer was presumably due to the Middle and Lower Siwaliks, deposited over a granitic basement. The present studies could not delineate the boundary between the Middle and Lower Siwaliks because both have similar resistivities. It is pertinent to note here that both these sedimentary layers were deposited under similar environmental conditions (brackish and freshwater environment, Krishnan, 1982). The results of the MT studies presented here are in good agreement with those reported from the stratigraphic studies (Prasad, 1976), sedimentological studies (Kumar and Nanda, 1989), palaeontological studies (Wadia, 1975) and aeromagnetic studies (Agocs, 1956).

As mentioned earlier, the resistive bottom layer (Fig. 3) could not be delineated below the stations GEN to SNH. It was also observed that at the stations KOT, KAN, AUG and BAN, the apparent resistivity curves show some 2-D effects at low frequencies with a strike direction along N 10° E. Although the 2-D behaviour was not sufficiently strong for a quantitative interpretation, it could be explained by assuming a deep-seated conductor aligned along N 10° E. A conductive zone has been reported in this region through MV studies (Arora, 1990). It was envisaged that the top of the conducting body is about 8 km below the surface and width of about 25 km on its west, whereas on the east, the top surface is at a depth of 12 km and a width of 38 km. The conductor coming as close as 8 km from the top reduces the thickness of the resistive bottom layer to about 5 km. The basement at depths of 5 km sandwiched between a conductive layer above and a massive conductor below does not provide adequate conductance contrast to appear as a separate layer and seems fortuitously joined to the conductor above. The present survey profile being almost perpendicular to the proposed conductive zone, the width of this zone was estimated to be 50-60 km (i.e., the distance between stations GEN and SNH). This observation gives an indirect support to the presence of the Trans - Himalayan conductor. However due to limited frequency band employed, it was not possible to carry out further quantitative interpretation.

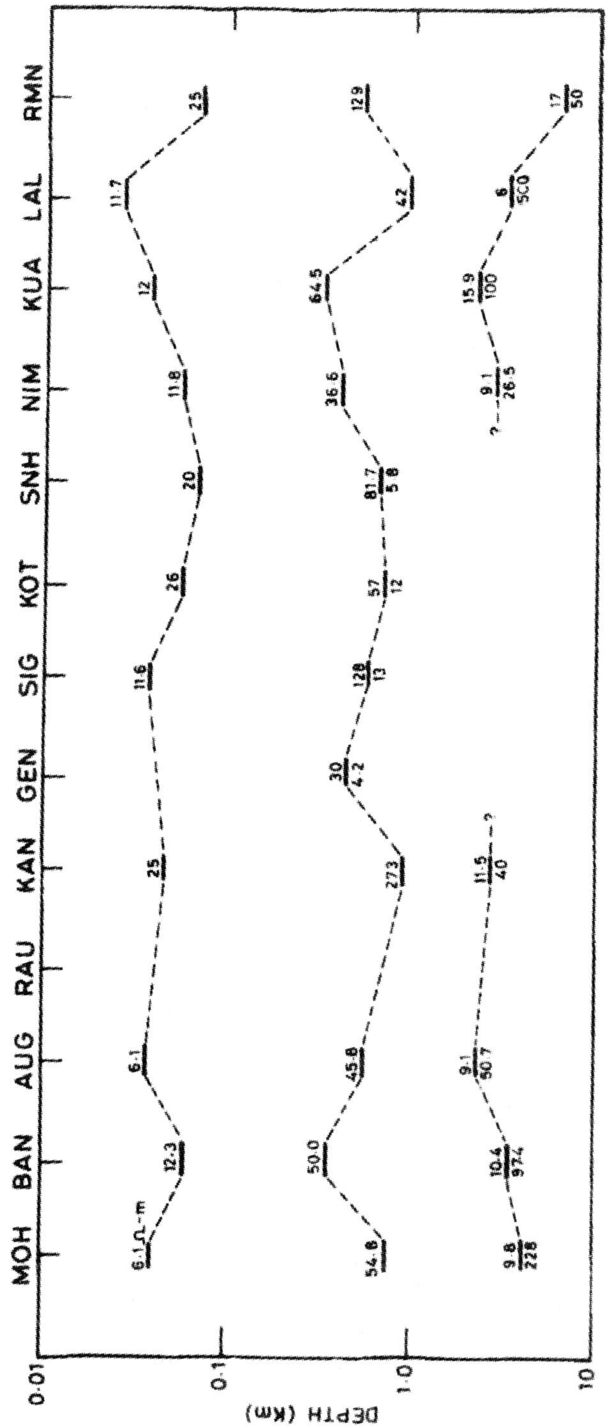

Figure 3: Geoelectric substructure of the Mohand-Ramnagar profile obtained from 1-D inversion.

Pang-Phobrang Profile

Magnetotelluric studies were conducted at 14 stations over the Pang-Upshi-Tangtse-Phobrang profile in the Ladakh-Karakoram mountains. The studies were conducted in the frequency range 1000-0.0005 Hz using the Phoenix V5 MT system. The survey profile passes from south to north through the Tso Morari Crystallines incorporating the Taglang La formation on the south, the Indus Tsangpo Suture Zone (ITSZ), Ladakh Batholith, Shyok Suture Zone (SSZ), Pangong metamorphics, and the associated Karakoram Batholith. The geological map of the survey region along with the magnetotelluric stations is shown in Fig. 4.

The study area is located in the high mountain regions with a rugged topography and thus, the station locations as well as the spacings were to a large extent governed by the availability and accessibility of the suitable survey sites. Of the MT data collected at 14 stations, data at one of the stations (Pang) were strongly contaminated by the possible topographic distortions and vehicular noise and hence was not considered for analysis purpose. The time series of the electric and magnetic field variations were analyzed using a combination of fast Fourier transform and Cascade decimation (Wight and Bostick, 1980) and the impedance tensor was obtained.

The rotation of the impedance tensor indicated a geoelectric strike direction of N 40° W and S 40° E. In view of the fact that most of the features including the ITSZ and the SSZ are coinciding with the N 40° W direction, this was assumed to be the geoelectric strike direction. For further analysis, the stations were projected on a straight line perpendicular to the strike (N 50° E). In such a projected map, station Dabruk (DBK) is located between Tangtse (TAN) and Lukung (LUK) and not between Tsoltak (TSO) and TAN as seen in the road maps. The apparent resistivities and phases in the E-polarization were inverted using an Occam 1-D inversion scheme (Constable et al., 1987) and the preliminary geoelectric cross-section was obtained by plotting together the depth-resistivity data on the projected line of stations. This was further refined using a 2-D inversion scheme (Rodi and Mackie, 2001) based on the finite element techniques. The geoelectric cross-section is shown in Fig. 5.

On the southwestern end, the profile passes through the domal structure of Tso Morari Crystallines (TMC). The resistivity structure shows a predominantly high resistive crust (> 1000 Ω-m). A vertical conductive feature was delineated on the south of TMC, which may coincide with the tectonic boundary with the overlying Tibetan Sedimentary Zone. The ITSZ, which appears as a vertical conductive feature at shallow depths, separates the TMC from the Ladakh Batholith on the NE. From the complex conductivity pattern associated at depth below the ITSZ and surrounding region, it is evident that the ITSZ is associated with strong crust- mantle interactions. This is also corroborated by the high heat flow here. From the geological and tectonic setting of this region, the high conductivity formations below the ITSZ may be closely related to the Puga geothermal regime, about 30 km SE of the MT survey profile. Further north, the ITSZ takes the form of a crustal detachment below the Ladakh Batholith. As expected, the resistivity of the Ladakh Batholith is more than 1000 Ω-m.

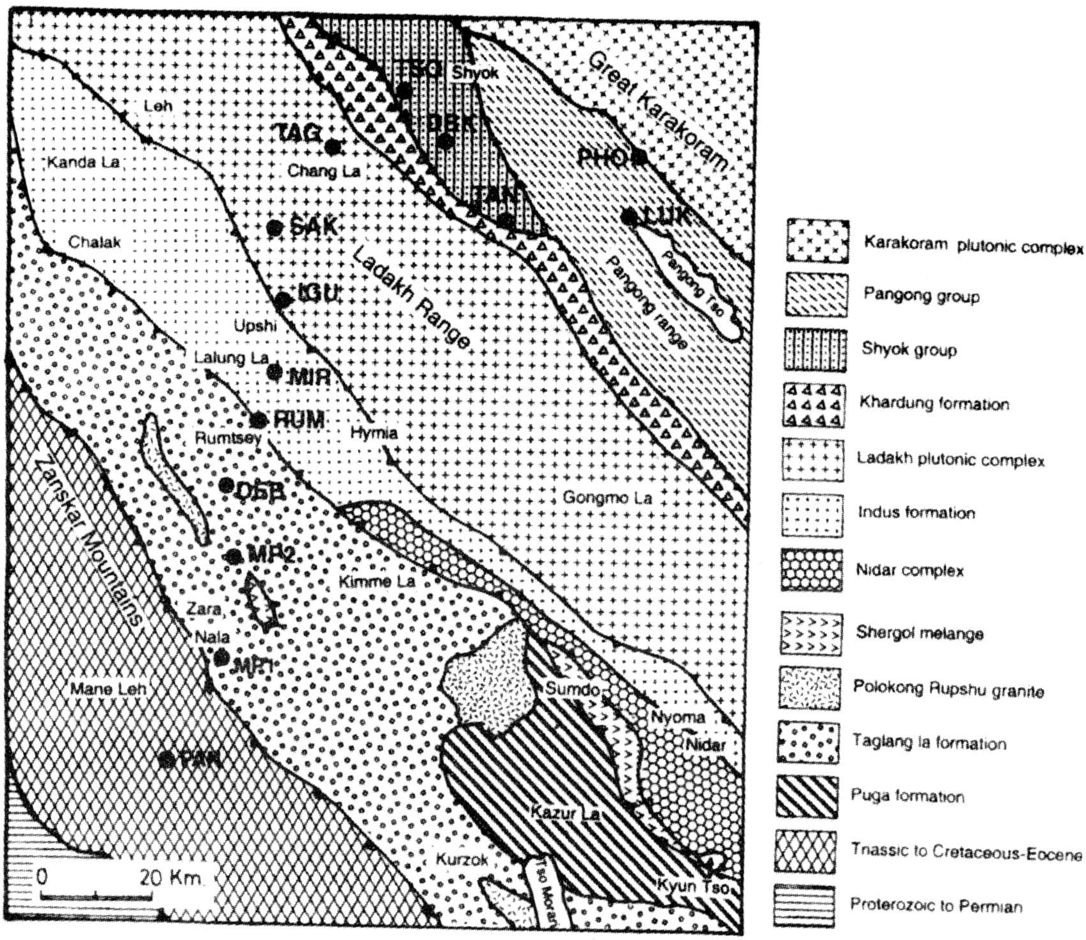

Figure 4: Geological map of Pang-Phobrang profile showing the magnetotelluric stations.

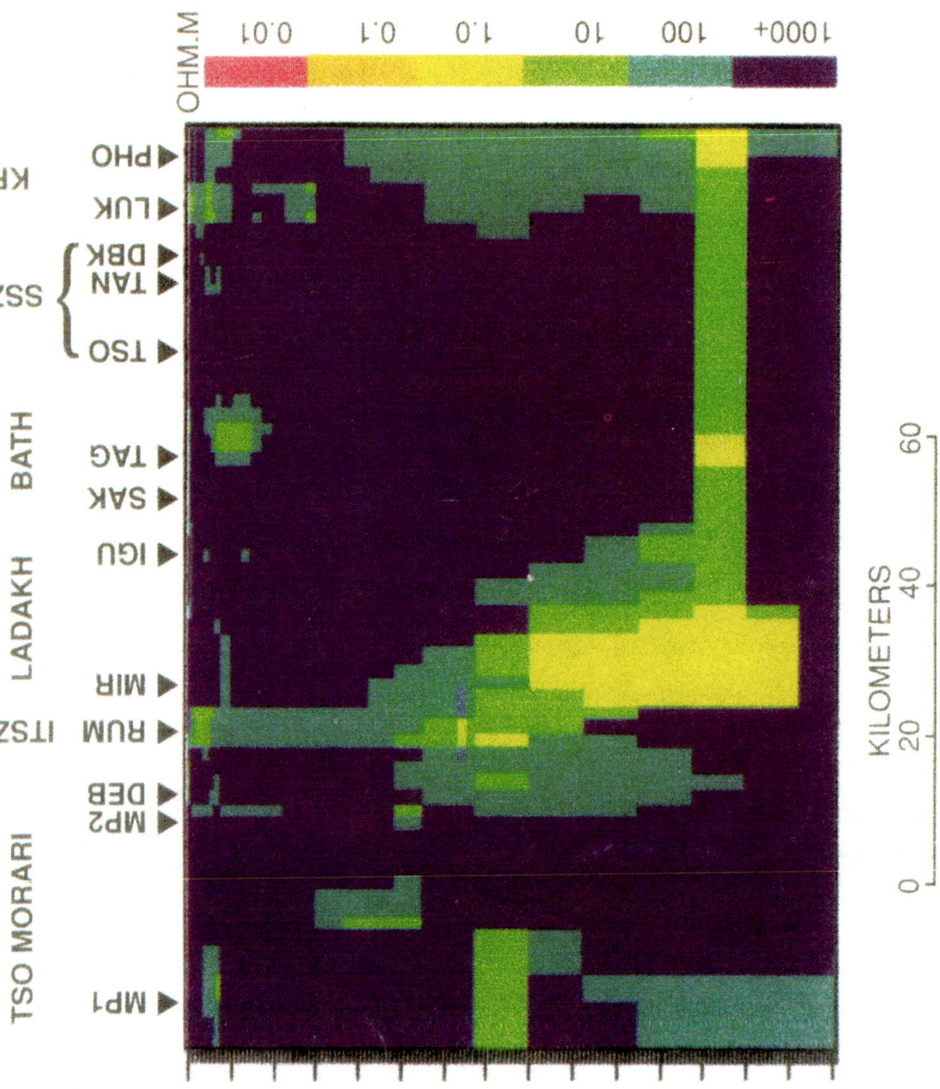
OHM.M
0.01
0.1
1.0
10
100
1000+
KF
PHO
LUK
SSZ
DBK
TAN
TSO
BATH
TAG
SAK
LADAKH
IGU
MIR
ITSZ
RUM
DEB
MP2
TSO MORARI
MP1
KILOMETERS
0
20
40
60
0
1
2
3
4
5
6
7
8
9
10
11
12
13
14
15

Further north, the profile terminates within the low-grade metamorphics of the Karakoram mountains near Phobrang. A fault zone was delineated below the stations LUK and PHO that is a relatively deep-rooted conductive feature, extending upto at least 12 km. This may coincide with the boundary of low-grade metamorphics near Phobrang and amphibolite facies rocks of Lukung. A well-constrained shallow conductive feature with resistivity of 10 W-m is delineated below the station TAG on the south of SSZ. Ophiolitic exposures are reported in the SSZ region (station TSO). However the association of the two in the conductivity framework is not apparent at present. The distortions due to the topography effects have not been considered in the present interpretation because the response functions at all the stations (except Pang, which was not included in the interpretation) were free from any significant phase distortions. All the stations were located in a narrow altitude range of 4.5 to 5 km and hence the topographic effects may not contribute to any major distortions. Furthermore, all stations are located in regions with slopes of less than 30 over a distance of more than 500 m from the stations.

SUMMARY

Magnetotelluric technique is a tool for exploring the deep interior of the earth. This is an interpretive method based on monitoring the components of the electromagnetic field on the earth's surface to obtain the information on the electrical resistivity of the various deep-seated formations in the earth's crust and mantle. The results of the magnetotelluric studies in the Mohand-Ramnagar region and the Pang-Phobrang region in the Himalaya have been discussed in this paper.

In the Mohand-Ramnagar region, the MT studies indicated a four-layered electrical substructure in this region except at stations between GEN and SNH where only three layers could be delineated. The observational results were understood as 50 m thick alluvial and post Siwalik sediments overlying an about 700-1000 m thick layer of Upper Siwalik sediments. The Middle and Lower Siwaliks together had a thickness varying between 3-4 km. The total thickness of the Siwalik sedimentary sequence was estimated to be in the range of 5-8 km. A 2-D behaviour was observed at KOT, KAN, AUG and BAN. Although this was not sufficiently strong for a quantitative interpretation, it could be explained on the basis of a 50 km wide conducting zone aligned parallel to N 10° E located between GEN and SNH. Similar observations have been reported from MV studies wherein the conductive zone has been explained due to the extension of the Aravalli formations in a north-northeasterly direction.

Over the Pang-Phobrang profile in the Ladakh and Karakoram mountains, the geoelectric structure shows a predominantly high resistive crust with resistivities more than 1000 Ω-m. A vertical conductive feature was delineated on the south of TMC. The ITSZ appears as a vertical conductive feature at shallow depths separating the TMC from the Ladakh Batholith. A deep-rooted fault zone was delineated below stations LUK and PHO, extending up to depths of 12 km. A 10 Ω-m conductive feature was delineated below station TAG on the south of the SSZ. These studies have established the utility of MT soundings in the Himalaya to supplement other geophysical studies.

REFERENCES

Agocs, W.B., 1956. Airborne magnetometer survey: 1. Indo-Gangetic Plains, 2. Rajasthan. Report submitted to Govt. of India, (Unpublished).

Arora, B.R., 1990. Magnetometer array studies in India: Present status, data interpretation and assessment of numerical modeling results. *Proc. Indian Acad. Sci. (Earth Planet. Sci.), 99*, 693-716.

Cagniard, L., 1953. Basic theory of the magnetotelluric method of geophysical prospecting. *Geophysics, 18*, 605-635.

Cerv, V. and Pek, J., 1981. Numerical solution of the two-dimensional inverse geomagnetic induction problem. *Stud. Geophys. Geod., 25*, 69-80.

Constable, S., Parker, R.L. and Constable, C., 1987. Occam's inversion: A practical algorithm for generating smooth models from electromagnetic sounding data. *Geophysics, 52*, 289-300.

Gansser, A., 1964. Geology of the Himalayas. Interscience Publishers, N.Y.

Groom, R.W. and Bailey, R.C., 1989. Decomposition of the magnetotelluric impedance tensor in the presence of local three-dimensional galvanic distortions. *J. Geophys. Res., 94*, 1913-1925.

Gupta, G., Gokarn, S.G. and Singh, B.P., 1994. Thickness of the Siwalik sediments in the Mohand-Ramnagar region using magnetotelluric studies. *Phys. Earth Planet. Inter., 83*, 217-224.

Jones, A.G., 1988. Static shift of magnetotelluric data and its removal in a sedimentary basin environment. *Geophysics, 51*, 967-978.

Kaila, K.L. and Narain, H., 1976. Evolution of the Himalaya based on seismotectonics and deep seismic soundings. *Geol. Surv. India, Misc. Publ., 41 (IV)*, 1-30.

Khattri, K.N. and Tyagi, A.R., 1983. The transverse tectonic features in the Himalaya. *Tectonophysics, 96*, 19-23.

Krishnan, M.S., 1982. Geology of India and Burma. 6th Edition, CBS, 422-438.

Kumar, R. and Nanda, A.C., 1989. Sedimentology of the Middle Siwalik Subgroup of Mohand area, Dehra Dun valley, India. *J. Geol. Soc. India, 34*, 597-616.

Marquardt, D.W., 1963. An algorithm for least square estimation of non-linear parameters. *J. Soc. Ind. Appl. Math., 11*, 431-441.

Ni, J. and Barazangi, M., 1984. Seismotectonics of the Himalayan collision zone: Geometry of the underthrusting Indian plate beneath the Himalaya. *J. Geophys. Res., 89, No. B2*, 1147-1163.

Prasad, C., 1976. Geomorphological and sedimentological studies in Doon valley, U.P. Ph.D. Thesis, Meerut University, India.

Qureshy, M.N., 1969. Thickening of a basalt layer as a possible cause for the upliftment of the Himalayas- a suggestion based on gravity data. *Tectonophysics, 7*, 137-157.

Rodi, W. and Mackie, R.L., 2001. Nonlinear conjugate gradients algorithm for 2D magnetotelluric inversions. *Geophysics, 66*, 174-187.

Valdiya, K.S., 1964. The tectonic design of the Himalaya: a survey of structure between Indus and Brahmaputra and their comparison with Alpine structure. *22nd Internat. Geol. Cong., Part II*, 283-315.

Vozoff, K., 1972. Magnetotelluric studies in the sedimentary basins. *Geophysics, 37*, 98-141.

Vozoff, K., 1984. Model study for the proposed magnetotelluric (MT) traverse in north India. *Tectonophysics, 105*, 399-411.

Wadia, D.N., 1975. Geology of India, 4th Edition, Tata McGraw-Hill, New Delhi.

Wight, D.E. and Bostick, F.X., 1980. Cascade decimation: A technique for real time estimation of power spectra. Proc. IEEE Internat. Conf. Acoustic, Speech Signal Processing, Denver, Colorado, April 9-11, 626-629.

SEDIMENTOLOGY OF THE PALAEOGENE SUCCESSION IN JAMMU REGION, JAMMU AND KASHMIR STATE, NW HIMALAYA.

B.P. Singh, D.S. Andotra and J.S. Pawar

P.G. Deptt. of Geology, University of Jammu, Jammu-180 006

ABSTRACT

Palaeogene sequences form significant units of the Himalayan foreland in the northwestern Himalaya. This paper presents a brief synthesis of the environments of deposition of individual facies and their provenance for the siliciclastic sediments of the palaeogecne times. The early Palaeogene Subathu facies consist of carbonaceous shale/ green/ greenish-yellow shale and yellow limestone of the coastal lagoon or bay. The conglomeratic limestone and oyster limestone form barrier bar, with laminated limestone, marl and purple shale deposited in a tidal flat environment. The overlying Murree Group facies are estuarine channels, along with levee in the intertidal secondary channels. The laminated mudstone was deposited in the supratidal zone of the estuary during slack-water period.

The chlorite and illite-bearing shale in the Subathu Formation, possibly were derived from the metamorphic terrain of the Himalayan Arc. The Murree sandstones containing quartz, feldspar, metamorphic and volcanic fragments were derived from the Higher Himalaya including the Indus-Suture Zone. The heavy Minerals such as epidote, garnet, staurolite, zircon and tourmaline further support their derivation form the Higher Himalayan metamorphic and igneous rocks. The sedimentary rock fragments suggest reworking of the earlier deposited sediments with shifting of the depositional sites.

INTRODUCTION

Palaeogene (late Palaeocene-early Miocene) successions are fairly exposed in the Sub-Himalayan zone of Jammu and Kashmir state and form a thick pile of argillaceous, arenaceous and calcareous facies (Fig. 1A). These together from an early Palaeogene Subathu Formation and a late Palaeogene Murree Group. The Subathu Formation crops out mainly as outliers whereas the Murree Group occurs in laterally continuing broad hill ranges continuing from Murree hills of Pakistan (Fig 1A & b). The northern limit of the Murree Group is marked by a thrust plane, i.e. the Main Boundary Thrust on Murree Thrust, and its southern limit is represented by the presence of Siwalik rocks (Fig. 1B). According to Pascoe (1964), there is a clear evidence of an unconformable contact or the Murree "series" with the Eocene on the northwestern foot of the Khairi-Murat ridge (Pakistan). However, palaeomagnetic studies carried out by Klootwijk et al. (1986) in the Jammu and by Najman et al. (1994) in Himachal Pradesh suggest a continuous sedimentation with no major hiatus between the Subathu Formation and the Murree Group/Dagshai-Kasauli formations, Karunakaran and Ranga Rao (1979) divided the Murree Group into lower Murree Formation

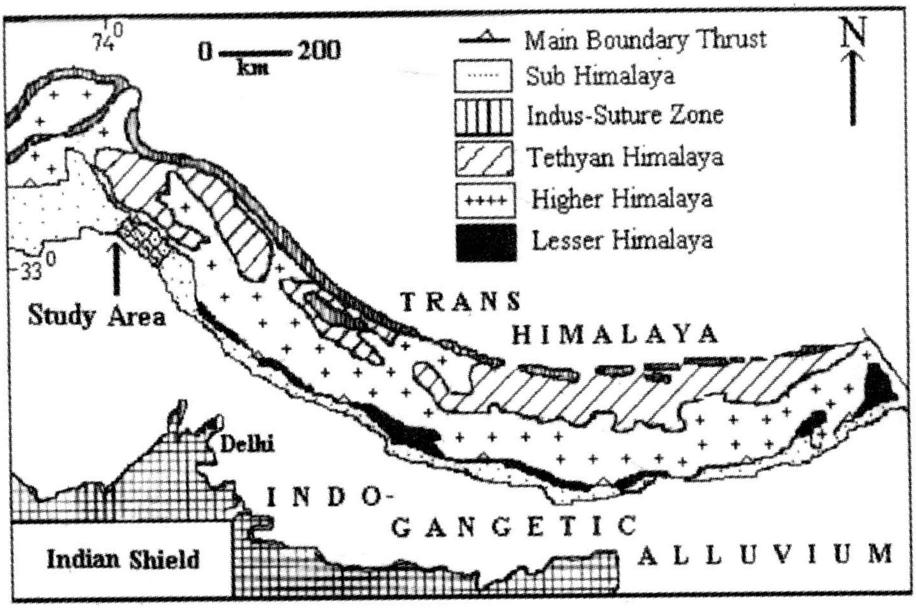

Figure 1: a Geological map showing Himalayan succession and the location of the study area in the Sub-Himalayan zone.

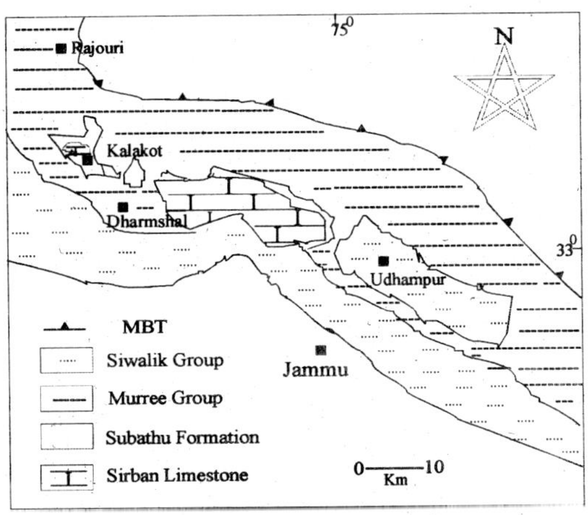

Figure 1: b Detailed geological map of the Sub-Himalaya in parts of the Jammu region exhibiting exposition of the different groups and formations.

and Upper Murree Formation (Table 1). Mud-pebble conglomerate, cross-bedded sandstone, planar-bedded sandstone, laminated siltstone, mudstone and calcrete are the important litho-types.

Middlemiss (1929) described the Subathu formation of the Jammu province about its economic significance. Singh (1970,73,80) discussed fossil zonation and stratigraphy of the Subathu formation of the Jammu and Kashmir State. Sahni and Khare (1973) and Sahni

Table1: Cenozoic stratiraphic succession (after Karunakaran and Ranga Rao. 1979)

Group	Subgrou/Formation	Age
Siwalik Group	Upper Siwalik Subgroup Middle Siwalik Subgroup Lower Siwalik Subgroup	Early Miocene-Pleistocene
Murree	Upper Muree Formation Lower Murree Formation	Mid. Eocene-Early Miocene
	Subathu Formation	Palaeocene-Mid Eocene
	Sirban LImestone	Precambrian (Basement)

et al. (1983) reported vertebrate fossils from the upper part of this formation. Palaeobiogeographic conclusions are drawn by Sahni et al. (1983), Mathur (1983, 1997) and Bhatia (1985,). Studies on the coal occurring in the basal part of the Subathu Formation were carried out by Singh (1988) and Singh and Singh (1995). Singh and Andotra (2000) carried out detailed facies analysis of the Subathu formation. Additionally, Singh et al. (2000) dealt with clay mineralogy and chemical composition of the Lower Tertiary mudrocks.

Oldham (1892) and Pilgrim (1910) carried out studies on depositional environment of the Murree rocks. Wadia (1928) described that the Murree Group comprised of shale, sandstone and pseudo-conglomerate (mud-pebble conglomerate) in the lower part and shale and sandstone in the upper part. Ranga Rao (1971) reported several non-marine vertebrates in rocks exposed near the contact of the Subathu Formation and the Murree Group. Sharda and Verma (1977) carried out textural analysis of the Murree sediments. Metha and Jolly (1989) discovered vertebrate fossils and suggested a fresh-water environment for the Murree sediments. Bossart and Ottiger (1989) reported lithofacies and palaeontological observations. Critelli and Garzanti (1994) presented detailed petrographic data of sandstones for determination of provenance of Murree Group from Hazara area in Pakistan. Singh and Singh (1995) did lithofacies analysis and reported variety of sedimentary structures such as herringbone cross-beds, flaser beds and tidal bundles. Singh (1996) presented petrographic and chemical composition data for interpretation of tectonic setting. Singh (2000) discussed the dispersal pattern and palaeogeographic configuration of the basin during sedimentation of the Murree Group. A brief compilation of the works carried out by us on depositional environments and provenance of the palaeogene rocks is being presented here.

SUBATHU FACIES (FIG. 2 & 3)

Carbonaceous shale and coal

This lithofacies with lateral extension occurs at the base of the Beragua member. Its total thickness is 12 to 18 m including three coal bands of approximately 3m in thickness and two coal seams of 1.1m and 1.2m in thickness occur within the carbonaceous shale at Kalakot and Mahogla whereas three seams of 1.0m, 0.7m and 0.6m thickness occur at Metka. The ash and sulphur (2-3%) contents are high in this coal and its rank varies from semibituminous to semianthracite. The macerals are vitrinite, inertinite and exinite with vitrinite being dominate (Singh 1988). This facies consists of hard platy black silty shale in the basal part grading upward to clay shale, dominated by laminae less than 5mm thick. Diagenetic pyrite and phosphate nodules are present at many horizons, together with calcite and dolomite

Green and greenish-yellow shale

These shales have extensive (many kilometers) lateral exposures and a few meters to tens of meters of vertical extent. The shales are laminated, hard and fissile with 1 to 2mm thick laminae. These shales are clay rich having textural similarities with the clay shales of Potter et al. (1980). The clay mineras include chlorite, illite and sepiolite. These shales are impregnated with calcareous matter including calcite and dolomite. Both these shales contain septarian nodules and elongated mud balls. Thin streaks of coal occur in the green shale at some levels. Phosphate nodules with thin concentric pyrite layers occur at some levels in the greenish-yellow shale.

Yellow limestone

This limestone, consists of thin beds (20 to 30cm thickness) and a few kilometers of lateral extent and contains foraminifers and lamellibranchs. It is composed of micro-spars of calcite and dolomite and classified as lime-mudstone following Dunham (1962). The pellets and welded faecal pellets are common in the micritic orthochem. The foraminiferal species like *Ranikothalia mittalli, Nummulites burdigalensis kuepperi, Assilina leymeriei, A. spinosa* and *Rotalia trochidiformis* are recovered from this lithofacies. The foraminifers are composed of subhedral and euhedral dolomite grains cemented by calcareous and argillaceous cement. The limestone facies also preserve well developed sediment-filled burrows with positive hyporelief such as *Imbrichmus* and Y-shaped irregular surface burrows as *Thalassinoides* along the bedding plane.

Conglomeratic limestone

An intraformational limestone conglomerate with erosional contact occurs above the yellow limestone. The conglomeratic limestone extends a few kilometers laterally, and shows up to 2m thick carbonate sequences. This limestone is clast supported and contains limestone clasts of 15 to 20cm diameter at basal part and 2to 10cm diameter toward the top part of beds. The clasts are encased in argillaceous lime-mud that contains micritic crystals of both calcite and dolomite. The pebbles are poor in fossil content but laminated argillaceous limestone which is rich in fossils, including *Nummulites cf. rotularius, N. cf. N. minervensis, Assilina spinosa, A. leymeriei, A. laminosa, A. dandotica* and *A. pustolosa*. Insoluble residues in this limestone vary from 17.35 to 21 and consists mostly of clay.

Oyster limestone

Limestones with abundant oyster shells along the lamina planes occur above the lag deposits. In many cases, the conglomeratic limestone facies grades upward in laminated limestone that is full of small and large foraminifers terminating into oyster rich limestone. Oyster limestones exhibit hummocky-cross stratification and symmetrical ripples. Petrographic study shows well-developed dolomite crystals. Dolomite crystals as large as 35μm size are present. The fossil oyster shells are composed of calcite. This limestone can be further classified into wackestone.

Yellow shale

The yellow shale is platy, with individual lamina thickness of 2 to 4mm. Silt sized grains of quartz are scattered throughout the clay matrix. This shale is mainly composed of illite, chlorite and sepiolite. Calcite is present as (under microscopic) in the laminated yellow mud shale. Ellipsoidal septarian nodules are also present. Srivastava & Kumar (1996) reported marine fish remains from thin calcareous hard bands present in this shale in the Mahogla area and they inferred as shallow marine environment for this deposition.

Laminated limestone

The laminated limestone is grey with thin bands alternating with mud flasers. It occurs at two levels in association with greenish- yellow shale and yellow shale. Finely laminated grey limestone is hard and devoid of fossils . Desiccation cracks are present on the upper surface at some locations. Planar - e and planar - s crystals of dolomite are present within the size range of 0.1 mm to 0.5mm along with calcite . The dolomite crystals show sucrose texture and cloudy centers. The dolomite limestone is comparatively clay free with a lower percentage (on an average 14.71%) of insoluble residue.

Limestone - marl laminites

Another important facies of the top portion of the Subathu Formation is white limestone with intercalated pink marl. In the lower part, the thickness of the beds is in centimeters whereas in the upper part the thickness decreases to millimeters. The limestone bands are thinner than the marl bands. The limestone is composed of silt and clay along with calcite. Some of the laminations are bioturbated. Upper surfaces of the marl preserve mud-cracks .

Purple shale

The purple shale is flaggy with lamina thickness of 4 to 6 mm . This shale is mud shale having clay and silt in subequal amounts. The clay minerals are mainly chlorite, illite and sepiolite. Haematite is the iron oxide imparting red pigmentation to this shale. Vertebrate fossils of fresh water and brackish water fishes have been reported by Ranga Rao (1971), Sahni & Khare (1973) and Sahni et al. (1983) from purple shale in Jammu region. Wells and Gingerich (1987) have described fresh water mammals, snails, crocodiles and turtles from adjoining Kotli area in Pakistan.

ENVIRONMENTS OF DEPOSITION (TABLE 2)

The marine faunal assemblage followed by brackish water assemblage suggests a major marine transgression in Palaeogene and a regression during middle Eocene in the Jammu region. Facies association (Fig. 2&3) indicates that along with the major transgression and regression, there were short - term transgressions and regressions also . The conglomeratic limestone and overlying planar - laminated to hummocky - bedded limestone deposited on a barrier showing existence of storms adjacent to gulf similar to the Persian Gulf. Fine - grained mud was deposited in quiet - water lagoons (e. g. Potter et al , 1980) above the carbonate barrier and the organic matter was supplied from the adjacent swamps. The change in colour of the shale from black to green and greenish - yellow is mainly because of varying organic matter supply and weakly oxidizing conditions. Abnormal salinity conditions in the vicinity of tidal inlets result in a high- diversity assemblage in which open marine and lagoonal species may be mixed (Parker, 1960 ; Rusnak, 1960) . The presence of a diversified fauna of

Table 2: Lithofacies and their Depositional Environment in the Subathu Formation

Lithofacies	Environment of Deposition
Carbonaceous Shale and Coal	Logoon
Green and Greenish-yellow SHale	Lagoon or bay
Yellow limestone	Shallow Lagoon
Conglomeratic limestone	Barrier
Oyster limestone	Storm-dominated, barrier
Yellow shale	Tidal flat
Laminated limestone	Tidal flat
Lamestone-marl laminities	Intertidal-Supratidal zone
Purple shale	Estuary and outer tidal flat

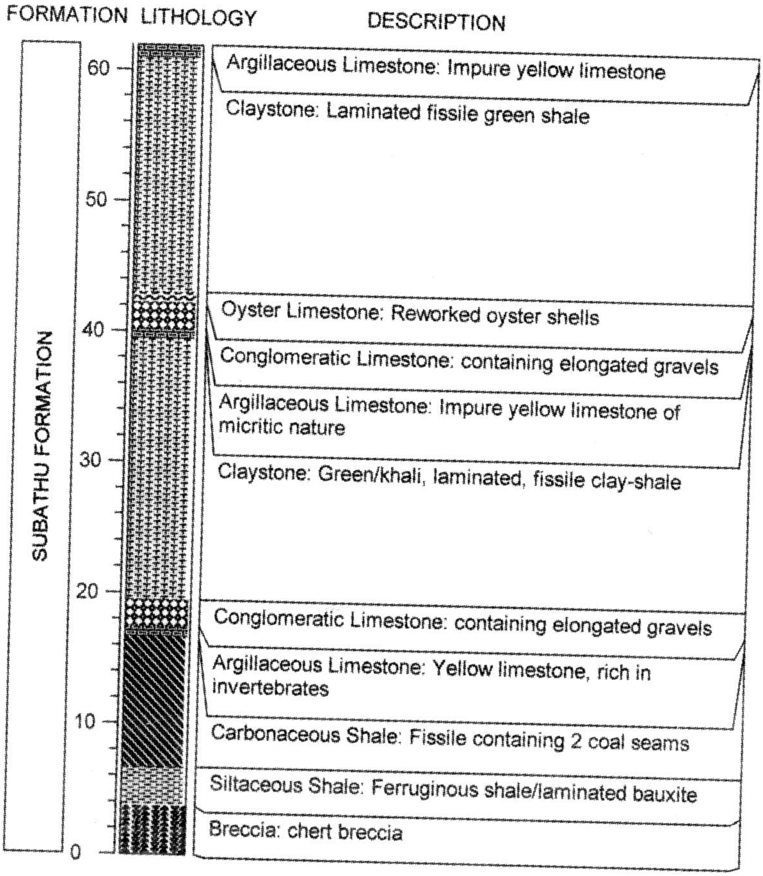

Figure 2: Lithocolumn showing facies association in the lower part of the Subathu Formation at Kalakot, district Rajouri, Jammu and Kashmir state, India.

foraminifers, ostracods, oysters and other lamellibranchs in the Subathu demonstrates connection between the enclosed water bodies and the sea either through storm cuts or tidal inlets. The laminated purple shale with high silt content and fresh water as well as brackish water fauna deposited in supratidal zone with definite contribution from river systems at this stage.

A comparison of the Subathu Formation of the Jammu region with that of the type area (Shimla hills) shows that the supply of siliciclastic mud was greater in the type area . The conglomeratic facies is poorly developed whereas oyster - bearing beds are well developed . This reveals that the gulf was protected and shallow in this part and open in the Shimla hills. Although the Tethys separated Indian and Eurasian plates, it became narrow enough for free migration of benthic fauna from Thanetian onwards and shallow marine conditions existed up to the Early Lutetian in the Indus - Suture, Zanskar, Tethyan and Lesser Himalayan zones (Mathur, 1997).

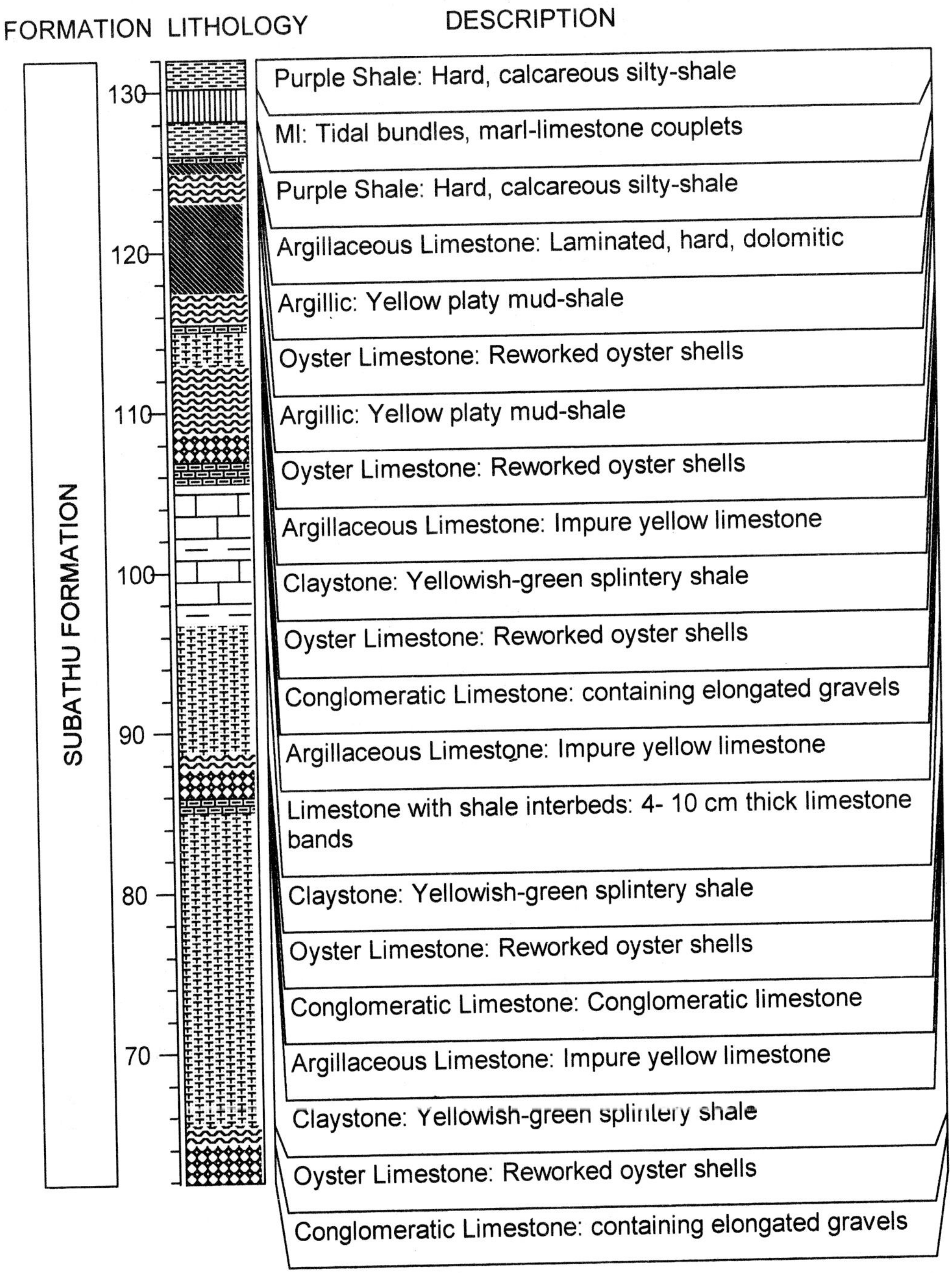

Figure 3: Lithocolumn showing facies assemblage the upper part of the Subathu Formation at Kalakot, district Rajouri, Jammu and Kashmir state, India.

Early Palaeogene rocks in Pakistan record rapid submergence of the northwestern Indian continental shelf in the palaeogene; followed by restriction and isolation of the former open marine Tethys sea to Persian gulf - like condition in early Eocene , and brief resubmergence of the basin in the middle Eocene (Pivnik & Wells, 1996) . The fine detrital siliciclastic sediments were deposited during the regressive phase and carbonates during transgressive phase. In adjoining Kotli area of Pakistan , Wells and Gingerich (1987) recognised ten units (A to J) in the Subathu Formation including one unit of sandstone . They concluded that the shales deposited either in offshore region or in coastal lagoons and the limestones developed during transgression in the coastal zones. In the Jammu region , the carbonate sequences represent barrier and shales as lagoonal deposit in the lower part. The higher - up, limestone and shale were deposited in intertidal - supratidal zones of a tidal flat.

During low supply of siliciclastic sediments a carbonates ramp deposited in a gulf. The carbonate ramp protected the lagoon in which siliciclastic sediments deposited during regression with added subsidence. After filling of the basin carbonate sedimentation started on the older lagoon . In general , the largest carbonate ramp developed along passive margins and in foreland basins where flexural subsidence dominated (Wright & Burchette, 1996). Subsidence followed by sedimentation and flexural bending originated cyclothems containing carbonate ramps in a foreland basin in the Subathu Formation of the Jammu and Kashmir state.

MURREE FACIES (FIG. 4 & 5)

Mud - pebble conglomerate

Mud - pebble conglomerate forms lowermost unit of individual cycles in the Murree Group. Elongated mud chips of 1.5 cm to 4.0 cm lie parallel to the bedding planes. In some cases, the lithofacies show arrangement of the pebbles in cross- bedded form. The mud clasts are present in the sandy matrix and can be categorized as intraformational clasts conglomerates. In general, the clast size increases from Lower Murree Formation to the Upper Murree Formation. There is a prominent conglomerate band (about 1.0m thick) 50 m above the contact between the Subathu Formation and the Murree Group, but this also contains intraformational clasts and exhibit primary sedimentary structures.

Cross-bedded sandstone

Cross-bedded sandstone occurs above the intraformational conglomerate in most of the cycles. The cross-beds are mainly tabular, but in few cases trough-shaped cross-beds occur. The basal part of the Murree Group also contains bidirectional cross-beds having fore-set directed towards NE and SW. At places multiple (composed of two or three sets) cosets occur in this facies. The cosets thickness varies from 0.2m - 1.2m. In general, the coset thicknesses is relatively more in the Upper Murree Formation than the lower Murree Formation. The particle size range varies form 0.5-1.4mm in this sandstone. The particles are subrounded to subangular showing composition of the lithic arentie sandstone.

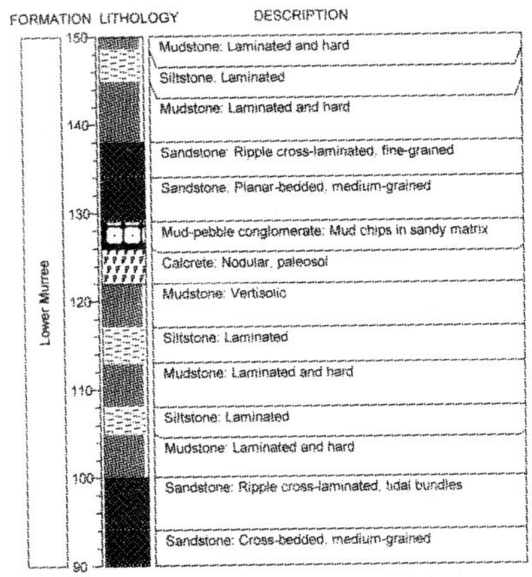

Figure 4: Lithocolumn displaying various facies occurring in the Lower Murree Formation near Kalakot

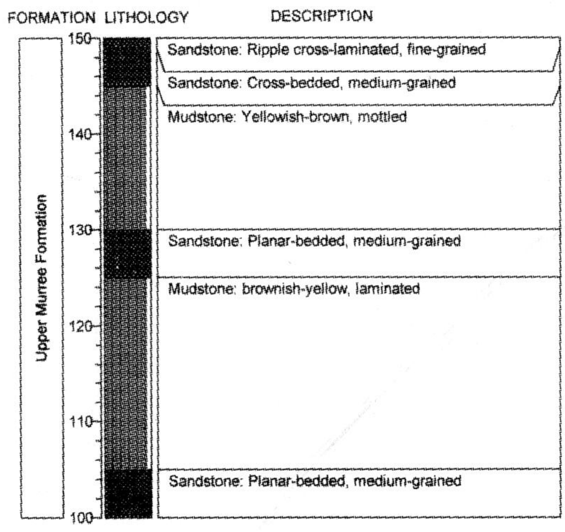

Figure 5: Lithocolumn of the Upper Murree formation showing various facies at Dharmshal, district Rajouri, Jammu and Kashmir state, India.

Planar-bedded sandstone

This sandstone unit is planar-bedded with a thickness of 0.5-2.0m in individual cycles. Flaser and wavy beds are also common in this facies. Thickness of individual beds ranges form 4-10 cm, and at places internal bedding planes are disrupted. The sandstone is grey and greenish-grey in the Lower Murree Formation, and yellow to yellowish-green in the Upper Murree Formation. Grain size varies from 0.2-1.2mm in this medium to coarse-grained sandstone. Subrounded and subangular grains are associated with a few angular ones. The lithic arenite fabric is mainly normal where by particles are cemented with carbonate cement and rock fragments dominate over feldspars. At places, large elongated fragments of mudstone are observed in thin-sections.

This facies is weakly bioturbated, only vertical burrows with wall lining at places are preserved.

Ripple cross-laminated sandstone

Ripple cross-laminated sandstone forms 1-2 m thick units in the lower and upper murree formations; the laminae range in thickness from 0.1-2.0 cm with couplets of thick and thin layers. In general, the thick layers are sandy and the thin layers are silty. These sandstones are of lithic wacke composition with abundant mica flakes on bedding planes. The upper surfaces record asymmetrical (bifurcated) straight-crested, undulatory and cuspate ripples of low ripple index (R.I.=4.0-8.0). The particles range in size from 0.25 to 0.1 mm. The particles also exhibit bimodality with fine and very fine sand size. This facies is moderately bioturbated by feeding-dwelling burrows of small tracts and grazing traces.

Laminated siltstone

Siltstone units of 1 - 3 m thickness occur sandwiched between mudstones with laminated silt and mud layers forming rhythmites. Thick (0.5 - 3.0 cm) silt laminae alternate with thin (0.1-0.5 cm) mud laminae. In a few cases, low angle (10^0-20^0) ripple cross-laminae produce cosetes with a thickness of 2 - 4 cm. In general, the ripple cross-laminae are overlain by drape laminae or planar laminae. The siltstone has a particle size range of 0.20-0.02 mm. Cuspate ripples of low ripple index (5-7) occur on the upper surface of the beds. These siltstones are rich in trace fossils and are moderately to strongly bioturbated, at places.

Mudstone

Brown to yellowish-brown and greenish-yellow laminated mudstones occur both in the Lower Murree and the Upper Murree Formations. The laminae planes are darker in colour because of iron oxide staining. Millimetre-thick mud layers are also present a thin interbeds in siltstones. The mudstones contain abundant angular silt-sized quartz grains. Singh et al. (2000) estimated the clay content of the Murree mudstones ranging form 54%-65% that mainly consists of illite, chlorite and sepiolite. At some levels, the mudstones are moderately bioturbated disturbing the normal fabric and lamina planes.

ENVIRONMENTS OF DEPOSITION (TABLE 3)

Macrotidal estuaries can be viewed within a deltaic estuarine coastal depositional settings influenced by riverine processes, wave regime and tidal energy (Wright and Coleman, 1973; Wright, 1985; Boyd et al., 1992). The mud-pebble conglomerate deposited on the floor of the channels and the mud-clasts were produced from the erosion of the channel walls. This suggests renewed activity after some degree of consolidation of the mud deposited around the channel walls. The bidirectional cross-beds present in the Murree Group formed within the channel similar to mesotidal channel complex as described by Nio and others (1980) from the Southwestern Netherlands. Channel sandstone is either cross-bedded or planar-bedded sandstone when the flow strength decreased in the channel. The flaser and

Table 3: LIthofacies and their Depositional Environment in Murree Group

Lithofacies	Environment of Deposition
Mud-pebble Conglomerate	Channel floor
Cross-bedded Sandstone	Estuarine channel
Planar-bedded Sandstone	Estuarine Channel
Ripple-laminated sandstone	Levee
Laminated Siltstone	Secondary Channel (Inter-tidal area)
Mudstone	Inter-tidal area

wavy beds are indicators of tidal influence (Reineck and Singh, 1980). Their presence suggests tidal influence during deposition of this sandstone unit. The subangular and subrounded nature of the grains indicate that the distance of transportation was limited. The mudstone fragments further indicate erosion of the pre-existing rock form a nearby area. The neap-spring tidal events are of tidal influence in the channel revealing estuarine sedimentation particularly in the eastern part of the study area. Channel deposits are generally followed by evenly laminated, fine-grained sand flat and mud flat deposits of the intertidal zone (Reineck and Singh, 1980). The ripple cross-laminated sandstone was deposited on the levee of the channels in the Murree Group. Increasing flow strength changed rippled bed forms from straight-crested into undulatory and eventually cuspate forms (e.g. Hayes and Kana, 1976). The occurrence of ripple laminae above climbing-ripple cross-laminae suggests decelerating flow strength along the crest of the levee. Sufficient sand for ripple migration was provided by a nearby river system during deposition of this litho-facies.

The siltstone unit of the Murree Group present within the thick (3-4 m) succession of mudstone above the fine-grained sandstone (Fig. 4 & 5) was probably deposited in between two slack-water periods. The rhythmic nature and the occurrence of silt and mud layers together suggest its deposition in the intertidal zone. The drape lamination and ripple cross-lamination in this facies are similar as reported by Lanier et al. (1993) from Carboniferous siltstone facies of Kansas, U.S.A. In Mont-Saint-Michel Bay, France, tidal rhythmites are observed within restricted secondary channels of the inner estuarine domain protected from ocean-wave reworking and fluvial energies (Tessier et al., 1995). Deposition of the laminated siltstone took place in a protected secondary tide-dominated channel. The occurrence of silt and mud layers together is possible in a flooded tidal channel. The strong bioturbation suggests sufficient faunal activity after sedimentation of this facies.

Deposition of the laminated mudstone suggests slack water conditions. The presence of angular silt grains is indicative of its local supply. Muddy drapes result from settling during the slack water period of the high tide (Tessier et al., 1995). The tidal currents deposited mud during flood tides, strong bioturbation indicates subaerial emergence and ploughing by the animals up to few centimeters below the surface. The upper intertidal zone is characterized by extended periods of subaerial emergence lasting at least for several days (Tessier et al., 1995). This indicates that the Murree mudstones were deposited in the upper intertidal zone.

The sedimentation was mainly governed by the strong tides annually and a comparatively high fluvial discharge during floods. The estuaries were tide-dominated in the basal and the middle parts whereas river-dominated in the upper part as reflected by the sedimentary structures and the nature of sediments fill.

A straight downstream funnel, a sinuous meandering channel and a straight downstream channel were recognized in the Fundy (Canada) and the Gironde (France) estuaries (Dalrymple et al., 1992; Allen and Posamentier, 1993). The minor variations in the directions of the cross-beds in the Murree group is primarily due to their formation in the meandering channels and broad funnel mouths. In the Gironde estuary, all the fluvially-derived sand and about 75% of the mud were deposited in the basin with only a minor seaward mud migration (Allen et al, 1980). The sources of sediments are Alpine in case of the present day Gironde estuary. Similarly, the source of the Murree sediments was the Himalaya in the north. According to Nichols and Allen (1992), the rhythmic alternations of sand and mud lamination are characteristic features of high energy tidal estuarine deposits. This feature is prominent in the sandstones and siltstones indicating their sedimentation in a high- energy estuary.

PROVENANCE

Provenance and sediments distribution paths are identified on the basis of light and heavy minerals along with rock fragments. The Subathu mudrocks are rich in detrital chlorite and illite that might have been derived from schists and gneisses occurring in the Himalayan Arc in the north. Singh and Andotra (2000) have suggested Ladakh-Kohistan Arc rock as the source for these mudrocks. The Murree sediments are composed of monocrystalline quartz, feldspar, muscovite, biotite, chlorite, iron minerals and lithic fragments. The feldpars are orthoclase as well as plagioclase and rock fragments are of phyllite, chert, siltstone, mudstone, quartzite, lime clast and volcanic clast. The significant aspect of the rock fragments distribution is the decreasing trend of volcanic and metamorphic rock fragments from the Lower Murree to the Upper Murree. The mudstones are also composed of detrital clay minerals within which scattered fine quartz grains occur.

Among the heavy minerals, zircon, tourmaline, garnet and epidote form major proportion (Singh et al., 1990). The zircon grains are subrounded to rounded and rod shaped and of finer size. The tourmaline grains are mostly blue and pink having subangular to subrounded shapes. Brownish red garnet with prismatic subrounded shape is present along with epidote and hornblende.

The heavy minerals of the Murree Group (tourmaline, zircon, garnet and epidote) suggest a gneiss and granite rich source (Higher Himalaya). The development of the foreland basin along the southern margin of the Himalaya and subsequent sedimentation of immature detritus created subsidence in the basin. The metamorphic rocks were exposed on the hanging wall of the MCT forming the Higher Himalayan Chain. The sediments were, thus, derived from hanging walls of this hinterland. The exhumation of the Higher Himalayan crystallines is interpreted to occur around 35-40 Ma based on cooling ages of mineral hornblende by Treloar and Rex (1990) and Sorkhabi et al. (1993). The phyllite and quartzite fragments present in the Murree rocks were contributed from the Himalayan crystallines. The volcanic rock fragments were either derived from the ophiolites or volcanics of the Indus-Suture zone. The sedimentary rock fragments indicate local erosion and shifting of the depositional basin towards south with the increasing basin load. The tourmaline and zircon were supplied by the granitic gneisses and granites present in the source terrain. Small proportion of garnet, epidote and hornblende points towards low to medium grade metamorphic source such as present in the Higher Himalaya. The southerly slope and palaeocurrents again indicate that the provenance was in the north and the contribution was mainly form the rising Himalaya. The thick skinned fold-thrust belt created due to collision of the two continental plates and deposited sediments were derived mainly from the sedimentary and partly metamorphosed rocks. Murree sandstones fall within the recycled orogenic provenance (Singh, 1996) and main petrographic parameters do not vary greatly where volcanic, sedimentary, low-grade metasedimentary and ophiolitic detritus persisted until the early Miocene (Critelli and Garzanti, 1994). The volcanic rock fragments present in the sandstones and chrome-spinel identified by Bossart and Ottiger (1989) substantially support the southerly derivation of the sediments due to unroofing of the ophiolite and other volcanic rocks.

CONCLUSIONS

The palaeogene succession in Jammu region indicates an overall regressive phase of a coastal regime where Subathu sediments were deposited in coastal lagoons or bay, forming barriers with much of storm activity and also, on tidal flats (Table 2). The Murree succession developed in estuarine realm depositing intraformational conglomerate, and cross-bedded and planar-bedded sandstones in channels (Table 3). The overlying laminated-and ripple cross-laminated sandstones deposited on the levee crest and the laminated siltstone-mudstone deposited in the intertidal zone.

The sandstones composition suggests that the detrital constituents were derived from the orogenic belts (recycled orogenic provenance). At the initial stages, the Palaeogene sediments were derived form the Himalayan arc followed by their derivation from the Indus-Suture Zone and from the Higher Himalayan uplifted mountains into the foreland basin.

REFERENCES

Allen, G.P. Salomon, J.C., Bassoullet, P. Dupenhoat, Y. and De Grandpre, C., 1980. Effects of tides on mixing and suspended transport in macrotidal estuaries. Sed. Geol. 26:69-90

Allen, G.P. and Posamentier, H.W., 1993 Sequence stratigraphy and facies model of an incised valley fill: the Gironde estuary, France, Jour. Sed. Petrol. 63: 387-391.

Bhatia, S.B. 1985. Some aspects of the Early Tertiary biostratigraphy and Palaeogeography of the Indian Subcontinent. Quarterly Journal Geological Mineralogical an Metallurgical Society of India. 57: 63-98.

Bossart, P. and Ottiger, R., 1989. Rocks of the Murree Formation in northern Pakistan: Indicators of descending foreland basin of late Paleocene to Mid. Eocene age. Eclog. Geol. Hevl. 82: 133-165.

Body R., Dalrymple, R.W. and Zaitlin, B.A. 1992. Classification of clastic coastal depositional environments. Sed. Geol. 80: 139-150.

Critelli, S. and Garzanti, E., 1994. Provenance of the lower Tertiary Murree redbeds (Hazara-Kashmir syntaxis, Pakistan) and initial rising of the Himalaya. Sed. Geol. 89:265-284.

Dalrymple, R.W. Zaitlin, B.A. and Body, R. 1992. Estuarine facies models: conceptual basis and stratigraphic implications. Jour, Sed. Petrol. 62: 1030-1043.

Dunham, R.J. 1962 Classification of carbonate rocks according to depositional texture. In: Classification of Carbonate Rocks (ed. W.E. Ham). Memoir American Association of Petroleum Geologists 1: 108-121.

Hayes, M.O. and Kana, T. W., 1976 Terrigenous/clastic depositional environments some modern examples. Tech. Rept. 11-CRD, Coast, Res. Div. Univ. South Carolina, II-184: 1-131.

Karunakaran, C. and Ranga Rao., 1979. Status of exploration for hydrocarbons in the Himalayan region-contribution to Stratigraphy and Structure. Geological Survey of India Miscellaneous Publication 41: 1-67.

Klootwijk, C.T. Sharma, M.L., Gergan. J., Shah, S.K. and Gupta, B.K. 1986. Rotational overthrusting of the northwestern Himalaya: further Palaeomagnetic evidence form the Riasi thrust sheet, Jammu foothills, India. Earth and Planet. Sci. Lett. 80: 375-393.

Lanier, W.P., Feldman, H.R. and Archer, A.W. 19936. Tidal system in a fluvial to estuarin transition, Douglas Group. Missourian-Virgilian, Kansas. Jour. Sediment. Petrol. 63: 860-873.

Mathur, N.S. 1983. Biostratigraphy of the Cretaceous-Eocene sequence in the Himalaya in relation of palaeobiogeography Himalayan Geology. 11:287-322.

Mathur, N.S. 1997. A comparative study of lower tertiary biostratigraphic sequences in the northwest Himalaya and their palaeogeographic significance. In: Geodynamic domains in Alpine-Himalaya Tethys (eds. A.K. Singh, F.P. Sassi and D.Papanikolaou). Oxford and IBH Publishers, Delhi, 125-158.

Metha, S.K. and Jolly. A. 1989. Leptomeryx, and Oligocene Artiodactyl form the Lower Murree of Sial-Sui (Kalakot tehsil) District Rajauri, Jammu and Kashmir Curr. Scien. 58: 625-627.

Middlemiss, C.S. 1929. Kalakot, Metka, Mahgola and other coalfields of Jammu Province. Mineral Survey Report, Jammu and Kashmir Government, 116 pp.

Najman, Y.M.R. Enkin. R.J. Johnson, R.W. Robertson, A.H.F. and Baker, J., 1994. Palaeomagnetic dating of the earliest continental Himalayan foredeep sediments: Implications for Himalayan evolution: Earth and Planetary Science Letters. 128: 713-718.

Nichols, M.M. and Allen , G.P. 1992. Comparison of facies and processes in macrotidal and microtidal estuaries. Tidal clastic 92 (Abstract Volume), 64-66.

Nio, S.D. Berg, J.H.V. Goesten, M. and Smulders, F., 1980. Dynamic and sequential analysis of a mesotidal shoal and intershoal channel complex in the eastern Scheldt (Southwestern Netherlands). Sed Geol. 26: 263-279.

Oldham, R.D., 1892. Report on the geology of Thal-Chotli and part of the Mari country. Rec. Geol. Surv. Ind. 25: 18-29.

Parker, R.H. 1960. Ecology and distribution pattern of marine macro-invertebrates, northern Gulf of Mexico, In: Recent Sediments of Northwest Gulf of Mexico (eds. F. P. Shepard, F.B. Phleger and T.H. Van Andel) American Association of Petroleum Geologists, Tulsa. 302-307.

Pascoe, E.H. 1964. A Manual of the Geology of India and Burma. 3rd ed. Geological Survey of India. Delhi. 3: 1345-2117.

Pecher, A., 1977. Geology of the Nepal Himalaya: deformation and Petrography in the Main Central Thrust Zone. Ecologie et Geologie de l' Himalaya,301-318.

Pilgrim, G.E., 1910. Preliminary Note on a revised classification of the Tertiary fresh water deposits of India. Geol. Surv. Ind. 40: 185-205.

Pivnik, D.A. and Wells, N.A., 1996. Transition from Tethys to the Himalaya as recorded in the Northwest Pakistan Geological Society of America Bulletin. 108: 1295-1313.

Potter. P.E., Maynard, J.B. and Pryor, A.W. 1980. Sedimentology of Shale. Springer-Verlag, New York, 310 pp.

Ranga Rao, A., 1971. New Mammals from Murree rocks of Kalakot Zone of the Himalayan foothills, Kalakot, J. & K State. Jour, Geol, Soc. Ind. 12: 126-134.

Reineck, H.E. and Singh I.B., 1980.Depositional Sedimentary Environments. Sec. Revised and update edition Springer, Verlag, New York, Heidelberg, Berlin.

Rusnak, G.A. 1960. Sediments of Laguna Madre, Texas. In: Recent sediments of the Northwest Gulf of Mexico (eds. F.P. Shepard, F.B. Phleger & T.H. Van Andel) American Association Petroleum Geologists, Tulsa. 153-196.

Sahni, A., Bhatia, S.B. and Kumar, K., 1983. Faunal evidence for the withdrawal of the Tethys in the Lesser Himalaya, Northwestern India. Bulletin Italian Society of Palaeontologists. 22: 76-86.

Sahni, A. and Khare, S.K. 1973. Additional Eocene mammals from the Subathu Formation of Jammu and Kashmir state. Journal Palaeontological Society of India. 17: 31-49.

Sharda, Y.P. and Verma, V.K. 1977. Palaeoenvironments during Murree and Siwalik sedimentation around Udhampur, Jammu Himalaya. Proc. Cent. Adv. Study Geol., Chandigarh, 23-35.

Singh, B.P. 2000. Sediment dispersal pattern in the Murree Group of Jammu area, India, NW Himalaya. Him. Geol., 21: 189-200.

Singh, B.P. Andotra, D.S., 2000. Barrier-lagoon and tidal cycles in Palaeocene to Middle Eocene Subathu Formation, NW Himalaya, India. Tertiary Res. 20: 65-78.

Singh B.P. and Singh H., 1995. Evidence of tidal influence in the Murree Group of rocks of the Jammu Himalaya. India. In: Tidal Signatures in Modern and Ancient Sediments. B.W. Flemming and A. Bartholoma (Eds). Inter. Assoc. Sediment. Spec. Publ. 24: 343-351.

Singh B.P. Fotedar, B.K. and Rao, A.S., 1990. Petrography and geochemistry of the sandstones of Murree Group around Laren, Udhampur, Jammu Himalaya. Jour. Geol. Soc. Ind. 36: 502-511.

Singh B.P., Andotra, D.S., Kumar, R., 2000 Provenance of the lower Tertiary mudrocks in the Jammu Sub-Himalayan Zone, Jammu and Kashmir State (India), NW Himalaya and its tectonic implications. Geosciences Jour. 4 (1): 173-186.

Singh, G.P. 1988. A preliminary study of the Tertiary coal deposits of Jammu Province (J&K). Journal of Science Research, Banaras Hindu University. 38 173-186

Singh, J. 1996. Biometric approach for determining younging direction in Subathu Formation in Dhundan area of Himachal Pradesh, India, In: Contributions to XV Indian Colloquium in Micropalaeontology and Stratigraphy, Dehra Dun (eds. J. Pandey, R.J. Azmi, A. Bhandari and A. Dave). 229-238.

Singh, M.P. Singh G.P., 1995. Petrological evolution of Palaeogene coal deposits of Jammu, Jammu and Kashmir, India. International Journal of Coal Geology. 27: 171-199.

Singh P., 1970. Larger Foraminifera form the Subathus of Beragua-Jangalgali area (Jammu and Kashmir state). Journal Geological Society of India. 11: 34-44.

Singh, P., 1973. Geology of the Subathu Group of Jammu and Kashmir state. Himalayan Geology. 3: 284-298.

Singh, P., 1980. Subathu Group of India. Navjyoti Scientific Publication Lucknow, India. (Professional paper) 1: 1-92.

Sorkhabi, R.B., Jain, A.K., Manikavasagam, R.M. and Stump, E. 1993. K/Ar hornblende ages from the Higher Himalaya: implications for the India-Asia Collision and Hiamalayan metamorphism. Current Science. 64: 40-42.

Srivastava, R. and Kumar, K. 1996. Taphonomy and palaeoenvironment of the middle Eocene rodent Localities of northwestern Himalaya, India. Palaeogeography Palaeoclimatology and Palaeoecolgy. 122: 185-211.

Tessier, B., Archer, A.W. Lanier, W.P. and Feldman, H.R., 1995. Comparison of ancient tidal rhythmites (Carboniferous of Kansas and Indiana, U.S.A.) with modern analogues (the bay of Mont-Saint-Michel, France). In: Tidal Signatures in Modern and Ancient sediments. B.W. Flemming and A. Bartholoma (Eds). Inter, Assoc. Sediment. Spec. Publ.24: 259-271.

Treloar, P.J. and Rex, D.C., 1990. Post-metamorphic cooling history of the Indian Plate crystalline thrust stack, Pakistan Himalaya. Jour. Geol. Soc. Lond. 147: 735-738.

Visor, M.J. 1980. Neap-spring cycles reflected in Holocene sub-tidal large scale bed form deposits: a preliminary note. Geology. 8: 543-546.

Wadia, D.N. 1928. Geology of Poonch State (Kishmir) and adjacent portion of the Punjab. Mem. Geol. Surv. Ind. 51 : 370.

Wadia, D.N. and Gingerich, P.D. 1987. Palaeoenvironmental Interpretation of Palaeogene strata near Kotli. Azad Kashmir, N.E. Pakistan. Kashmir Journal of Geology. 5: 23-41.

Wright, L.D. 1985. River Deltas, In: Coastal sedimentary Environments R.D. DAVIS (Ed.) Springer-Verlag, 1-76.

Wright, L.D., and Coleman, J.M., 1973. Variations in Morphology of Major river deltas as functions of ocean wave and river discharge regimes, Amer. Assoc. Petrol. Geol. Bull. 57: 370-398.

Wright, V.P. and Burchette, T.P. 1996. Shallow water carbonate environments in: Sedimentary Environments: Processes, Facies and Stratigraphy (ed. H.G. Reading). Blackwell Science. 325-394.

BIOTITE FROM NEOPROTEROZOIC GRANITE GNEISS, CAMBRO-ORDOVICIAN GRANITOIDS AND NEOGENE LEUCOGRANITE OF KINNAUR DISTRICT, HIMACHAL PRADESH

Santosh Kumar and Brajesh Singh

Department of Geology, Kumaun University, Nainital - 263 002

ABSTRACT

The rocks exposed along the Satluj and Baspa and its tributary- valleys belong to the Central Crystalline Zone in the southern part of Himachal Higher Himalaya and are mainly represented by granite gneiss (601±19 *Ma*), migmatites and intercalated metasediments intimately associated with Rakcham Granitoids (453±9*Ma*), belonging to the rocks of Vaikrita Group whereas tourmaline bearing leucogranite (14-28 *Ma*), is considered younger lithounit of Haimanta Group. Electron microprobe analyzed biotites, barring a few exceptions, from granite gneiss (GGn), Rakcham Granitoids (RG) and tourmaline bearing leucogranite (TLg) lie in the field of igneous biotites and are considered primary in nature. Biotites from GGn, RG and TLg can be classified as Fe^{2+} biotites while biotites from TLg are slightly enriched in octahedral Al+Fe^{t}+Ti+Mn. The biotites from these lithotypes have chemical affinity with the biotites coexisting with aluminosilicates and muscovite, which typically crystallized in the peraluminous (S-type) granite melts. Biotites crystallized in GGn and TLg melts exhibit negative correlations between Al_2O_3-MgO and Al_2O_3-FeO^t suggesting pronounced 3(Mg,Fe) = 2Al substitution during evolution of peraluminous granitoid melts. However, RG biotites show only 3(Fe) = 2Al substitution. The GGn, RG and TLg biotites show distinct compositional trends in terms of Fe^{2+}-Fe^{3+}-Mg mostly buffered between FMQ and HM through NNO, though GGn biotites even evolve into HM buffer field. Using T-X-fO_2 plane of biotite stability at 2070 bars, the deduced $fO_2 = 10^{-16.68}$ to $10^{-13.26}$ bars and T = 860 to 520° C for GGn melt, $fO_2 = 10^{-14.50}$ to $10^{-12.50}$ bars and T = 795 to 780° C for RG melt, and $fO_2 = 10^{-16.07}$ to $10^{-15.39}$ bars and T = 810 to 730° C for TLg melt. The observed reducing evolutionary trends of GGn, RG and TLg biotites are typically differentiated from ilmenite (monazite) series of granites derived from metasedimentary sources, as similarly reflected by magnetic susceptibility (÷) values of GGn (÷ = 0.162×10^{-3}SI), RG (÷ = 0.144×10^{-3}SI) and TLg (÷ = 0.051×10^{-3}SI). However post-magmatic modification of few biotites particularly noted for GGn resulted in high Fe^{3+}/Fe^{2+} ratio of biotite composition buffered above HM. The observed biotite compositions from GGn, RG and TLg can be intrinsic to the source region during melting events and/or their subsequent crystallization of episodic peraluminous granitic melts.

INTRODUCTION

The Central Crystalline Zone (CCZ) forms an integral part of the Tethys Himalayan Tectogen (Srikantia and Bhargava, 1998). The lithounits exposed along the Satluj and Sangla river valleys and its tributaries belong to the CCZ located in the southern portion of Himachal Higher Himalayas. These lithounits are granite gneiss (601±19 Ma; Mehta, 1977), migmatites and intercalated metasediments associated with granitoids (453±9 Ma; Kwatra et al., 1999) belonging to the Vaikrita Group of rocks exposed in and around Reckong Peo, Ribba, Karcham, Akpa and Rakcham areas and beyond. However tourmaline bearing leucogranite (14-28 Ma; Srikantia and Bhargava, 1998; p. 187) is thought to be formed during post-Himalayan Neogene time belonging to Haimanta Group of rocks.

The regional geology of the studied area was first discussed by Stoliczka (1866) followed by a detailed work of Hayden (1904). Berthelsen (1951) gave a preliminary account of various rocks exposed in the Kinnaur Basin. Iyenger (1949) carried out a survey of the mineral resources of Kinnaur district. In the recent years many workers have described the geology of different parts of the Kinnaur Basin (e.g. Srikantia et al., 1975; Sharma, 1977, 1983; Tewari et al., 1978; Bhargava, 1980; Srikantia, 1981; and Prashra and Rapa, 1982). Bassi and Chopra (1983) were first to give a detailed litho-stratigraphy of the Kinnaur Basin. Kakar (1988) described the geology of the Central Crystalline Rocks of the Himachal Higher Himalaya. Detailed exploration for the polymetallic mineralisation in the Kinnaur district was taken up by the Geological Survey of India, to assess the economic potential of the area. This involved reconnaissance survey, detail mapping and bedrock sampling. Kumar and Jain (1993) have reported the occurrence of scheelite, pyrite, chalcopyrite, pyrrhotite and bornite at Purbani in Kinnaur district. Sharma (1995) suggested that magmatic solutions responsible for mineralisation hosted in granite gneiss at Purbani were probably related to the mobilization of Rakcham granitoids. Tungsten mineralisation has also been discovered in the adjoining areas (Dorka and Gupta, 1993; Ruby and Singh, 1994; Sharma, 1995). More recently Srikantia and Bhargava (1998) have published a memoir on geology of Himachal Pradesh.

Numerous studies were carried out on biotites and it was noted that the primary and hydrothermally altered biotites from felsic igneous rocks could provide valuable petrogenetic informations and condition of ore-mineralisation, if any, respectively (e.g. Heinrich, 1946; Nockolds, 1947; Foster, 1960; Wones and Eugster, 1965; Rowins et al, 1980; Wones, 1972, 1981; Wones and Eugster, 1965; Abdel-Rahman, 1994). Biotite is a significant ferromagnesian mineral present in GGn, RG and TLg. In this paper we present electron microprobe analyzed composition of biotites from representative GGn, RG and TLg samples of Kinnaur district of Himachal Higher Himalaya , and discuss the physical conditions, source, nature and tectonic environment of biotites and their host felsic magmas.

GEOLOGICAL SETTING

The rocks of the studied area comprise of metamorphosed lithounits (meta-sediments) of the Vaikrita Group (Sharma, 1977; Kumar and Jain, 1993) and are represented by psammitic gneiss, banded gneiss, granite gneiss, coarse psammitic gneiss (collectively referred to as granite gneiss, GGn), calcareous bands and pegmatites. Rakcham Granitoids (RG) collectively referred to as Cambro-Ordovician (453±9 Ma) granitoids are best exposed in and around Ribba, Akpa and Rakcham villages along the Satluj and Baspa river valleys, which have sharp to gradational contacts with the coarse psammitic gneiss variety of GGn of the Vaikrita Group near Ribba village in the north (Fig. 1) and between Sangla and Rakcham in the south.

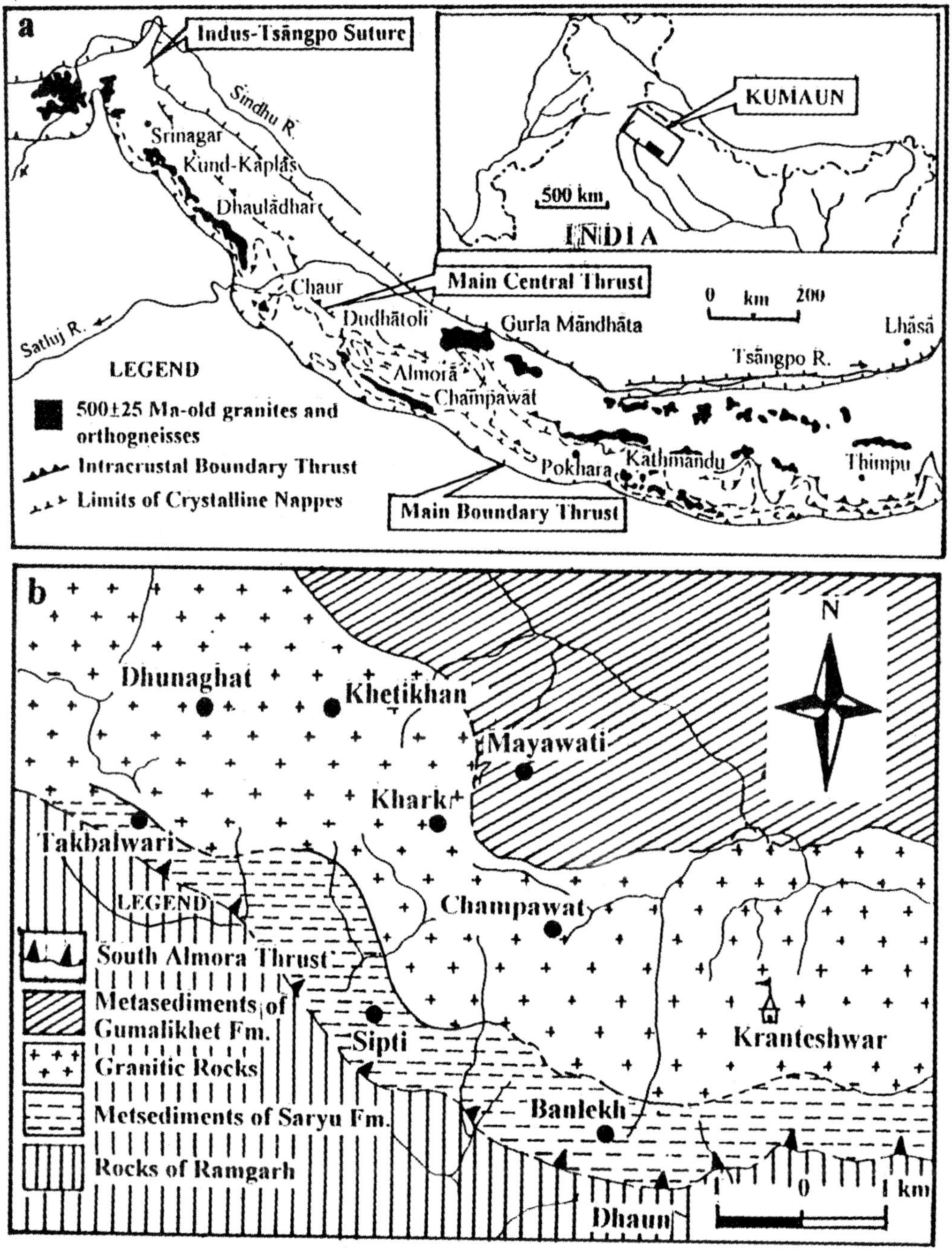

Figure 1: Geological map of the area (after Kakar, 1988).

Succession of Rocks of the Study Area

Simplified lithological succession of rocks of the study area as evident from field relationships is given below (compiled and simplified from Bartarya et al., 1996; Srikantia and Bhargava, 1998; Gupta et al., 2001).

	Tourmaline leucogranite, Pegmatite / Quartz Vein / scheelite / skarn
Haimanta Group	(14-28 Ma)
	Schist, Phyllite and Quartzite
	TETHYAN THRUST
	Rakcham Granitoid
	(453±9 Ma)
Vaikrita Group	Granite Gneiss (Psammitic Gneiss, Banded Gneiss, Granite Gneiss, Coarse Psammitic Gneiss), / Calcareous Bands, Migmatites (601±19 Ma)
	VAIKRITA THRUST
Karcham Group	Schist, Gneiss and Quartzite
	KARCHAM THRUST
	Wangtu Gneissic Complex (Jutogh Group)

Vaikrita Group mainly consists of various types of Neoproterozoic granite gneisses and Cambro-Ordovician Granitoids. However quartz veins, tourmaline bearing leucogranite and pegmatites are younger lithounits of Haimanta Group of rocks.

GENERAL PETROLOGY

Granite Gneiss (GGn) is medium- to coarse-grained, moderately foliated rock characterised by felsic (quartz and feldspar) and mafic (ferro-magnesian) minerals. The variety of GGn is represented by psammitic gneiss, banded gneiss and coarse psammitic gneiss, showing varied structure. Despite the varied textures and structures the bulk modal and chemical compositions of these gneisses do not vary significantly. Modal compositions of GGn collectively correspond to monzogranite (*sensu stricto*) representing anatectic granite melt formed by crustal fusion (Laymere and Bowden, 1982). The mineral assemblage, whole rock data and magnetic susceptibility (Singh and Kumar, 2002) of GGn typically represent peraluminous (S-type) granite (mus+bt+tour+ap; molar A/CNK=1.1; SiO_2=67.77-72.95 wt. %, Al_2O_3=12.36-15.73 wt. %, K_2O=1.78-5.23 wt.%, CaO=0.78-2.19 wt.%, TiO_2=0.07-0.72 wt.%; magnetic susceptibility (÷) ranging from 0.129 to 0.344 × 10^{-3} SI; CIPW corundum 2.42 to 6.41 wt.%; initial $^{87}Sr/^{86}Sr$=0.711, Kwatra et al., 1999; tungsten mineralisation with pegmatite).

Rakcham Granitoids (RG), named after Rakcham village where they are best exposed, intrude the psammitic gneiss (variety of GGn) of Vaikrita Group near Ribba village showing sharp to gradational contact. The RG are fine- to coarse-grained, meso- to leucocratic and non-foliated but enclose abundant small-spotted surmicaceous enclaves and xenoliths of country-rocks (metasediments and granite-gneiss). The xenoliths and surmicaceous enclaves are variable in shape and size. Frequency of the former is more near the contact. The RG are modally two-mica (biotite-muscovite) leuco-monzogranite belonging to the field of anatectic granite melt formed by crustal fusion (Laymere and Bowden, 1982). At places, due to the presence of quartz and feldspar phenocrysts and megacrysts, RG become porphyritic in nature. Frequently veins, patches and clots of tourmaline bearing leucogranite (TLg) and mica pegmatites are found closely associated with RG. The whole-rock geochemistry of RG (Singh and Kumar, 2002) characterizes its nature similar to syncollisional, peraluminous (S-type) and ilmenite series of granitoids (mus+bt+tour+ap; abundant meta-sedimentary enclaves; molar A/CNK=1.1; SiO_2=70.75-72.95 wt.%, Al_2O_3=14.33-15.94 wt.%, K_2O=3.53-6.52 wt.%, CaO=0.51-1.99wt.%, TiO_2=0.02-0.17 wt.%; magnetic susceptibility (÷) ranging from 0.087 to 0.165×10^{-3} SI; CIPW corundum 0.05 to 3.79 wt.%; initial $^{87}Sr/^{86}Sr$=0.737±0.002, Kwatra et al., 1999; showing association of tourmaline and tungsten mineralization with pegmatite).

Tourmaline bearing leucogranite (TLg) is medium- to coarse-grained non-foliated rock occurring as veins, pockets, lenses within RG, and is considered magmatic product of Himalayan collisional tectonics (Kwatra et al., 1999). TLg contains radiating needle-shaped to well developed tourmaline crystals and is thought to be formed due to remobilisation of RG and probably the older metamorphites during Himalayan orogenesis. Although such a huge quantity of TLg produced by remobilization is highly questionable. The modal mineralogy of TLg corresponds to monzogranite (*sensu stricto*) having character similar to anatectic granite formed by crustal fusion (Laymere and Bowden, 1982). The mineral assemblage and whole rock composition of TLg (Singh and Kumar, 2002) typically represent peraluminous (S-type) granites (mus+bt+tour+ap; molar A/CNK=1.1; SiO_2=67.77-71.64 wt.%, Al_2O_3=13.61-16.03 wt.%, K_2O=1.60-9.90 wt.%, CaO=0.16-2.19 wt.%, TiO_2=0.02-0.65 wt.%; magnetic susceptibility (÷) ranging from 0.045 to 0.056×10^{-3} SI; CIPW corundum 1.15 to 4.34 wt.%). Quartz and mica-pegmatite veins of variable dimensions also cut across the RG and GGn lithounits, but are oftenly devoid of mineralisation.

Skarns (chiefly calc-silicates constituted by plagioclase, clinopyroxene, and garnet etc.) are developed along shear zones in calcareous bands occurring within GGn unit, and act as host for tungsten (scheelite) mineralisation. Scheelite occurs sporadically as specks and stringers in the skarns forming lenticular zones (Sharma, 1995). Pyrrhotite, pyrite (± chalcopyrite) and bornite are also associated with these skarns.

OCCURRENCE AND TEXTURAL CHARACTERISTICS OF BIOTITE

GGn shows porphyritic texture with megacrysts of quartz and feldspar, and preferred orientation of flaky minerals such as biotite and muscovite. Euhedral to subhedral flakes of strongly pleochroic (dark brown to pale yellowish brown colour) biotite of primary nature are noted (Fig. 2a). Common accessories include euhedral apatite, zircon, tourmaline and traces of iron-titanium oxides occurring either as inclusions in felsic minerals or as components of the interstitial ground mass. The biotite along with muscovite shows orientation along the schistosity plane, wherever suffered foliation and gneissosity. Poor alteration of biotite to chlorite along cleavages is due to hydration with the liberation of secondary Fe-Ti oxides forming minute granules arranged in clusters along fractures and cleavages. Secondary biotite is noted where secondary muscovite is found replacing the plagioclase (Fig. 2b).

RG overall shows fine- to coarse-grained hypidiomorphic texture and even attains to granitic pegmatitic variety. It mainly consists of quartz, K-feldspar (± perthite), sodic plagioclase (± antiperthite), biotite and muscovite and accessories such as tourmaline, apatite and occasional zircon. Biotite and muscovite are common ferromagnesian minerals in all varieties of GGn, RG and TLg. The modal constituent of biotite in RG always dominates over muscovite, and both occur as randomly oriented flakes (Fig. 2c). Primary biotite is strongly pleochroic (pale yellowish brown to deep brown) whereas secondary biotite partly replacing the primary biotite shows pale reddish to dark reddish pleochroism (Fig. 2d).

TLg shows medium- to coarse-grained hypidiomorphic texture. It mainly consists of quartz, K-feldspar (± perthite), sodic plagioclase (± antiperthite), biotite, muscovite and tourmaline and accessories such as zircon, apatite and traces of iron-titanium oxides. The micas occur as randomly oriented flakes and modal content of muscovite dominates over biotite (Fig. 2e). Biotite is associated with muscovite showing its replacement with muscovite. Primary biotite shows yellowish brown to dark brown pleochroism while the secondary biotite shows pleochroism from light green to yellow. Tourmaline, homogeneous and zoned, occurs in medium- to coarse-grains bearing prominent cracks but occasionally as minute needles showing pleochroism from light brown to shades of green. It has inclusions of quartz and feldspars.

MINERAL CHEMISTRY OF BIOTITE

Analytical Technique

Biotites from GGn, RG and TLg were analysed by wavelength dispersive electron microprobe analyzer (JEOL - JXA–8600M) at Institute Instrumentation Centre, Indian Institute of Technology, Roorkee. In most cases accelerating voltage was 15 KV and electron beam was kept at 2μm at a probe current of 2×10^{-8} amp. Since total iron was determined as FeO, hence ferric (Fe^{3+}) and ferrous (Fe^{2+}) iron were divided using the method outlined by Dymek (1983), which proposes the procedures of iron partitioning based on vacancies associated with Al substitution by means of the dioctahedral component, which eliminates the calculated charge excess in all instances and permits an estimate of Fe^{3+} content. For calculating the structural formulae and other manupulations of microprobe analysis of biotite, DOS-based software *bioterm* (Yavuz and Öztas, 1997) was used. The electron microprobe data of biotites and their recalculated cations from representative GGn, RG and TLg are given in tables 1, 2 and 3 respectively.

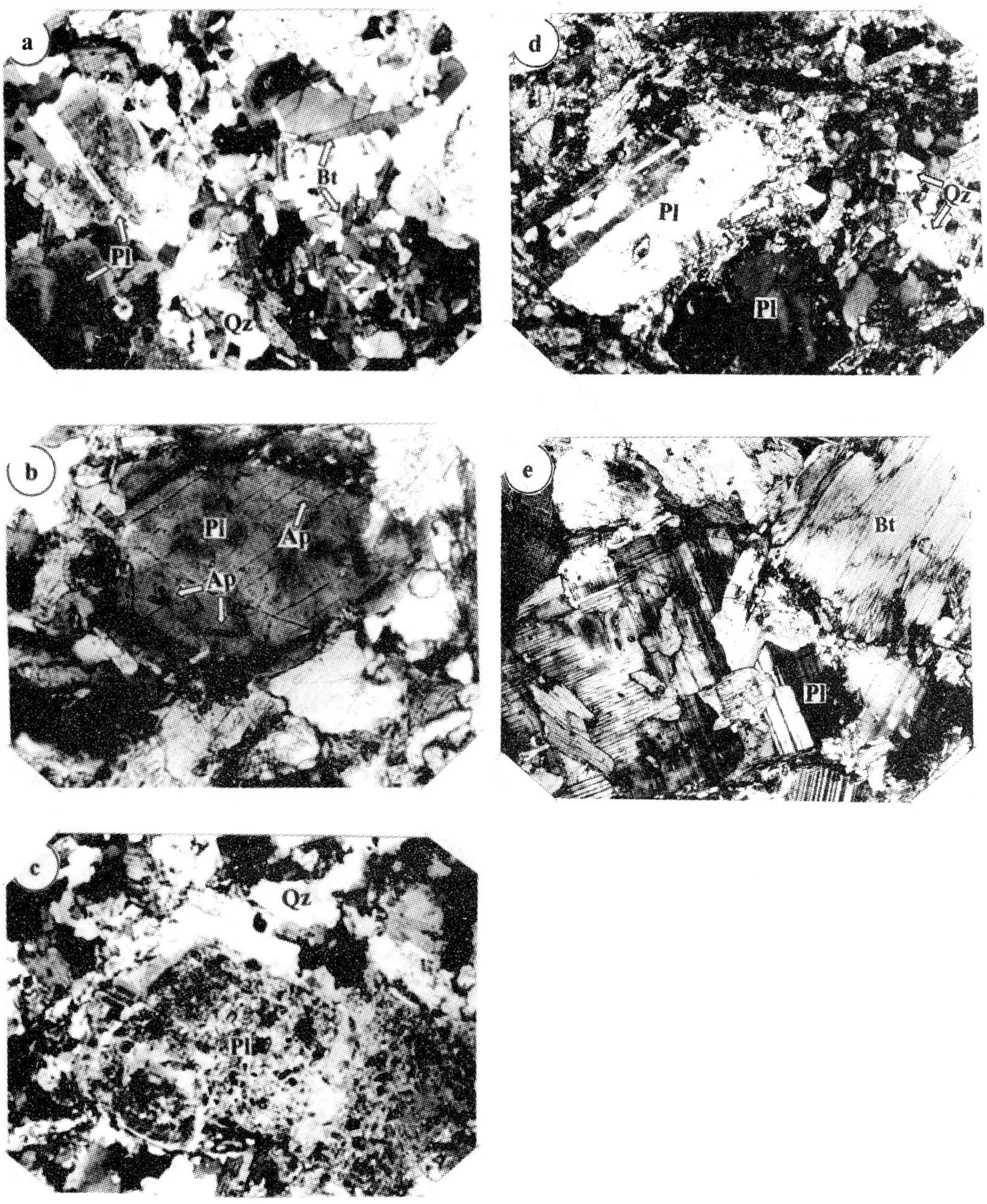

Figure 2 a-e : (a) Euhedral to subhedral flakes of strongly pleochroic primary biotite (Bt) in granite gneiss (GGn). **(b)** Secondary muscovite (Ms) replacing the plagioclase (Pl) in association with primary biotite and quartz. **(c)** Randomly oriented biotite and muscovite flakes in Rakcham Granitoids (RG). **(d)** Secondary biotite partly replacing along the cleavages in the core of muscovite of Rakcham Granitoids (RG). **(e)** Randomly oriented biotite (Bt) and muscovite flakes in tourmaline bearing leucogranite (TLg). All photomicrographs are taken under cross polars. Bar given in each figure equals 0.5mm.

Classification, Nomenclature and Compositional Variation

Primary and secondary hydrothermal biotites can be distinguished using ternary composition annite-phlogopite-oxyannite (proton deficient) as suggested by Beane (1974), which is actually modified from Heinrich (1946) and Foster (1960). The biotite compositions from GGn, RG and TLg plot in the field of igneous biotite except a few biotites, which were probably affected by magmatic and post-magmatic oxidation leading to enrichment of Fe^{+3}

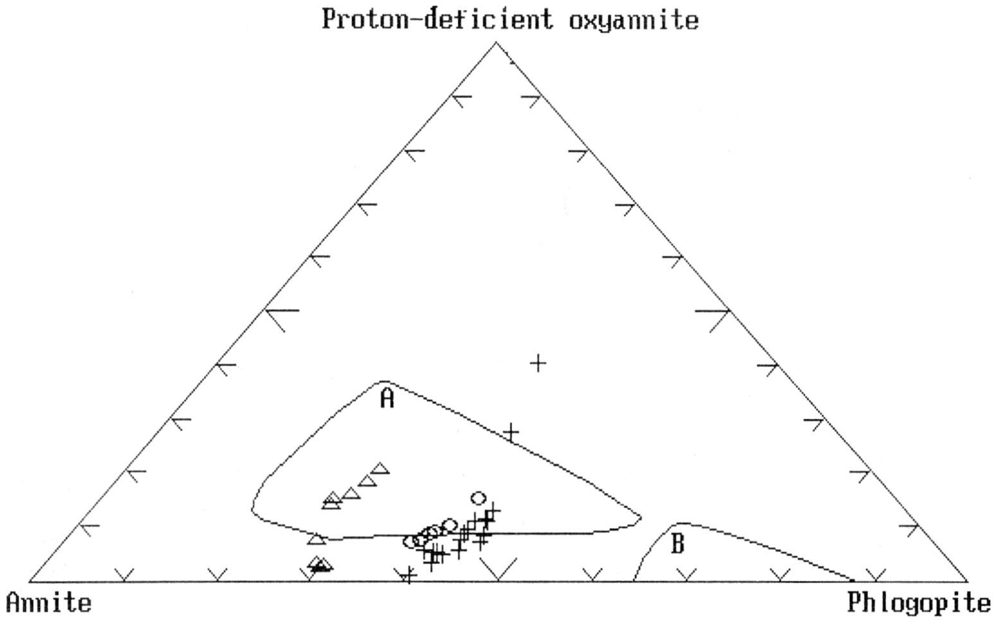

Figure 3: Annite-phlogopite-oxyannite (proton deficient) traingular diagram after Beane (1974) showing fields of primary and secondary hydrothermal biotites. Field A after Foster (1960) and field C after Heinrich (1946) represent primary biotites in granite-granodiorite rocks. Field B is taken after Beane (1974), which represents hydrothermally altered biotite. Biotite compositions from GGn(+), RG(o) and TLg(Ä) mostly plot in the field of igneous biotite except a few noted for the GGn biotites.

ion particularly noted in the case of biotite hosted in GGn (Fig 3). Therefore, all analysed biotite can be considered primary in origin, which in most cases has not been affected by hydrothermal or post-magmatic alteration except for a few with high Fe^{3+}/Fe^{2+} ratio hosted in GGn.

The composition of biotite was compared directly to the petrography of its host rock. Biotite from GGn is more Mg-rich compared to that of RG and TLg (Tables 1, 2 and 3), but they show low- to moderate compositional range TLg ($Mg^{\#}$ = 0.30 to 0.25) to RG ($Mg^{\#}$ = 0.40 to 0.37) and GGn ($Mg^{\#}$ = 0.45 to 0.34). Foster (1960) proposed a ternary diagram of the octahedral cations in terms of Mg^{2+}-[Al^{VI}+Fe^{+3}+Ti^{4+}]–[Fe^{+2}+Mn] and delineated the biotite fields of major igneous rock types, although these fields show a considerable overlap. The biotites from TLg, RG and GGn can be equivocally classified as Fe^{+2} biotites however biotites from TLg are slightly Fe-enriched plotting towards siderophyllite and lepidomelane (Fig. 4).

Table 1: Electron Microprobe Analysis of representative biotite from selected Granite Gneiss (GGn) samples.

	1	2	3	4	5	6	7	8	9
SiO_2	34.45	34.87	34.90	34.44	33.13	34.63	35.06	32.71	34.14
Al_2O_3	17.90	17.81	18.69	18.23	17.98	18.39	18.84	17.64	18.87
TiO_2	5.19	5.14	4.59	4.77	4.16	4.26	4.48	5.34	4.05
FeO	20.45	20.06	20.67	20.88	20.01	19.92	20.41	23.82	20.32
MnO	0.05	0.06	0.11	0.08	0.00	0.00	0.04	0.05	0.03
MgO	8.67	8.66	8.49	8.49	8.51	8.95	7.96	8.07	8.30
CaO	0.01	0.01	0.01	0.00	0.00	0.01	0.01	0.00	0.07
Na_2O	0.32	0.20	0.13	0.24	0.08	0.16	0.00	0.00	0.04
K_2O	9.53	9.54	9.62	9.67	9.20	9.51	9.74	8.97	9.13
TOTAL	96.57	96.35	97.21	96.80	93.07	95.83	96.54	96.60	94.95
Structural Formulae based on 11 oxygen basis									
Si	2.62	2.65	2.63	2.62	2.61	2.64	2.65	2.53	2.63
Al^{IV}	1.38	1.35	1.37	1.38	1.39	1.36	1.35	1.47	1.37
Z = 4	4.00	4.00	4.00	4.00	4.00	4.00	4.00	4.00	4.00
Al^{VI}	0.22	0.24	0.29	0.25	0.28	0.29	0.33	0.13	0.34
Ti	0.30	0.29	0.26	0.27	0.25	0.24	0.25	0.31	0.23
Fe^{3+}	0.29	0.24	0.21	0.25	0.26	0.19	0.11	0.68	0.19
Fe^{2+}	1.01	1.03	1.10	1.08	1.06	1.08	1.18	0.86	1.12
Mn	0.00	0.00	0.01	0.01	0.00	0.00	0.00	0.00	0.00
Mg	0.98	0.98	0.95	0.96	1.00	1.02	0.90	0.93	0.95
$\Sigma Y \approx$ 2.5 to 3	2.80	2.78	2.82	2.82	2.85	2.82	2.77	2.91	2.83
Ca	0.00	0.00	0.00	0.00	0.00	0.00	0.00	0.00	0.01
Na	0.05	0.03	0.02	0.04	0.01	0.02	0.00	0.00	0.01
K	0.92	0.92	0.92	0.94	0.92	0.92	0.94	0.88	0.90
$\Sigma X \approx 1$	0.97	0.95	0.94	0.98	0.93	0.94	0.94	0.88	0.92
Fe#	0.57	0.57	0.58	0.58	0.57	0.56	0.59	0.62	0.58
Mg#	0.43	0.43	0.42	0.42	0.43	0.44	0.41	0.38	0.42

Fe# = Fe^t/Fe^t+Mg Mg# = $Mg/Mg+Fe^t$

	10	11	12	13	14	15	16	17	18
SiO_2	34.94	35.76	35.15	35.41	35.59	34.86	35.37	37.38	34.58
Al_2O_3	18.53	19.60	19.68	18.94	18.67	18.75	19.42	20.82	18.61
TiO_2	4.84	4.19	4.63	4.50	4.62	4.67	4.21	4.21	4.78
FeO	19.87	17.92	19.79	19.94	20.22	19.87	19.63	17.26	20.50
MnO	0.02	0.02	0.07	0.04	0.05	0.09	0.01	0.07	0.07
MgO	8.16	8.12	8.37	7.46	7.96	7.99	7.50	5.04	7.50
CaO	0.02	0.00	0.00	0.00	0.03	0.01	0.02	0.03	0.00
Na_2O	0.05	0.16	0.26	0.21	0.09	0.15	0.25	0.27	0.23
K_2O	9.42	9.81	9.57	9.60	9.69	9.65	9.97	9.83	9.52
TOTAL	95.85	95.58	97.52	96.10	96.92	96.04	96.38	94.91	95.79
Structural Formulae based on 11 oxygen basis									
Si	2.65	2.69	2.62	2.68	2.68	2.65	2.67	2.81	2.64
Al^{IV}	1.35	1.31	1.38	1.32	1.32	1.35	1.33	1.19	1.36
Z = 4	4.00	4.00	4.00	4.00	4.00	4.00	4.00	4.00	4.00
Al^{VI}	0.31	0.44	0.350	.37	0.33	0.33	0.40	0.65	0.32
Ti	0.28	0.24	0.26	0.26	0.26	0.27	0.24	0.24	0.27
Fe^{3+}	0.16	0.14	0.12	0.02	0.07	0.10	0.11	0.66	0.11
Fe^{2+}	1.10	0.99	1.12	1.24	1.21	1.16	1.13	0.42	1.20
Mn	0.00	0.00	0.00	0.00	0.00	0.01	0.00	0.00	0.00
Mg	0.92	0.91	0.93	0.84	0.89	0.90	0.84	0.56	0.85
ΣY≈ 2.5 to 3	2.77	2.72	2.78	2.91	2.76	2.77	2.72	2.53	2.75
Ca	0.00	0.00	0.00	0.00	0.00	0.00	0.00	0.00	0.00
Na	0.01	0.02	0.04	0.03	0.01	0.02	0.04	0.04	0.03
K	0.91	0.94	0.91	0.93	0.93	0.94	0.96	0.94	0.93
ΣX≈ 1	0.92	0.96	0.95	0.96	0.94	0.96	1.00	0.98	0.96
Fe#	0.58	0.55	0.57	0.60	0.59	0.58	0.59	0.66	0.61
Mg#	0.42	0.45	0.43	0.40	0.41	0.42	0.41	0.34	0.39

Fe# = Fe^t/Fe^t+Mg Mg# = $Mg/Mg+Fe^t$

Table 2: Electron Microprobe Analysis of representative biotite from selected Rakcham Granitoid (RG) samples.

	1	2	3	4	5	6
SiO_2	36.66	34.08	34.56	35.07	35.35	35.27
Al_2O_3	19.17	18.35	19.25	18.74	18.28	18.91
TiO_2	3.21	2.76	2.79	2.80	3.20	2.98
FeO	19.17	19.52	20.11	20.33	21.15	20.85
MnO	0.17	0.14	0.15	0.10	0.12	0.16
MgO	7.25	7.22	7.06	7.26	7.04	7.17
CaO	0.00	0.14	0.01	0.04	0.04	0.05
Na_2O	0.07	0.24	0.00	0.13	0.05	0.18
K_2O	10.02	9.22	9.53	9.42	9.78	9.50
TOTAL	95.72	91.67	93.46	93.89	95.01	95.05
Structural Formulae based on 11 oxygen basis						
Si	2.75	2.71	2.70	2.73	2.73	2.71
Al^{IV}	1.25	1.29	1.30	1.27	1.27	1.29
Z = 4	4.00	4.00	4.00	4.00	4.00	4.00
Al^{VI}	0.47	0.43	0.47	0.44	0.39	0.43
Ti	0.18	0.17	0.16	0.16	0.19	0.17
Fe^{3+}	0.31	0.22	0.18	0.19	0.15	0.16
Fe^{2+}	0.91	1.08	1.13	1.13	1.21	1.18
Mn	0.01	0.01	0.01	0.01	0.01	0.01
Mg	0.82	0.86	0.82	0.84	0.81	0.82
$\Sigma Y \approx$ 2.5 to 3	2.70	2.77	2.77	2.77	2.76	2.77
Ca	0.00	0.01	0.00	0.00	0.00	0.00
Na	0.01	0.04	0.00	0.02	0.01	0.03
K	0.97	0.94	0.95	0.93	0.96	0.93
$\Sigma X \approx 1$	0.98	0.99	0.95	0.95	0.97	0.96
Fe#	0.60	0.60	0.62	0.61	0.63	0.62
Mg#	0.40	0.40	0.38	0.39	0.37	0.38

Fe# = Fet/Fet+Mg   Mg# = Mg/Mg+Fet

Table 3: Electron Microprobe Analysis of representative biotite from selected Tourmaline bearing Leucogranite (TLg) samples.

	1	2	3	4	5	6	7	8	9
SiO_2	36.38	34.58	35.81	33.83	34.33	33.35	33.35	34.33	33.70
Al_2O_3	20.35	20.61	19.53	18.53	18.77	18.93	19.18	19.89	20.15
TiO_2	1.61	1.87	1.80	1.81	1.77	2.05	1.97	1.96	2.02
FeO	24.00	22.95	24.42	24.77	24.89	23.99	24.20	24.20	23.65
MnO	0.37	0.44	0.41	0.54	0.53	0.71	0.68	0.40	0.39
MgO	5.05	4.78	4.93	5.95	6.01	5.01	5.62	4.49	4.51
CaO	0.00	0.01	0.01	0.05	0.03	0.05	0.04	0.02	0.06
Na_2O	0.18	0.01	0.14	0.00	0.08	0.00	0.00	0.05	0.01
K_2O	9.77	9.84	9.53	9.52	9.35	9.60	9.56	9.85	9.70
TOTAL	97.71	95.09	96.58	95.00	95.76	93.69	94.60	95.19	94.19
Structural Formulae based on 11 oxygen basis									
Si	2.75	2.69	2.75	2.67	2.68	2.67	2.64	2.69	2.66
AlIV	1.25	1.31	1.25	1.33	1.32	1.33	1.36	1.31	1.34
Z = 4	4.00	4.00	4.00	4.00	4.00	4.00	4.00	4.00	4.00
AlVI	0.56	0.58	0.52	0.40	0.41	0.45	0.43	0.53	0.54
Ti	0.09	0.11	0.10	0.11	0.10	0.12	0.12	0.12	0.12
Fe3+	0.43	0.38	0.34	0.05	0.06	0.16	0.07	0.32	0.29
Fe2+	1.09	1.12	1.23	1.58	1.56	1.45	1.53	1.27	1.27
Mn	0.02	0.03	0.03	0.04	0.04	0.05	0.05	0.03	0.03
Mg	0.57	0.55	0.56	0.70	0.70	0.60	0.66	0.52	0.53
$\Sigma Y \approx$ 2.5 to 3	2.76	2.77	2.78	2.88	2.87	2.83	2.86	2.79	2.78
Ca	0.00	0.00	0.00	0.00	0.00	0.00	0.00	0.00	0.00
Na	0.03	0.00	0.02	0.00	0.01	0.00	0.00	0.01	0.00
K	0.94	0.98	0.93	0.96	0.93	0.98	0.97	0.98	0.98
$\Sigma X \approx 1$	0.97	0.98	0.95	0.96	0.94	0.98	0.97	0.99	0.98
Fe#	0.73	0.73	0.74	0.70	0.70	0.73	0.71	0.75	0.75
Mg#	0.27	0.27	0.26	0.30	0.30	0.27	0.29	0.25	0.25

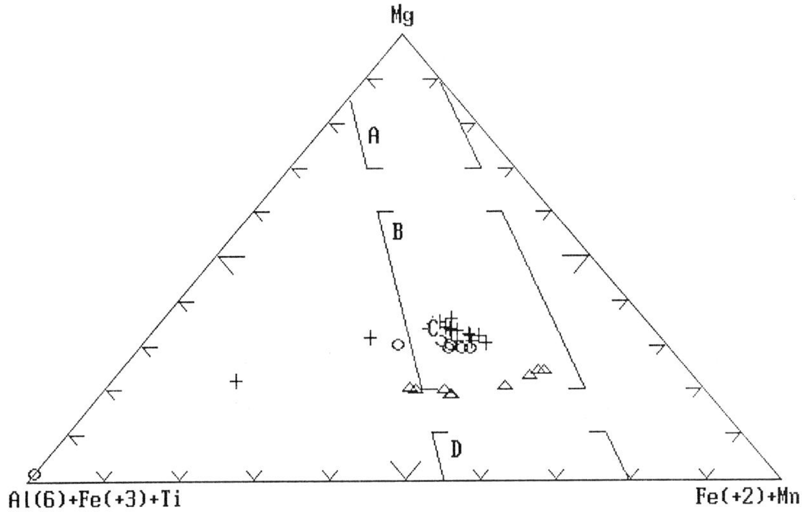

Figure 4: Classification of biotites from major igneous rock types in terms of Mg^{+2}-[Al^{VI}+Fe^{+3}+Ti^{+4}] [Fe^{+2}+Mn] components (Foster, 1960). Fields A- Phlogopite; B- Mg^{+2} biotite; C- Fe^{+2} biotite; and D- Siderophyllite and Lepidomelane. Biotite compositions from GGn, RG and TLg mostly belong to Fe^{+2} biotite. Note that TLg biotites show distinct composition more towards siderophyllite and lepidomelane compositions. Symbols are the same as used in figure 3.

A triangular Al-Mg-Fe^t diagram (not shown) after Deer et al. (1963) is incapable to discriminate the biotite compositions from GGn, RG and TLg, showing clustering in the field of biotite. However, ternary compositions MgO-FeO^t-Al_2O_3 proposed by Albuquerque (1973) clearly illustrate the position of biotites from these lithounits similar to the biotites coexisting with aluminosilicates and muscovites (Fig. 5a,b), though the majority of biotites from TLg has compositional affinity with biotite coexisting with aluminosilicate. The observed compositional affinity of biotites is petrographically consistent, and is strong indication of biotite stability in the crustally-derived granite melts probably equilibrated with restite assemblages of aluminosilicates (garnet, kyanite, sillimanite) induced by ultrametamorphism, a condition required for generation of anatectic granite.

Chemical zonation within interstitial grains was not observed, however, octahedral site occupancy between 2.5-3.0 atom per formula unit is indicative of trioctahedral micas. The phlogopite-annite-eastonite-siderophyllite end–members in terms of Fe/Fe+Mg against Al^{IV} (Fig. 6) are capable to display the condition of magmatic oxidation during primary crystallization. Biotites from GGn, RG and TLg display negative trends (Fig. 6), increasing Fe/Fe+Mg with decreasing Al^{IV}, which are compatible with decreasing fO_2 during its primary crystallisation reported for biotites of Shonkin Sag laccolith (e.g. Nash and Wilkinson, 1970 in Rowins et al., 1991). The observed biotite compositional path(s) from GGn, RG and TLg is opposite to the biotites from many igneous complexes shown by Rowins et al. (1991), which are believed to decreasing fugacity of oxygen during its crystallisation or melting (intrinsic to source) events.

Nature of the Host Magmas

Based on Al_2O_3, FeO^t and MgO contents of biotite, Abdel-Rahman (1994) proposed series of diagrams to discriminate that the biotite crystallised from alkaline, calc-alkaline and peraluminous magmas. Biotites crystallized in peraluminous melts mostly exhibit negative

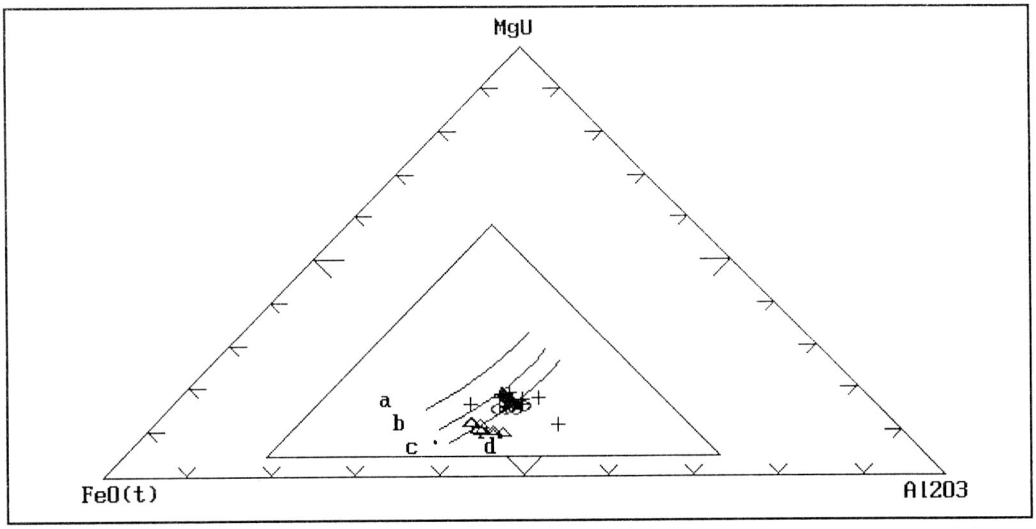

Figure 5a: FeO^t-MgO-Al_2O_3 triangular diagram after Albuquerque (1973) showing fields of biotite coexisting with different mineral association. Field a- biotite coexisting with amphibole; b- biotite coexisting with other ferromagnesian minerals; c- biotite coexisting with muscovite; and d- biotite coexisting with aluminosilicates. Biotite compositions from GGn, RG and TLg mostly cluster in the fields c and d suggesting their coexistence with muscovite and aluminosilicates.Symbols are the same as used in figure 3.

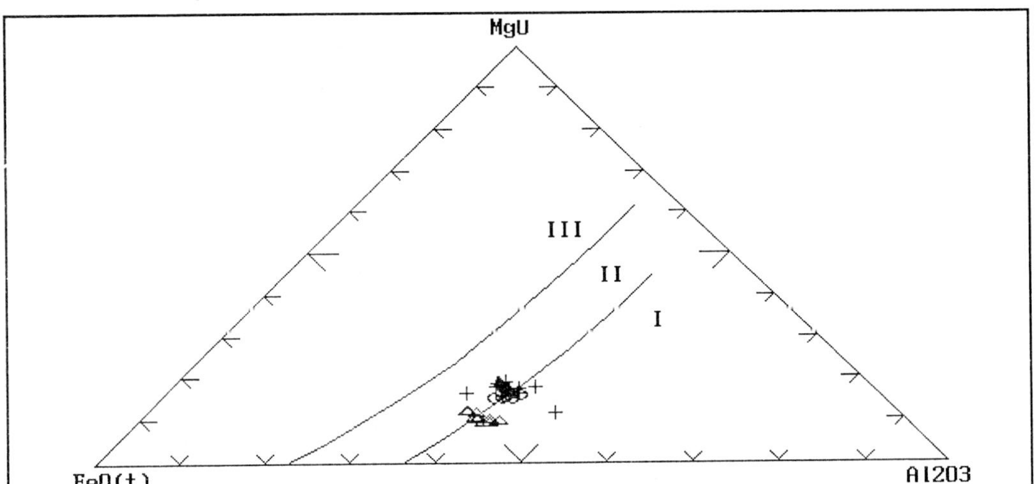

Figure 5b: FeO^t-MgO-Al_2O_3 triangular diagram after Albuquerque (1973) showing fields of biotite coexisting with different mineral association. Field I- biotite associated with muscovite or topaz; II- biotite associated with unaccompanied other mafic minerals; and III- biotite associated with hornblende, pyroxene or olivine. Biotite compositions from GGn, RG and TLg mostly cluster between fields I and II suggesting their association with muscovite and other mafic minerals. Symbols are the same as used in figure 3.

$MgO-Al_2O_3$ correlation, suggesting pronounced 3Mg=2Al substitution producing Al-rich biotites in peraluminous melts (Abdel-Rahman, 1994). FeO^t seems to behave in a similar manner but $FeO^t-Al_2O_3$ correlation is not as pronounced as noted for $MgO-Al_2O_3$ correlations. The biotites from GGn, RG and TLg can be discriminated as biotites typically crystallized in peraluminous (S-type) granite melts (Figs. 7a,b,c). TLg biotites are more enriched in total iron content compared to RG and GGn biotites, but a limited negative correlation between FeO^t and Al_2O_3 and to a certain extent between MgO and Al_2O_3 can be established (Fig. 7a,b) for GGn and TLg biotites, suggesting 3(Mg,Fe)=2Al substitution in biotites during evolution of GGn and TLg peraluminous melts. However $MgO-Al_2O_3$ correlation in RG biotites is not apparent as they tightly cluster, hence for RG biotites only a limited 3Fe=2Al substitution is suggested. The aspect of substitution relation can be explained more clearly if we analyze more biotites from RG.

Estimation of Oxygen Fugacity (fO_2) and Temperature

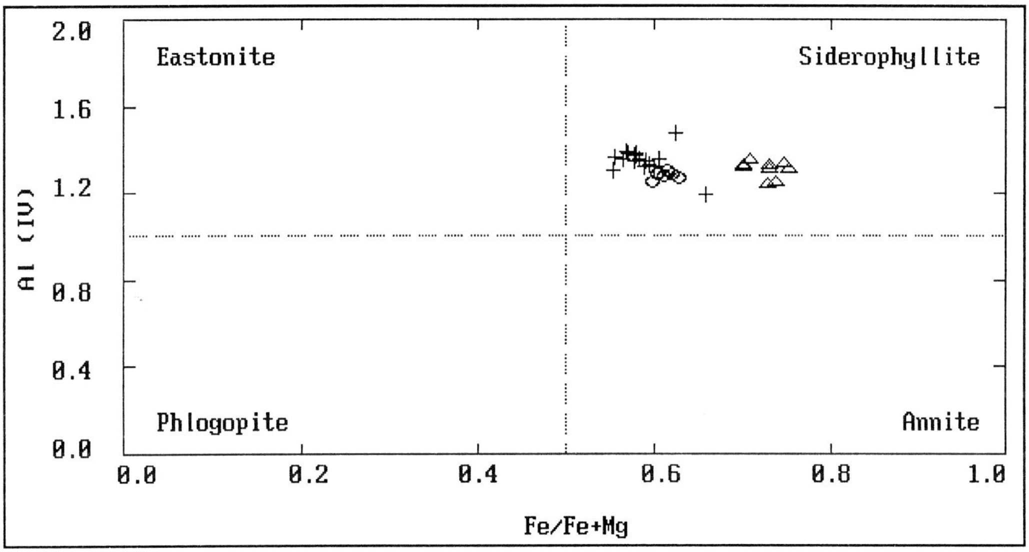

Figure 6: Plot of Fe/Fe+Mg against Al^{IV} showing phlogopite-annite-siderophyllite-eastonite end members of biotite (Deer et al., 1963). Biotite compositions from GGn, RG and TLg plot in the siderophyllite field showing increasing Fe/Fe+Mg with decreasing Al^{IV}. Symbols are the same as used in figure 3.

The condition (intensive parameters viz. T, fO_2 and fH_2O) of biotite crystallized in host felsic magmas can be deduced using biotite equilibria (Wones and Eugster, 1965; Wones, 1972) if biotite is found in cogenetic mineral assemblage of magnetite and K-feldspar. Biotite in GGn, RG and TLg co-exists with abundant K-feldspar and quartz but magnetite is either absent or present in trace amounts. This is consistent with the measurement of magnetic susceptibility (MS) values, which were noted in the range of 0.087 to 0.165×10^{-3} SI for RG, 0.344 to 0.129×10^{-3} SI for GGn and 0.056 to 0.045×10^{-3} SI for TLg. The observed MS values of these lithotypes suggest their nature similar to ilmenite series of granitoids (Ishihara, 1977; Ishihara et al., 2002). Nevertheless, with existing limitations of poor presence of magnetite, ternary compositions annite (Fe^{2+}) – phlogopite (Mg) – oxyannite (Fe^{3+}) of biotites can be applied to estimate the fO_2 conditions during biotite crystallisation in GGn, RG, TLg melts with various oxygen fugacities (Wones and Eugster, 1965, Fig.8)

Figure 8 shows FMQ, NNO and HM buffer equilibria where biotites from GGn, RG and TLg follow clearly distinct evolutionary trends and mostly lie between FMQ and HM buffers crossing through NNO. The compositional curvature of biotites from GGn, RG and TLg appear more related to the magmatic differentiation (crystallisation and/or partial melting of the source) perhaps least influenced by post-magmatic alteration but it was more pronounced in case of biotite hosted in GGn, which evolved above the HM buffer. A better evaluation of fO_2 and thermal stability of biotite can be obtained using contoured [Fe/Fe+Mg] × 100 ratio of biotite projected on to a calibration curve of Wones and Eugsters (1965) in fO_2-T space at 2070 bars. Biotites from GGn, RG and TLg overall show reducing trends

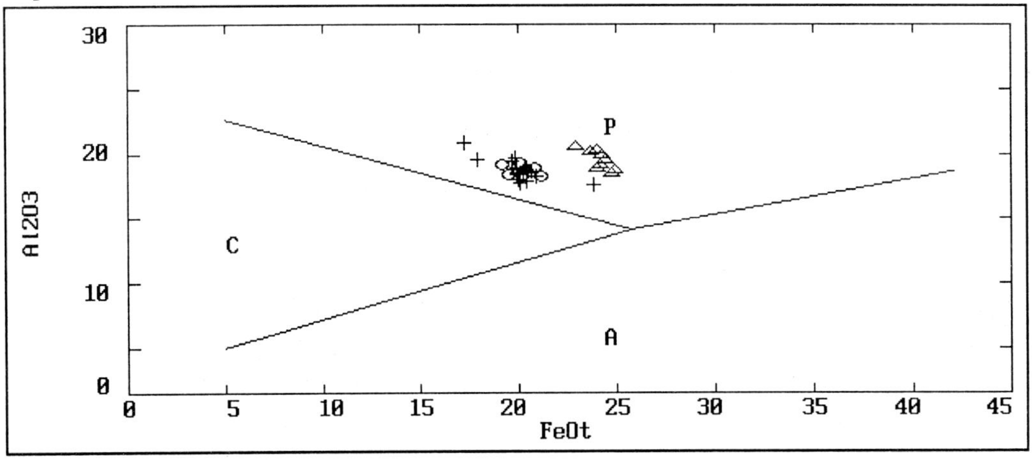

Figure 7a: Al_2O_3 vs. FeO^t plot for GGn, RG and TLg biotites showing negative correlation. Fields A, C and P are taken after Abdel-Rahman (1994). A- biotites in anorogenic alkaline suites; C- biotites in calc-alkaline orogenic suites; and P- biotites in peraluminous (including S-type) suites. Note that biotites from GGn, RG and TLg typically plot in the biotites crystallised in peraluminous melt. Symbols are the same as used in figure 3.

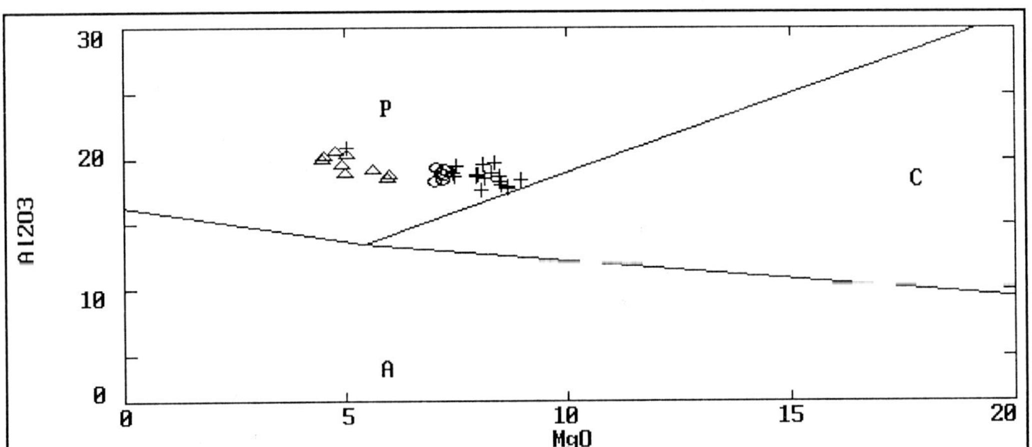

Figure 7b:Al_2O_3 vs. MgO plot for GGn, Rg and TLg biotites showing negative correlation for GGn and TLg whereas RG biotites tightly cluster and do not show variation. Note that all biotites plot in peraluminous field. The explanation of the figures A, C and P are the same as given for figures 7a. Symbols are the same as used in figure 3.

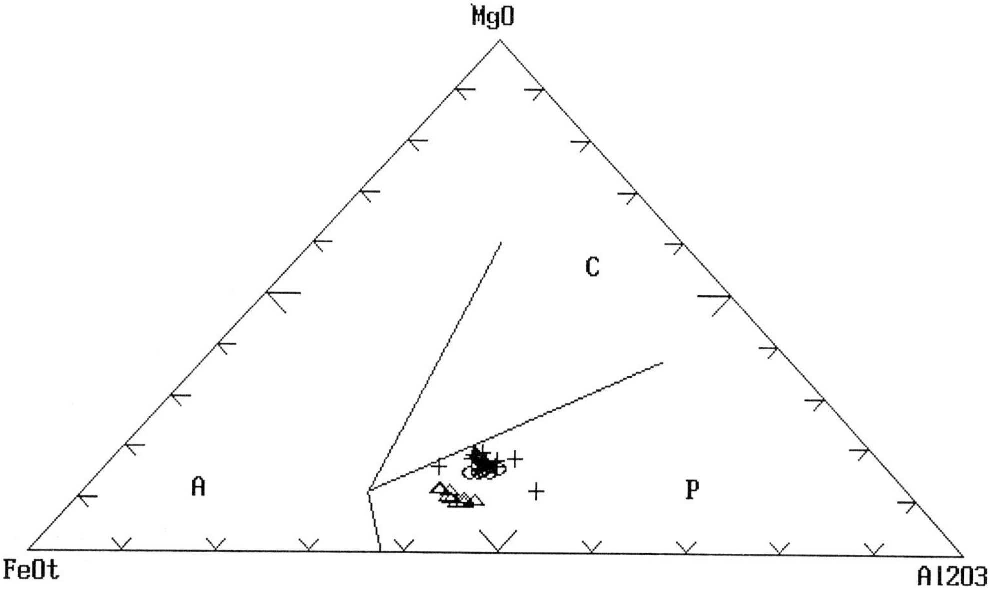

Figrue 7c: FeO^t-MgO-Al_2O_3 triangular plot suggesting the nature of biotites crystallised in peraluminous melts. The explanation of the figures A, C and P are the same as given for figures 7a,b. Symbols are the same as used in figure 3.

with decreasing temperature at their respective FMQ, NNO and HM buffers. The $Fe^{\#}$ (=Fe/Fe+Mg) of biotite ranges from 0.66 to 0.55, T= 860 to 520° C, and fO_2 $10^{-16.68}$ to $10^{-13.26}$ bars for GGn, $Fe^{\#}$ = 0.63 to 0.60, T = 795 to 780° C, and fO_2 = $10^{-14.50}$ to $10^{-12.50}$ bars for RG, and $Fe^{\#}$ = 0.75 to 0.70, T = 810 to 730° C, and fO_2 = $10^{-16.07}$ to $10^{-15.39}$ bars for TLg. These values suggest an overall decrease in fO_2 (reducing trend; ilmenite series) with the falling temperature. It should be noted that primary biotite crystallisation in all these rock types (irrespective of their ages) broadly had taken place between 860°C - 725°C at P_{total} =2070 bars which is typical of primary biotite crystallising at peraluminous granite melts generated at different thermal episodes ranging from Pre-Himalayan (Neoproterozoic and Cambro-Ordovician) to Post-Himalayan time (Neogene) but occurred in a close spatial proximity. However, a few GGn biotites was also stabilized at sub-solidus temperature around 530° C.

A small increase or decrease in pressure will shift the biotite isopleth to higher or lower temperature respectively. Thus all biotites in studied rocks appear to have been in equilibrium with peraluminous magma experiencing lowering of fO_2 (10^{-14}-10^{-16} bars), a condition typical of ilmenite series (Ishihara, 1977) or reduced-type (Takagi and Tsukimura, 1997), peraluminous (S-type; Chappell and White, 1974) granitoids analogous to monazite type of granite melt (Petrik and Broska, 1994). The essential difference between oxidized and reduced type granite magmas is the redox state of volatile components, which indeed

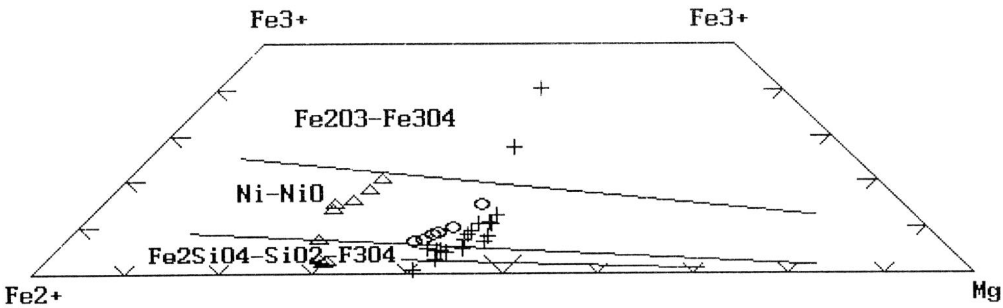

Figure 8: FMQ, NNO and HM buffer equilibria in terms of Fe^{+2}-Fe^{+3+}-Mg (Wones and Eugster, 1965). Biotites from GGn, RG and TLg follow clearly distinct evolutionary trends and mostly lie between FMQ and HM buffers crossing through NNO. Note that GGn biotites also plot in the HM (Fe_2O_3-Fe_3O_4) field. Symbols are the same as used in figure 3.

influences the crystallisation of hydrous phases. Due to the lack of modal primary magnetite in the studied lithotypes no further attempt was made to calculate the water fugacity (fH_2O).

CONCLUSIONS

Based on electron microprobe analyses of biotites from GGn, RG and TLg and their interpretations, the following conclusions can be drawn out:

1. Biotite is ubiquitous ferromagnesian mineral found in Neoproterozoic granite gneiss, Cambro-Ordovician Granitoids and Neogene tourmaline bearing leucogranites of Kinnaur district, Himachal Higher Himalaya. These granitoids are indeed two-mica (biotite-muscovite) granites (*sensu stricto*) but biotite dominates over muscovite in Cambro-Ordovician Granitoids.

2. Biotites from granite gneiss (GGn), Rakcham Granitoids (RG) and tourmaline bearing leucogranites (TLg) can be equivocally classified as Fe^{+2} biotites, however biotites from TLg are slightly enriched in octahedral components compared to those of GGn and RG.

3. Substitution relations of 3 (Mg, Fe) =2Al in biotites of GGn and TLg during evolution of peraluminous melts were recognized, although, substitution of Mg to Al of GGn and TLg biotites was more pronounced compared to Fe to Al. Biotites of RG show a restricted Fe to Al substitution but no Mg to Al substitution.

4. The range of thermal (T) and oxybarometer (fO_2) of biotites has been deduced; T = 860 to 520° C, fO_2 = $10^{-16.68}$ to $10^{-13.26}$ bars buffered at FMQ, NNO and HM buffers for GGn biotites, T = 795 to 780° C, fO_2 = $10^{-14.50}$ to $10^{-12.50}$ bars buffered at FMQ, NNO and HM for RG biotites, T = 810 to 730° C, fO_2 = $10^{-16.07}$ to $10^{-15.39}$ bars

buffered between FMQ, and HM for TLg biotites. The reduced trends of biotite evolution in peraluminous melts may be due to the fractionation of Fe-Ti oxide (magnetite-ilmenite) and/or intrinsic to source region during melting events. The former is more plausible as magnetite and ilmenite were not observed to be present in trace amounts in these rock types. However, a post-magmatic modification of biotite by percolating fluids via granites, particularly for GGn, cannot be fully ignored. Granitic (peraluminous) melts formed from meta-sedimentary rock had very reduced oxygen fugacities whereas intrinsicially oxidised granites came from a oxidized source (Wones, 1981). The metasedimentary protolith seems to be the best candidate for the generation of GGn, RG and TLg melts governed by varying degrees of partial melting. These melts were later experienced Fe-Ti oxide (magnetite and ilmenite) fractionation at intermediate or late stage of crystallisation leading to Fe/Fe+Mg enrichment of biotite, consistent with changing fO_2 with falling temperature.

5. Tourmaline and possibly the scheelite (W) mineralization in the area appears to be more related to the RG melt, generated during Neogene-remobilization of RG. This is consistent with the observed higher tungsten content (av. 42.29ppm) of RG compared to W content (av. 11.30ppm) of GGn (Singh and Kumar, 2002). Experimental results (Štemprok, 1990) predict that the solubility of SnO_2, WO_3 and MoO_3 in granitic melts was maximum (= 1000 ppm) at 750° C - 800° C, and highly evolved granitic melts were possibly undersaturated in these metals. This suggests that the scheelite (and cassiterite) can neither be liquidus minerals of ore-bearing granites nor restite minerals from partial melting events (Štemprok, 1990). Therefore, it is more plausible that tourmaline bearing leucogranite and scheelite of the studied area were probably formed by mobilization of RG during Himalayan orogen.

ACKNOWLEDGEMENTS

Part of the present work was supported by DST- grant. Thanks are due to Dr. Tamal K. Ghosh, for the electron microprobe analyses at Institute Instrumentation Centre, IITR, Roorkee. Drs. F. Yavus and T. Öztas are specially thanked for providing original bioterm program. Profuse thanks are due to Prof. P.S. Saklani, who improved an earlier version of the manuscript.

REFERENCES

Abdel-Rahman, A.M., 1994. Nature of biotites from alkaline, calc-alkaline and peraluminous magmas. J. Petrol., V. 35, Pt. 2, pp. 525-541.

Albuquerque, C.A.R. de, 1973. Geochemistry of biotites from granitic rocks, northern Portugal. Geochim. Cosmochim. Acta, V. 37, pp. 1779-1802.

Bassi, U.K. and Chopra, S., 1983. A contribution to the geology of Kinnaur Himalaya, Himachal Pradesh. Ind. Jour. Earth Sci., V. 10, pp. 96-99.

Bartarya, S.K., Virdi, N.S. and Sah, M.P., 1996. Landslides hazards: Some case studies from the Satluj valley, Himachal Pradesh. Him. Geol., V. 17, pp. 193-207.

Beane, R.E., 1974. Biotite stability in the porphyry copper environment. Econ. Geol., V. 69, pp. 241-256.

Berthelsen, A., 1951. A geological section through the Himalaya: A preliminary report. Bull. Geol. Soc. Denmark, V. 12, pp. 102-104.

Bhargava, O.N., 1980. Outline of the stratigraphy of Eastern Himachal Pradesh, with special reference to the Jutogh Group. *In*: K.S. Valdiya (*Ed.*). Stratigraphy and correlation of Lesser Himalayan Formations. Hindustan Publishing Corporation (India), Delhi, pp. 117-125.

Chappell, B.W. and White, A.J.R., 1974. Two contrasting granite types. Pacific Geology, V. 8, pp. 173-174.

Deer, W.A., Howie, R.A. and Zussman, J., 1963. Rock-forming minerals, Sheet Silicates, V. 3, London, Longman Green and Co., 270p.

Dorka, B.S. and Gupta, S.K., 1993. Study of granites and pegmatites of Himachal Pradesh and search for tin-tungsten and lithium mineralisation in Jeori-Wangtu Gneissic Complex, Kinnaur District, Himachal Pradesh. Geol. Surv. India Records, V. 126, pp. 112-114.

Dymek, R.F., 1983. Titanium, aluminium and interlayer cation substitutions in biotite from high-grade gneisses, West Greenland. Am. Min., V. 68, pp. 880-889.

Foster, M.D., 1960. Interpretation of the composition of trioctahedral micas. U.S. Geol. Surv. Prof. Paper, V. 354-B, pp. 11-49.

Gupta, V., Virdi, N.S. and Prakash, S., 2001. Morphometric assessment of active landslides in the higher Himalayan Crystallines, Satluj valley, Himachal Pradesh (India). Him. Geol., V. 22, pp. 99-107.

Hayden, H.H., 1904. The geology of Spiti with parts of Bashar and Rupshu. Mem. Geol. Surv. India., V. 36, pp. 1-129.

Heinrich, E. Wm., 1946. Studies in the mica group: the biotite-phlogopite series. Am. J. Sci., V. 244, pp. 836-853.

Ishihara, S., 1977. The magnetite-series and ilmenite-series granitic rocks. Min. Geol., V. 27, pp. 293-305.

Ishihara, S., Robb, L.J., Anhaeusser, C.R. and Imai, A., 2002. Granitoid series in terms of magnetic susceptibility: A case study from the Barberton Region, South Africa. Gond. Res., V. 5, No. 3, pp. 581-589.

Iyengar, B.R.C., 1949. Report on mineral investigations in Bushahr State, H.P. (Unpublished report) Geol. Surv. India, pp. 1-7.

Kakar, Ramesh K., 1988. Geology and tectonic setting of Central Crystalline rocks of southern part of Higher Himachal Himalaya. J. Geol. Soc. India, V. 31, pp. 243-250.

Kumar, Prem and Jain, A.K., 1993. Detailed investigation of the polymetallic mineralisation at Purbani, Kinnaur District, Himachal Pradesh. Geol. Surv. India Records, V. 126, pp. 109-111.

Kwatra, S.K., Sandeep Singh, Singh, V.P., Sharma, R.K., Bimal Rai and Naval Kishor, 1999. Geochemical and geochronological characteristics of the Early Paleozoic granitoids from Sutlej-Baspa Valleys, Himachal Himalayas. *In*: (A.K. Jain and R.M. Manickavasagam *Eds.*) Geodynamics of the NW Himalaya. Gondwana Research Group Memoir No. 6, pp. 145-158.

Laymere, J. and Bowden, P., 1982. Plutonic rock type series: discrimination of various granitoid series and related rocks. J. Volcanol. Geotherm. Res., V. 14, pp. 169-186.

Mehta, P.K., 1977. Rb/Sr geochronology of the Kulu-Mandi belt; its implication for the Himalayan tectonogenesis: a reply. Geol. Rundhsch., V. 68, pp.383-392.

Nash, W.P. and Wilkinson, J.G.F., 1970. Shonkin Sag Laccolith, Montana. Part I. mafic minerals and estimates of temperature, pressure, oxygen fugacity and silica activity. Contrib. Mineral. Petrol., V. 25, pp. 241-269.

Nockolds, S.R., 1947. The relation between chemical composition and paragenesis in the biotite micas of igneous rocks. Am. J. Sci., V. 245, pp. 401-420.

Petrik, I and Broska, I., 1994. Petrology of two granite types from Tribec Mountains, Western Carpathians: an example of allanite-magnetite vs. monazite-ilmenite dichotomy. Geol. J., V. 29, pp. 59-78.

Prashra, K.C. and Rapa, D.A., 1982. Recent observations on Mandhali Formation and Simla Slate -- Tertiary sequence in the Tons region of Garhwal Himalaya. Himalayan Geol. Seminar 1976. Misc. Pub. Geol. Surv. India, No. 41(II).

Rowins, S.M., Lalonde, A.E. and Cameron, E.M., 1991. Magmatic oxidation in the syenitic Murdock Creek Intrusion, Kirkland Lake, Ontario: evidence from the ferromagnesian silicates. J. Geol., V. 99, pp. 395-414.

Ruby, S. Balvinder and Singh, J., 1994. Detailed investigation for polymetallic mineralisation at Purbani, Kinnaur District, Himachal Pradesh. Geol. Surv. India Records, V. 127, pp. 88-90.

Sharma, K.K., 1977. A contribution to the geology of Satluj valley, Kinnaur, Himachal Pradesh, India. *In*: Colloq. Inter. CNRS: Himalaya Science de la Terre, CNRS, Paris, V. 268, pp. 369-378.

Sharma, K.K., 1983. Granitoid belts of the Himalayas. *In*: Shams, F.A. (*Ed.*) Granites of Himalayas, Karakorum and Hindu Kush, Punjab University, Lahore, pp. 341-354.

Sharma, V.K., 1995. Tungsten mineralisation in parts of Himachal Pradesh. Symposium on recent advances in geological studies on the NW Himalayas and the foredeep. Organised by GSI, Lucknow, 21-23 Feb.'95, pp. 218-219.

Singh, B. and Kumar, Santosh, 2002. Geochemistry of Cambro-Ordovician Rakcham Granitoids and associated late Proterozoic granite gneiss, Himachal higher Himalaya: Implication on petrogenesis and tungsten mineralisation. Extended Abstract, 17th Himalaya-Karakorum-Tibet Workshop, 25-27 March, 2002, Gangtok, pp. 13-14.

Srikantia, S.V., 1981. The lithostratigraphy, sedimentation and structure of the Proterozoic-Phanerozoic Formations of Spiti Basin in the Higher Himalaya of Himachal Pradesh, India. *In*: A.K. Sinha (*Ed.*), Contemporary Geoscientific Research in Himalaya, Bishen Singh Mahindra Pal, Dehra Dun, V. 1, pp. 31-48.

Srikantia, S.V. and Bhargava, O.N., 1998. Geology of Himachal Pradesh. Geol. Soc. Ind. Mem. No. 9, Bangalore, 406p.

Srikantia, S.V., Jangi, B.L. and Reddy, K.P., 1975. Lithostratigraphic classification of the Jutogh Formation of the Chaur Mountain area, Simla Himalaya. Seminar on Recent Geological studies in the Himalaya at Calcutta. Misc. Geol. Surv. India, No. 24, pp. 1-10.

Štemprok, M., 1990. Solubility of tin, tungsten and molybdenum oxides in felsic magmas. Mineral. Deposita, V. 25, pp. 205-212.

Stoliczka, F., 1866. Geological section across the Himalayan mountains from Wangtu bridge on the river Sutlej to Sungdeo on the Indus, with an account of the formations in Spiti, accompanied by a revision of all known fossils from that district. Mem. Geol. Surv. India., V. 5, 153p.

Takagi, T. and Tsukimura, K., 1997. Genesis of oxidised- and reduced-type granites. Econ. Geol., V. 92, pp. 81-86.

Tewari, A.P., Gaur, R.K. and Ameta, S.S., 1978. A note on the geology of a part of Kinnaur District, Himachal Pradesh. Him. Geol., V. 8, pp. 574-582.

Wones, D.R. and Eugster, H.P., 1965. Stability of biotite: experiment, theory and application. Am. Min., V. 50, pp. 1228-1272.

Wones, D.R., 1972. Stability of biotite: a reply. Am. Min., V. 57(1-2), pp. 316-317.

Wones, D.R., 1981. Mafic silicates as indicators of intensive variables in granitic magma. Min. Geol., V. 31, pp. 191-212.

Wones, D.R., 1989. Significance of the assemblage titanite+magnetite+quartz in granitic rocks. Am. Min., V. 74, pp. 744-749.

Yavuz, F. and Öztas, T., 1997. BIOTERM – A programme for evaluating and plotting microprobe analysis of biotite from barren and mineralised magmatic suites. Comp. Geosci., V. 23, pp. 897-907.

RELATION BETWEEN MINERALOGICAL AND PHYSICAL PROPERTIES OF COARSE AND FINE AGGREGATES AND INFLUENCE OF TEXTURAL, CHEMICAL AND DIMENSIONAL DIFFERENCES ON STRENGTH OF MORTAR AND CONCRETE-MANALI (H.P.)

S.C. Bhatt[1] and Rajeev Kumar[2]

1. Department of Geology, Bundelkhand University, Jhansi, U.P.
2. Central Soil and Materials Research Station, New Delhi

ABSTRACT

A detailed mineralogical and textural study of coarse and fine aggregates (gneiss, quartzite, schist and sand) samples collected from the Rohtang Pass Highway Project, Manali (H.P.) were conducted under petrological microscope and the results were correlated with physical and mechanical properties of these aggregates. It is inferred that the differences examined in crushing strength, impact value and compressive strength of coarse aggregate and corresponding concrete were due to variation in mineralogical characters such as deformational bands, undulose extinction, kinking, alteration, twinning and rotation of asymmetric porphyroblasts, pressure shadows and S-C fabric, observed in deformed quartz, feldspar and flaky minerals of each rock aggregate.

The presence of higher amount of strained quartz (60-70%), altered or twinned feldspar (30-40%), elastic (micaceous) minerals associated with high grade clayey and carbonaceous matrix in few samples of quartzite, gneiss, biotite and mica-schist may cause deleterious alkali silica reaction in concrete and subsequently may produce expansion, swelling and distress in the concrete structures in the long run. Contrary to this, few samples of granitic gneiss (A-144 and A-145) and quartzite (A-146A) composed of hard, strong interlocked, globular grains of quartz and feldspar with low fraction of clay, micaceous and ferromagnesium matrix may show higher strength than that of mica-schist and mylonitised gneiss.

On the other hand, the rock aggregates (A-145-B-quartzite, A-152-gneiss) showing lower water absorption values (0.68 to 0.50) composed of hard rounded size of quartz and feldspar grains with few voids and interstices between the interlocked mineral grains may produce concrete of higher strength than that produced by aggregates showing higher water absorption values. The presence of excess amount of strained quartz and other deleterious reactive minerals show adverse effects on strength, hardness and chemical stability of aggregate and concrete.

INTRODUCTION

It is universally accepted that the aggregate constituting about 60 to 80% of concrete may always have strong effects on the physical, engineering and chemical behaviour of concrete. It was also emphasized that under certain conditions the strength of aggregate may be controlled by the strength of concrete whereas the water absorption property of an aggregate may have different effects on corresponding concrete property only to limited conditions (Popovics, 1979). In the last decade petrography has played a vital role in identification of suitable (non-reactive) aggregates for sound concrete as well as in the identification of reactive aggregates and reaction rims in the expanded/distressed and deteriorated concrete cores. Therefore, in these days the geologists and engineers have quite similar opinion and observations in selecting the qualitative aggregates for concrete purpose. The aggregates used for concreting must have low degree of weathering i.e. both physical and chemical and must be quite strong (undeformed, well bonded and hard) to resist abrasion and degrading in a Portland cement mixer and also able to withstand forces applied by traffic, weather and reservoir and running water during service life of the pavements, dams, bridges and other concrete structures. (Baalbaki, et al., 1992 and PCCM Manual, 1989).

Most of the crushed stone, gravel and sand are used in the concrete industry as aggregate and generally are composed of 90% of silicate minerals (except few calcareous rocks) and therefore, they may be considered very prone to alkali silica/silicate reaction rather than carbonate and sulphate reactions.

Recently, optical, electron and transmission microscopy has become indispensable for examining the deformation features like extension/ healed fractures etc. in the crystal faces of aggregate and reaction rims, gel morphology and cement paste interface in the concrete cores. On other hand, x-ray diffraction, texiogoniometry and microprobe analysis would have significant importance in the study of crystal lattice and crystal bonding and in finding the rare organic and inorganic traces of elements in the aggregate and other admixtures.

Geologically the area consists of highly deformed and metamorphosed rocks which are exposed in a overturned limb of Central Crystalline Nape of Higher Himalaya (Fig. 1). Kumar (1981) mapped in detail the Manali and Rohtang Pass area and described various litho units of metapsammites, metasemipelites, migmatites and gneiss and granites (Fig. 1). In the present study attempts were made to correlate the physical and petrographical characters of coarse (crushed rocks) and fine aggregates and light has also been thrown on differences of shape, size, texture, microstructural orientation, fineness modulus and density with reference to their influence on the strength (crushing and compressive both) of aggregate and concrete.

PETROGRAPHIC AND MICROSTRUCTURAL CHARACTERS

About five specimens of gneiss, two of mica-schist, one of biotite-schist and two specimens of quartzite collected as coarse aggregate (Table-1) from selected quarries and two specimens of sand (fine aggregates) collected from the shoal of river bed at Rohtang Pass Highway, Tunnel Project, Manali, H.P. were examined under polarizing microscope as per IS: 2386 (Part-VIII) – 1963 (Table-1 and Table-2) for petrographical analysis.

Quartzite

The rock (A-145B) was massive, medium to coarse grained and consisted mostly of euhedral to subhedral, granoblastic to polygonal, tabular to elongated porphyroclasts of quartz-

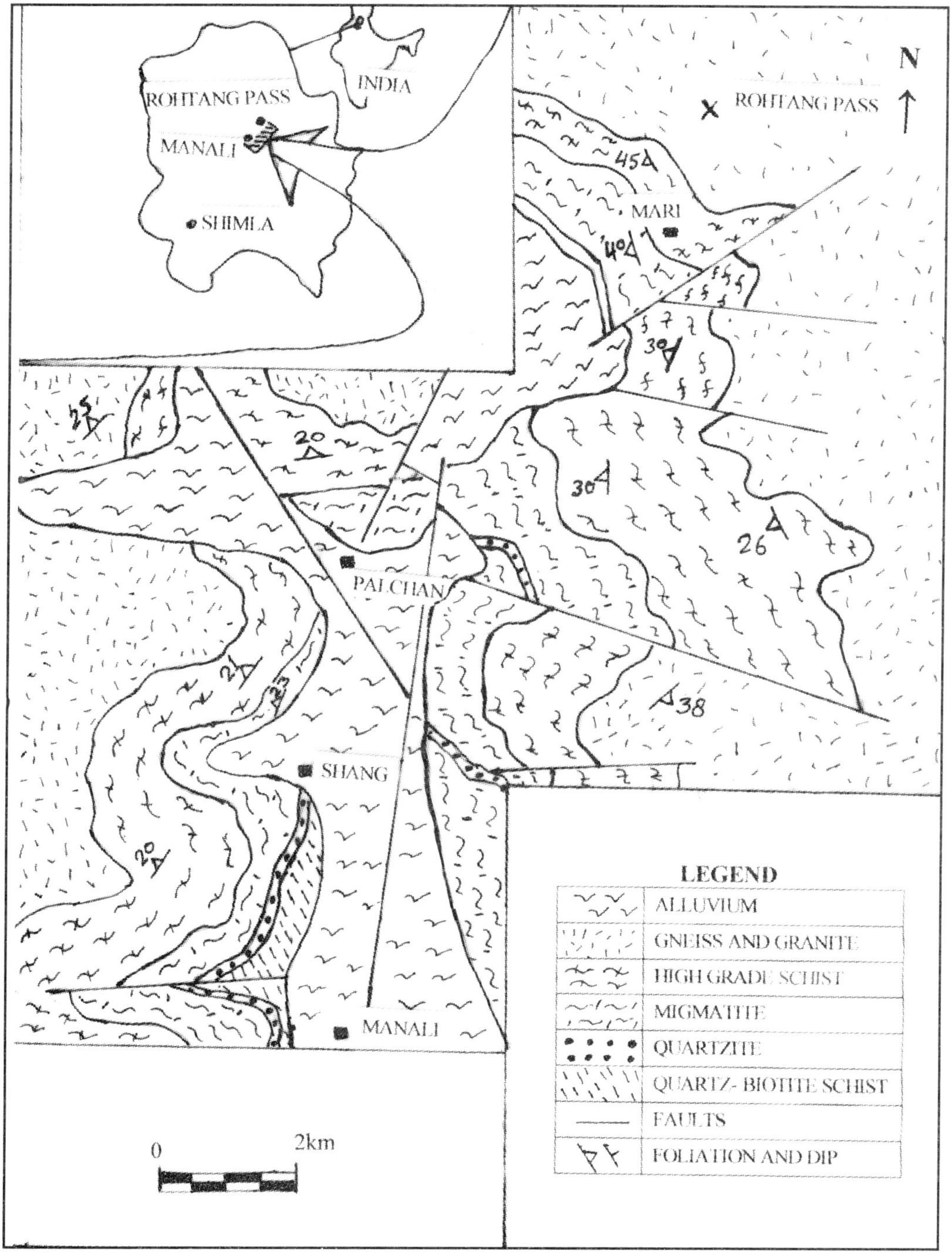

Figure 1: Geological Map of Manali and Rohtang Pass Area, Himachal Himalaya. (After Kumar, 1981).

grains (98-99%) with traces of chlorite, sericite and biotite (Fig.2a). About 75 to 85% grains of total quartz were found to be strained with an undulatory extinction angle varying from 23 to 30°. The average grain size of quartz varied from 0.20mm to 0.50mm. Most of the strained quartz grains were recrystallized and fractured. In few thin sections the main foliation was defined by elongated/tabular porphyroclasts of quartz and feldspars associated with fluxion bands of chlorite, biotite, muscovite and sericite.

The rock (A-146A) examined under polarizing microscope was micaceous quartzite. Its principal mineral constituents were quartz (68-75%) biotite (20-25%) and albite (2-4%) and accessories were muscovite, chlorite and magnetite (Fig.2b). The quartz grains were globular to subrounded, interlocked and tightly packed. Nearly 25 to 35% of the quartz exhibited strain effects with undulatory extinction angle varying from 20 to 32°.

Gneiss

The rock was observed to be medium to fine grained, gneissose and composed of quartz (30-35%), feldspar (40-50%; mainly plagioclase) and muscovite (15-20%) as main constituents with accessories of biotite (1-2%) and tourmaline (2-3%) and traces of chlorite, magnetite and zircon (Fig.3a & 3b). The quartz grains (A-144B & A-145-A) were fine to coarse grained, granoblastic, polygonal to elongated and recystallized.

About 60% to 75% of total quartz grains were examined to be strained with UE angle varying from 26 to 32° but in few specimens nearly 90-95% (A-150 and A-152) and 55-60% (A-146A) quartz grains showed strain effects with UE angle varying from 28 to 35°. These grains were characterized by deformation bands (Bohem's lamellae, strong undulose extinction and extension fractures. The grain size of quartz varied from 0.15mm to 0.52mm.

In all specimens except A-150 the plagioclase was found as main feldspar with rare microcline and orthoclase. The preferentially oriented porphyroclasts of quartz and feldspar formed the main foliation along with flakes of muscovite, biotite and chlorite. At places the inclusion of minute sericite and quartz were seen in the plagioclase grains which resulted due to alteration of muscovite and quartz. Few feldspar grains had myrmekitic texture with perthitic intergrowth and twinned lamellae of sub-idioblastic and tabular plagioclase (A-150 and A-152).

The fine and coarse flakes of muscovite and biotite were seen to be preferentially oriented and parallel or sub parallel to main foliation except in few thin sections (A-150 to A-152) they were characterized by lepidoblastic texture. Chlorite was found as alteration product of biotite. The presence of gneissose texture, two foliation planes, and recrystallized quartz grains and alteration of muscovite and biotite into sericite and chlorite indicate that the rock has experienced more than two phases of deformation.

Schist

a. **Mica-schist**: The light grey schistose rock (A-141 and A-144A) was composed of quartz, plagioclase and muscovite with rare biotite, magnetite, illmenite and chlorite (Fig.4a & 4b). The quartz grains were fine to medium grained, granoblastic and

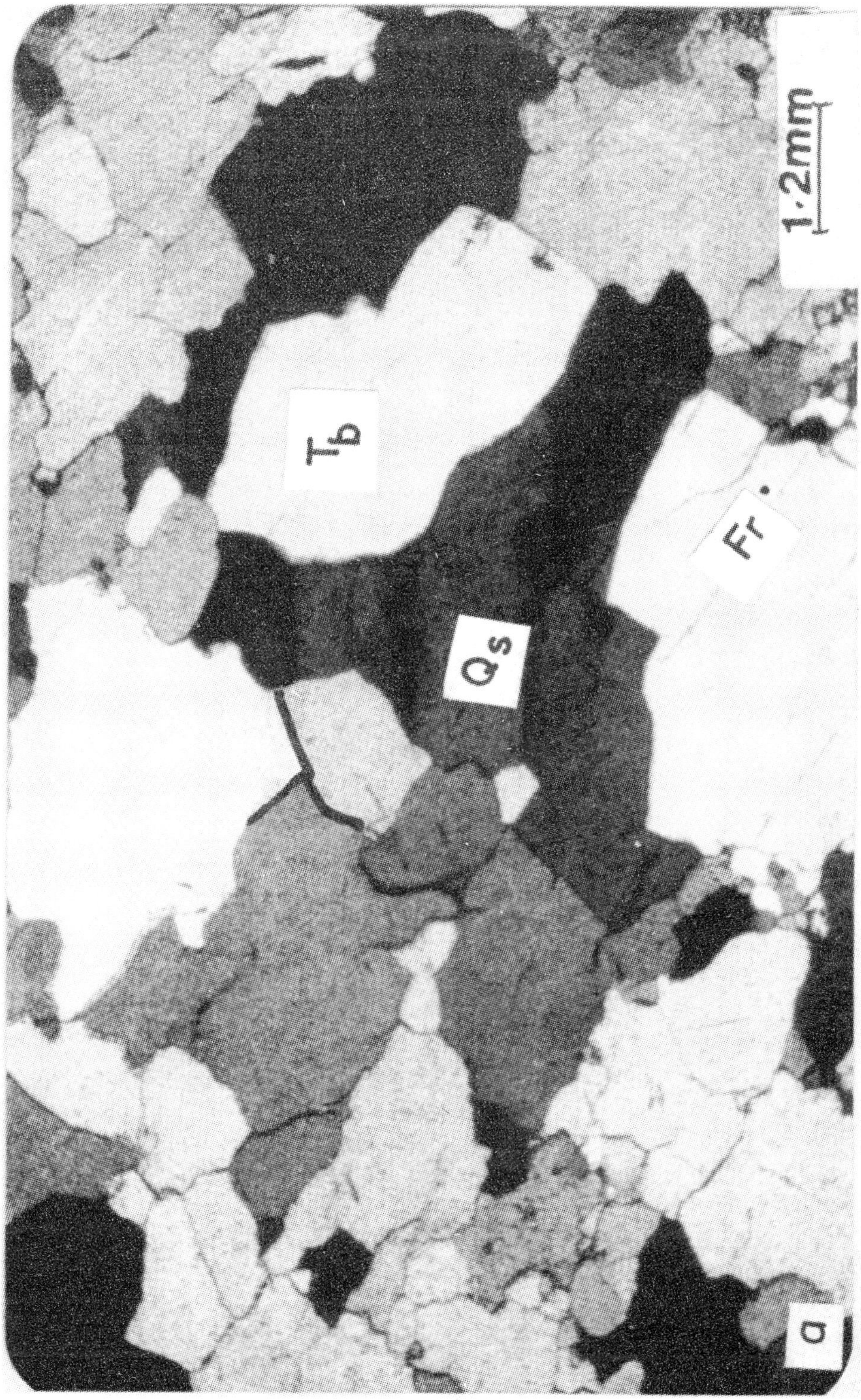

Figure 2.a: Photomicrograph of quartzite showing anhedral and tabular (Tb) grains of strained quartz (Qs) marked by extensional fractures and deformation bands. Crossed Nicols, A-145-B.

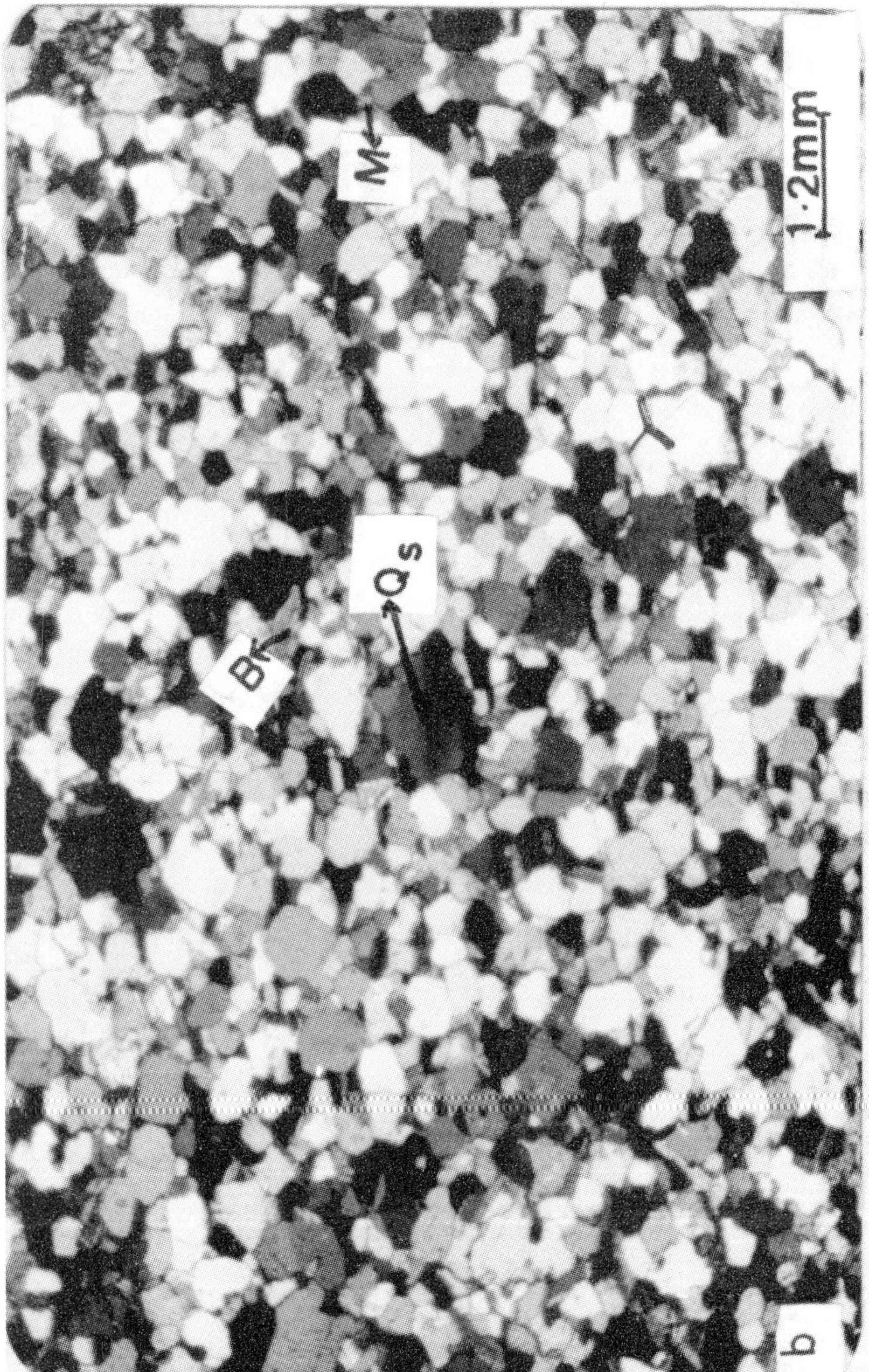

Figure 2.b: Randomly oriented globular and tabular shaped quartz grains (Qz &Qs) surrounded by flakes of muscovite (M), biotite (B) and few recrystallized quartz grains; Crossed Nicols, A-146-A.

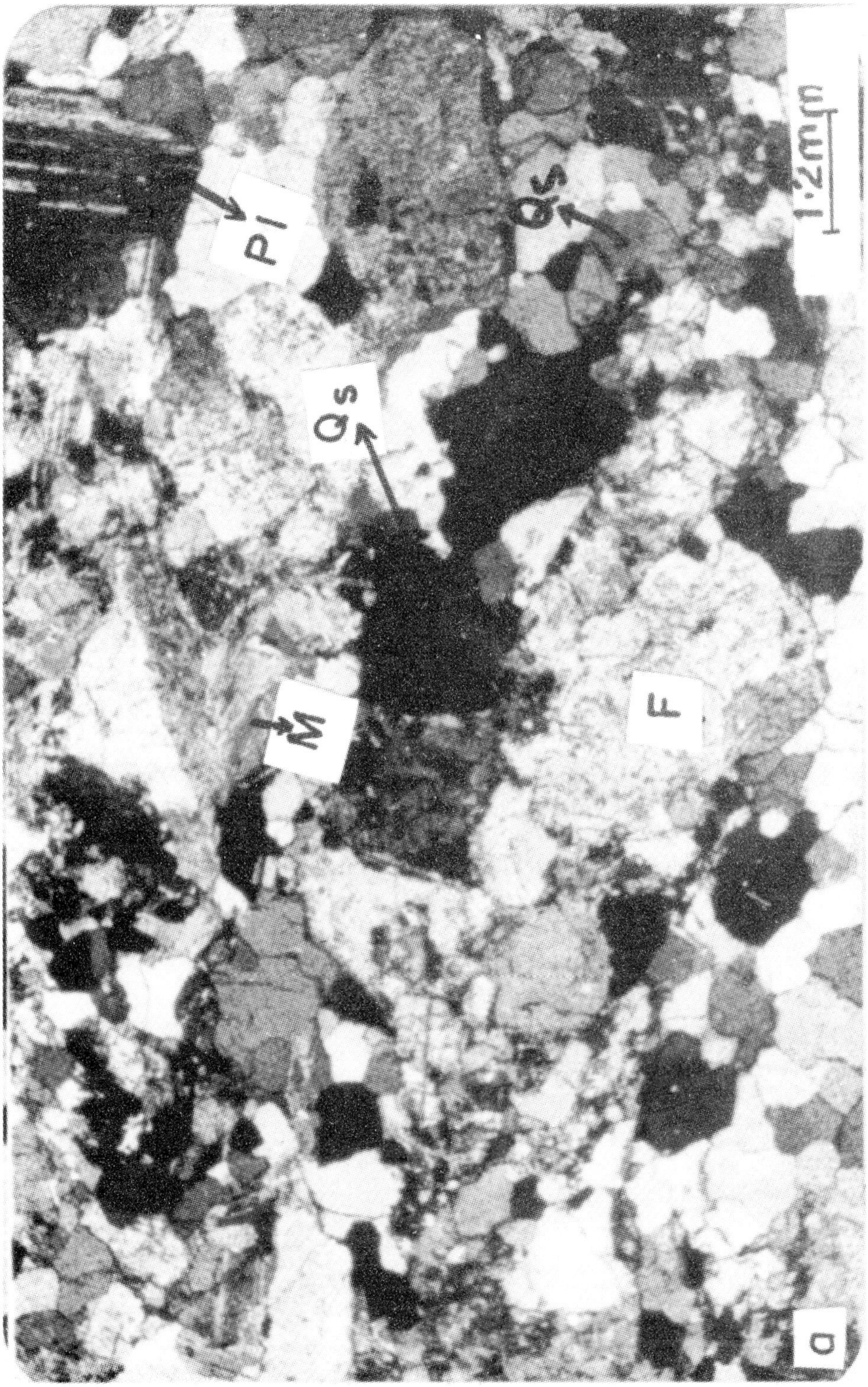

Figure 3.a: Gneiss showing altered and cleaved grains of feldspar (Mainly Plagioclase; Pl) and fractured recrystallized grains of strained quartz (Qs) are associated with stretched mica flakes of muscovite (M) and biotite (B); Crossed Nicols, A-146-B.

Figure 3.b: The altered pleochroic grains of plagioclase (Pl) and strained quartz (Qs) are associated with stretched mica flakes; muscovite (M), biotite (B) and anhedral tourmaline grains (T); crossed nicols, A-152.

polygonal to elongated in outlines. The main foliation was defined by porphyroclasts of quartz and feldspar (mainly albite) along with flakes of mica and recrystallized quartz grains. About 80-90% quartz grains exhibited strain effects with UE angle varying from 35 to 40°. In the coarse laths of biotite the inclusion of quartz were also seen. The fine to medium grained and at places subidioblastic laths of plagioclase (mainly albite) oriented in the direction of main schistosity were also noticed. The traces of chlorite were found in the coarse laths of biotite. The occurrence of two types of schistosity, altered minerals and recrystallized quartz grains suggest that the rock has suffered more than two phases of deformation.

b. **Biotite-schist:** It was medium to coarse grained and light to dark grey schistose rock. Its prinicipal mineral constituents were quartz (55-60%), biotite (20-25%) and albite (18-25%). The accessories were chlorite, magnetite and tourmaline (Fig.4C). The porphyroblasts of quartz were medium to coarse grained, xenoblastic and tabular to elongated in shape. The strain effects were shown by 78-85% of quartz grains. These grains were characterized by strong undulose extinction (>28-32°) and deformation bands and dynamic recrystallization. The albite grains were examined to be fine to medium grained, subidioblastic and oriented parallel to main schistosity direction along with quartz and micaceous minerals. Biotite having lepidoblastic texture showed inclusion of recrystallized quartz grains. The fine traces of chlorite, magnetite and tourmaline showing idioblastic texture were also seen.

Sand

The collected material (A-147 and A-151) consisted of subangular to subrounded, fine to coarse grained gneissose, schistose, quartzitic and feldspathic grains with accessories of biotite and muscovite grains (Fig. 5a & 5b; Table-2).

The gneissose grains were found to be subrounded to rounded consisting of xenoblastic to granoblastic quartz (40-45%) feldspar (15-18%), lepidoblastic biotite (4-6%) and muscovite/sericite (8-10%) often associated with inclusion of magnetite and tourmaline. In few gneissose grains the elongated and platy grains of pyroxene associated with the laths of muscovite and bands of quartz and feldspar were also noticed. The schistose bands were composed of polygonal to elongated/tabular quartz (30-35%), albite (15-20%) biotite (2-3%) and muscovite/sericite (25-45%) grains. In few thin sections the schistose bands had abundant carbonaceous material (10-15%). Quartz (90-95%) with rare flakes of muscovite and biotite was found as principal mineral constituent of quartzitic material whereas the sub-angular to subrounded laths of altered plagioclase and microcline with perthitic texture constituted main composition of feldspathic grains. Basaltic material had small laths of plagioclase associated with glassy material (35-40%), magnetite and rare pyroxene. In few thin sections of specimen A-151 the schistose grains had composition of subangular to elongated quartz (40-45%), muscovite/sericite (25-30%) along with preferentially oriented grains of albite, biotite and magnetite. Most of the (micaceous) minerals were characterized by lepidoblastic texture. Nearly 78-86% quartz grains showed strain effects which were characterized by strong undulose extinction, deformation bands and fractures.

Table 2: Mineralogical composition of sand samples collected from Rohtang pass project, Manali (H.P.)

	S.No	Gneissose Grains %	Schistose Grains %	Quartzitic Grains %	Feldspar Grains %	Musco-vite%	Biotite %	Basaltic Grains%	Strained Quartz %
	1	30-35	20-22	15-18	15-18	8-10	-	4-5	80-85
A-147	2	30-32	20-25	18-22	13-15	8-10	-	2-3	78-80
	3	30-35	20-22	15-18	15-18	7-9	-	1-2	75-80
70 mesh									
A-147	1	20-22	8-10	35-40	12-15	2-3	4-6	6-8	75-85
	2	18-20	7-9	38-42	10-15	1-2	6-7	5-7	75-80
	3	20-22	8-10	35-40	14-16	1-2	5-7	5-7	80-85
A-147	1	15-18	30-35	25-30	15-18	5-7	-	1-2	70-85
	2	15-18	35-40	25-30	12-15	8-10	-	1-2	75-85
	3	12-15	35-40	30-32	15-18	6-8	-	1-2	78-86
70 mesh									
A-147	1	10-12	20-25	35-40	10-12	4-5	5-6	4-5	72-85
	2	10-15	20-25	35-40	10-12	8-10	3-4	1-2	75-82
	3	10-12	20-22	35-40	12-15	5-6	6-8	1-2	78-85

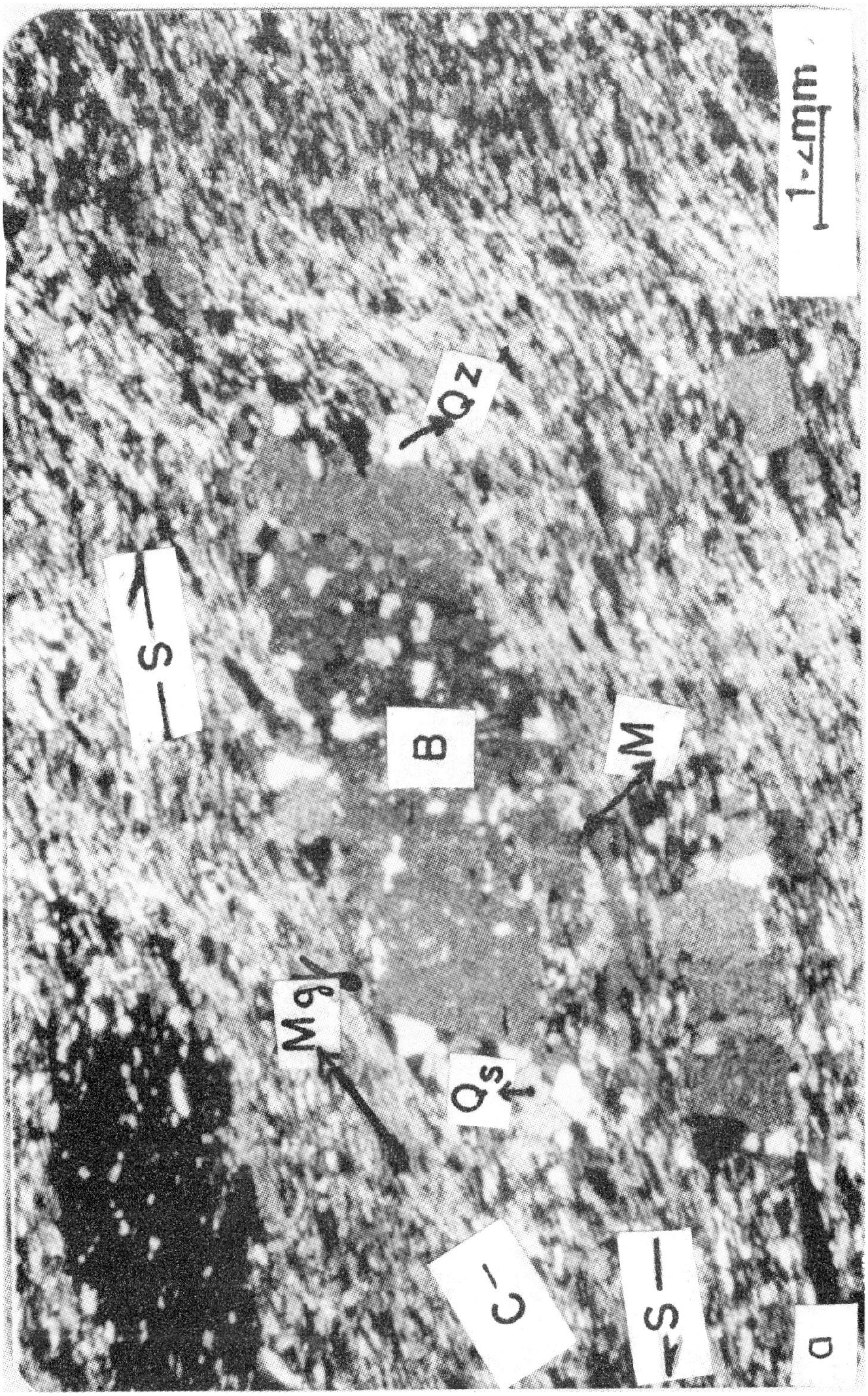

Figure 4.a: Mica-fish consisting of coarse laths of biotite (B), muscovite (M) with inclusion of recrystallized quartz (Q) wrapped by micaceous c-bands and intersected by earlier schistose bands (s-); Crossed Nicols, A-141.

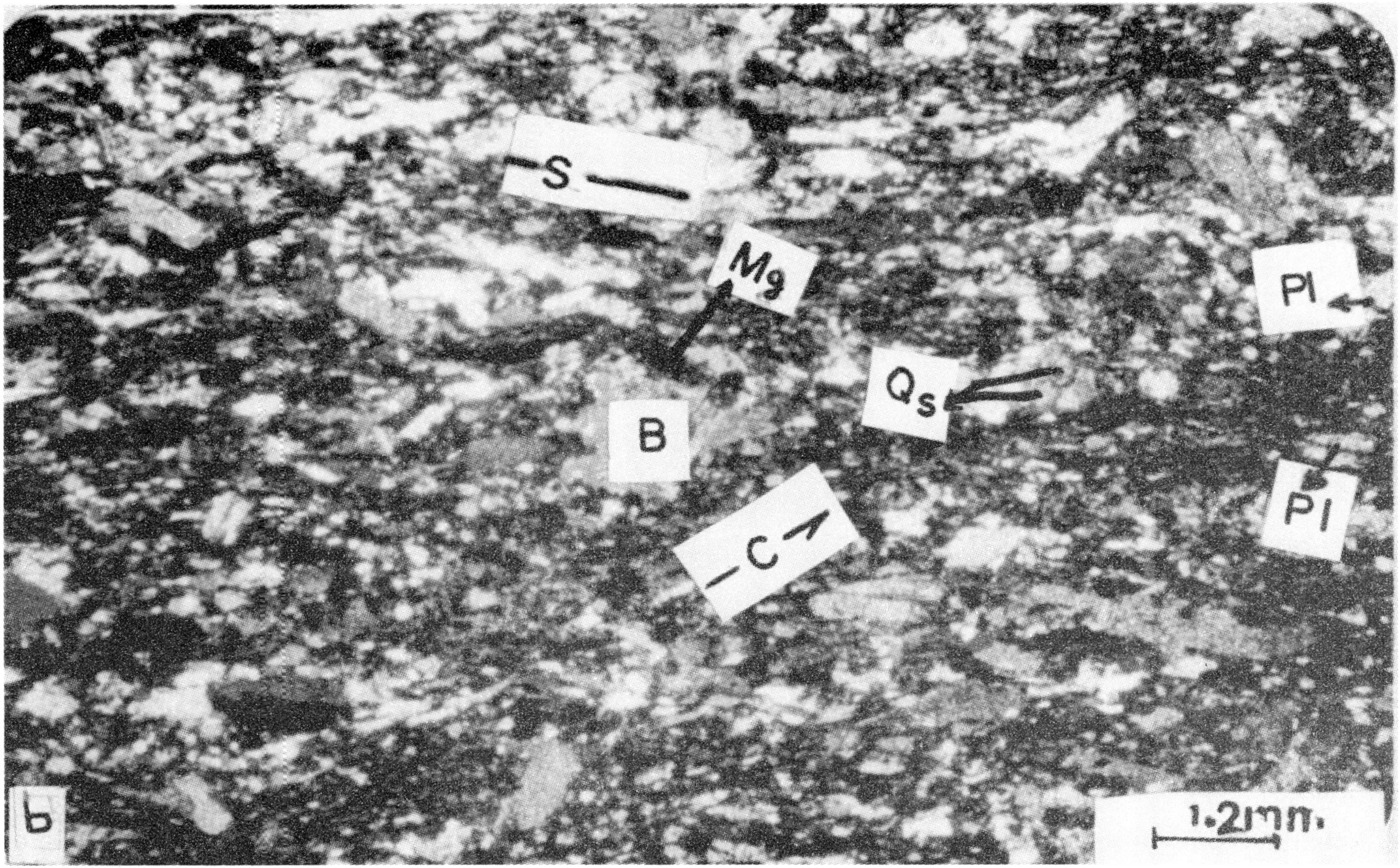

Figure 4.b: Photomicrograph of mica-schist showing two types of schistosity; The predominant schistosity (s-bands) is represented by parallel elongated and recrystallised quartz (Qs) and feldspar (Pl) grains with flakes of mica; secondary schistosity (c-b ands) are defined by recrystallized, equant and elongated quartz (Qs) with flakes of biotite (B) and muscovite (M); Crossed Nicols, A-144A.

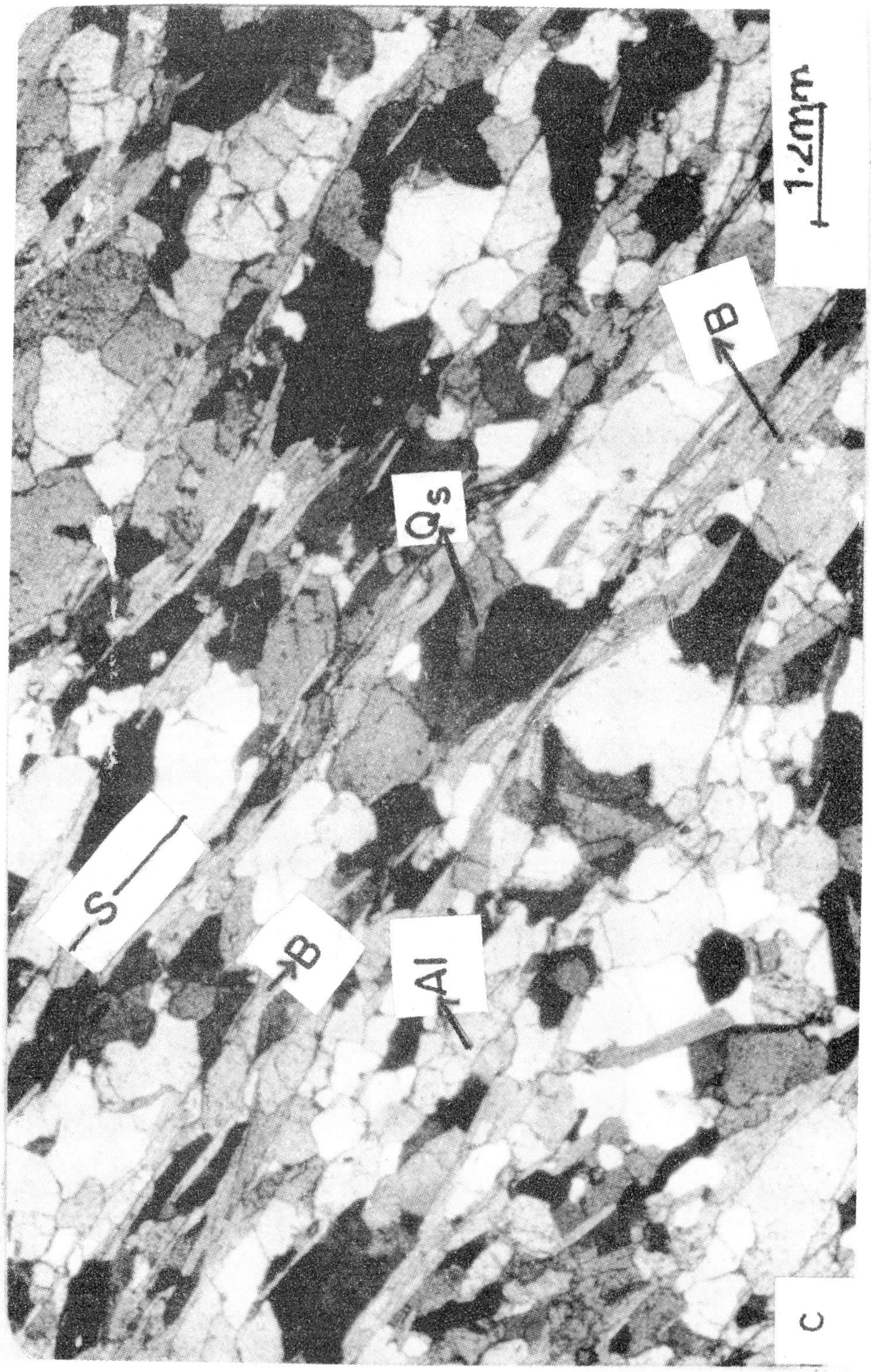

Figure 4.c: Biotite-schist showing prominent schistosity defined by parallel laths of biotite (B), feldspar (Al=albite) and phenocrysts of strained quartz (Qs); Crossed Nicols, A-149.

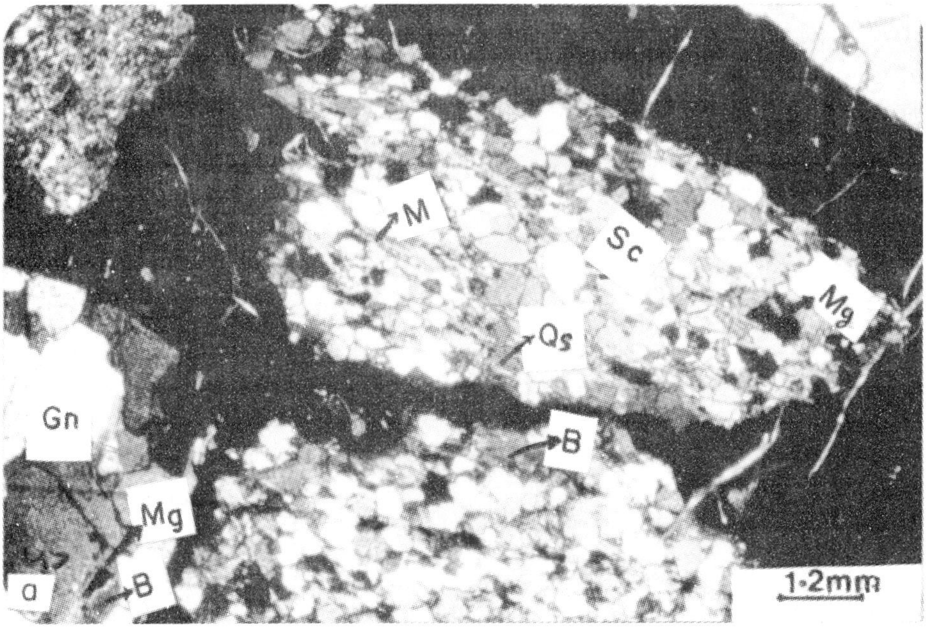

Figure 5.a: Sand showing showing schistose (Sc) and gneissose structures (Gn) consist of tabular and elliptical grains of strained quartz (Qs), stretched laths of biotite (B) and muscovite (M) and magnetite (Mg).

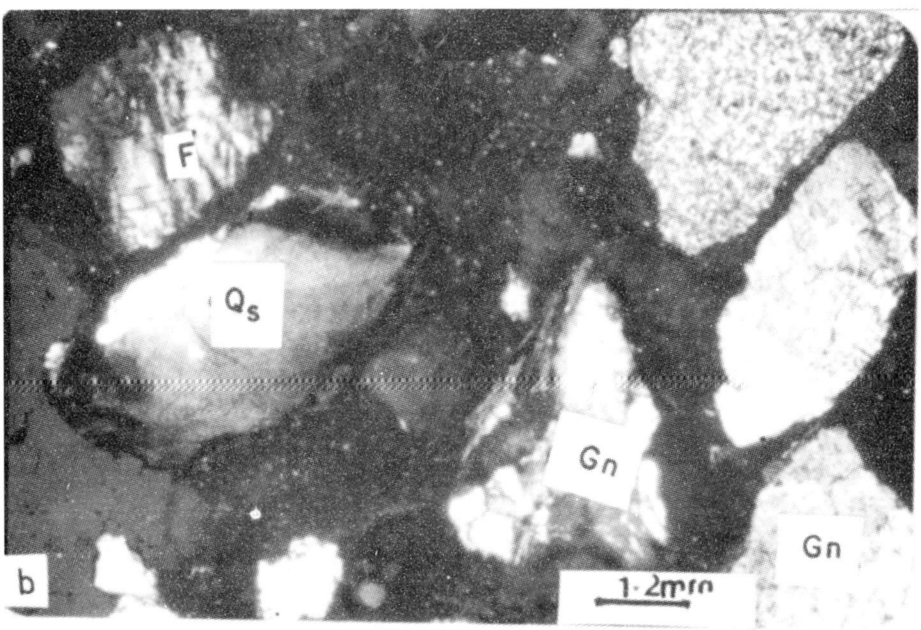

Figure 5.b: Photomicrograph of coarse sand showing gneissose structures (Gn) composed of phenocrysts of strained quartz (Qs) and feldspar (F).

PHYSICAL PROPERTIES OF AGGREGATES AND THEIR INTERPRETATION WITH PETROGRAPHIC PROPERTIES

Physical properties viz specific gravity, water absorption, aggregate impact value, aggregate crushing value, abrasion loss, soundness loss, fineness modulus and compressive strength tested as per IS; 2386 (part III and IV) for coarse and fine aggregates (collected from Rohtang Pass Highway project, Manali) were recorded in Table-1a & b; and 3.

The highest water absorption value (1.70%) was recorded for biotite-schist (A-149). It was pointed out that the amount of absorbed water depends on the abundance and continuity of the pores in the particle, whereas the rate of absorption depends on the size and continuity of these pores (Kucynski, 1974). The textural and mcrostructural features observed in biotite-schist indicate that the rock constituting low proportion of large quartz and feldspar grains than that of smaller flakes of mica, chlorite, magnetite and tourmaline always have grater number of pore spaces. The gneiss composed by grater amount of large porphyroclasts of quartz and feldspar showed low porosity i.e. low water absorption value.

The highest crushing and impact value % (Table-1a) was exhibited by gneiss (A-146) consisting of lowest proportion of strained quartz and rare traces of micaceous, carbonaceous and clayey matrix. Its quartz and feldspar grains were observed to be well packed, interlocked and well cemented with negligible number of pore spaces. The lowest crushing and impact values were shown by mylonitised mica – schist (A-144) and gneiss (A-152). These rocks were found to be composed of greater amount of strained quartz, elastic (mica), altered and clayey minerals. The porphyroclasts of quartz and feldspar in these rocks were often marked by weak foliated, cleaved deformed and fractured planes.

The soundness, toughness and hardness of an aggregate may be defined as its ability to resist abrasion, degradation, impact and grinding actions and natural forces applied by various agencies on concrete structures (PCCM manual, 1981 and ACI, 1967). The maximum abrasion loss % was noticed in biotite schist (A-149) and minimum was in gneiss (A-152). Accessing these examinations it may be suggested that the gneiss (A-150) showing lowest soundness loss value may be considered strong rock while the mica-schist (A-141) exhibiting maximum soundness loss may be observed as weak rock. Moreover, these differences in strength and hardness and soundness of rock are due to variation in mineralogical, textural and crystal bonding of each rock.

The data pertaining to physical properties and main mineralogical constituents of rock aggregates collected from the selected quarries (Rohtang Pass Highway Project, Manali) were graphically correlated in figures 6 to 9. The crushing and impact values of rock aggregates plotted against the percentage of strained quartz constituents indicate that the average strength of crushing and impact of rock is decreasing with consistent enhancement of strained quartz (Fig.6, Table 1a). However, this relationship was not reflected by a few specimens of gneiss constituting higher proportion of unaltered, undeformed feldspar and quartz grains with low percentage of elastic and clayey minerals. Contrary to this the adverse relationship between abrasion/soundness loss % and percentage of strained quartz was shown in fig. 6. The

Tble 1.a: Physical Properties and mineralogical contituents of rocks used as coarse aggregates in Rohtang- Pass High Way Project, Manali (H.P.)

Sample No.	Specific gravity	Water Absorption %	Aggregate Impact Value %	Aggregate Crushing value %	Los Angeles Abrasion Loss %	Soundness Loss %
A-146B	2.74	0.80	47.74	41.74	49.80	2.80
A-148	2.66	1.00	48.54	44.45	63.60	1.33
A-149	2.61	1.70	45.08	41.30	64.02	1.34
A-150	2.64	0.80	34.94	27.55	45.00	0.64
A-152	2.66	33.07	33.89	40.44	0.81	
A-141	2.73	0.80	41.74	36.56	49.80	5.00
A-144A	2.71	0.80	34.00	32.27	40.86	3.32
A-145A	2.61	0.80	37.40	34.16	49.60	1.60
A-145B	2.63	0.68	36.00	34.40	44.20	1.20
A-144B	2.66	0.87	38.20	39.50	41.00	0.92
A-146B	2.64	0.78	42.00	45.20	39.00	0.81

Table 1b: Mineralogical composition of sand samples collected from Rohtang Pass project, Manali (H.P.)

Sample No.	Quattz %	Feldspar %	Sericite or Muscovite %	Biotite %	Chlorite %&Magentite %	Tourmaline Quartz %	Strai-ned	Rock Type
A-146-B	45-60	30-40	10-20	1-2	1-2	-	55-56	Gneiss
A-148	40-50	35-45	10-20	1-2	1	-	85	Gneiss
A-149	50-60	18-20 (Albite)	-	18-25	-	-	78-85	Biotite-schist
A-150	32-42	40-50	6-10	5-8	-	2-4 Tourmaline	65-75	Gneiss
A-152	48-58	33-38	5-10	1-2	-	3-4 Magnetite	70-75	Gneiss
A-141	50-62	3-5 (Albite)	15-30	5-9	3-4	-	80-90	Mica- schist
A-144-A	50-55	25-35	- Plagioclase	5-6	-	1-2	80-85 Tourmaline	Mica- schist
A-145-A	38-45	32-40	12-20	2-4	-	-	70-75	Gneiss
A-145-B	98-99	-	-	-	1-2	Tourmaline	75-85 & Magnetite	Quartzite
A-144-B	30-35	40-50	15-20	1-3	-	-	65-72	Gneiss
A-146-A	68-75	2-4 (Albite)	1-2	18-25	1-2	Magnetite	25-35	Micaceous Quartzite

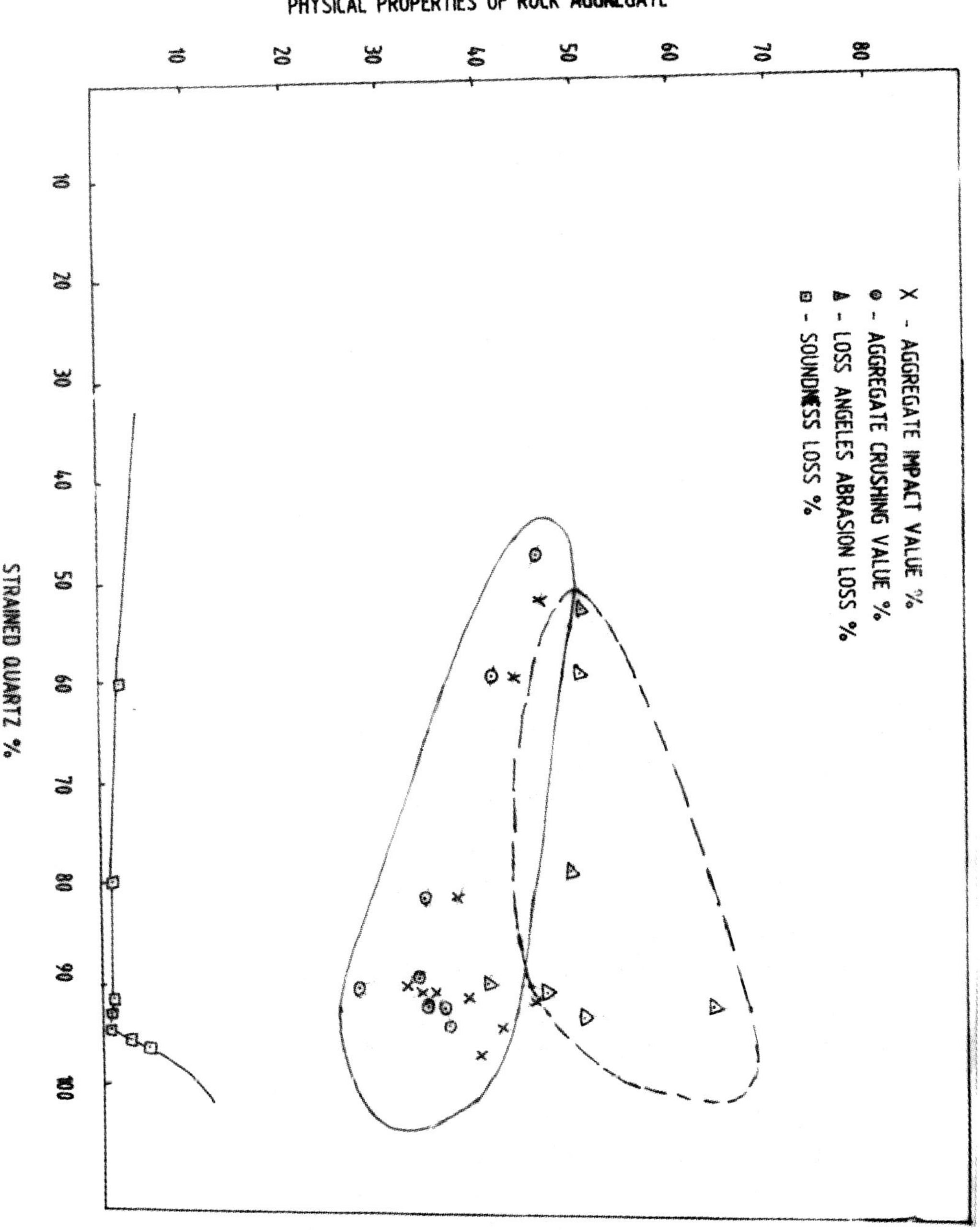

Figure 6: Graphical representation showing correlation between percentile values of strained quartz and physical properties rock aggregates.

average values of abrasion and soundness loss percentage is decreasing as the percentage of strain quartz is increasing.

Figure 7 shows the relationship between the physical properties of aggregate and percentage of quartz. The average percentage of crushing and impact values is considerably increasing as the amount of undeformed quartz increases whereas the average percentage of abrasion and soundness loss declines as the fraction of strained quartz percentage increases.

The slight enhancement in impact and crushing values with consistent increment of feldspar percentage is shown in figure 8. Apart from it a gradual decline in the percentage of abrasion and soundness loss values with considerable increment of percentage of feldspar is also demonstrated in figure 8. The increasing order of strength (crushing and impact value %) and abrasion and soundness loss % with remarkable enhancement in micaceous (muscovite/sericite and biotite) minerals is also seen in figure 9.

Because of differences in the mineralogical and textural components of various types of rocks or even in similar rocks a direct or consistent relationship between these properties was not found. It was also observed that gneiss having similar mineralogical composition but showing differences in textural and micro structural features possesses different range of crushing and impact values. These investigations reveal that a strong rock like gneiss can not be so sound than that of slightly weak quartzite as its principal minerals (quartz and feldspar) are quite harder than they are in gneiss and schist. Thus it can be inferred that a strong rock can not be a sound rock and strong aggregates may not able to produce strong concrete.

COMPRESSIVE STRENGTH AND FINENESS MODULUS

The fine aggregates of sand (A-147 and A-151) of –150 micron to 2.36 mm sizes were tested under 2000- tonne universal testing machine as per IS: 2386 (part-IV) – 1963. The samples of sand collected from the bed of river shoal were furnished to be graded sand and selected for test specimens. These test specimens were dried for four hours in the oven at a temperature of 100 to 110°C and cooled before the test. The specimens were placed centrally between the steel plates without packing. Each was conducted by applying 5 tonnes load/per minute to the ends of cylindrical test specimens and the stress (compressive strength) was calculated in kg./cm^2. The data of compressive strength along with fineness modulus, specific gravity and silt and clay contents are given in the Table-3.

The mortar specimens prepared under the peculiar conditions were tested over the range of compressive strength for 7 and 28 days. The mortar (A-151) of 1 cement : 2.75 sand ratio constituted of higher fraction of quartzitic, schistose material with low amount of gneissose, basaltic and micaceous minerals showed an increasing order of compressive strength than that of the mortar (1cement : 2.8 sand) having higher percentage of gneissose and feldspar material and low fraction of quartzitic and schistose components when tested for 7 and 28 days (Figure-10, Table-3). These both mortar specimens (A-147 to A-151) showing differentiation in textural, mineralogical and cement admixures are marked by a uniform increasing order of compressive strength (Figure-10). Therefore the experimental curve (fig.10) indicates a very intense increase of compressive strength of both specimens for 28 days.

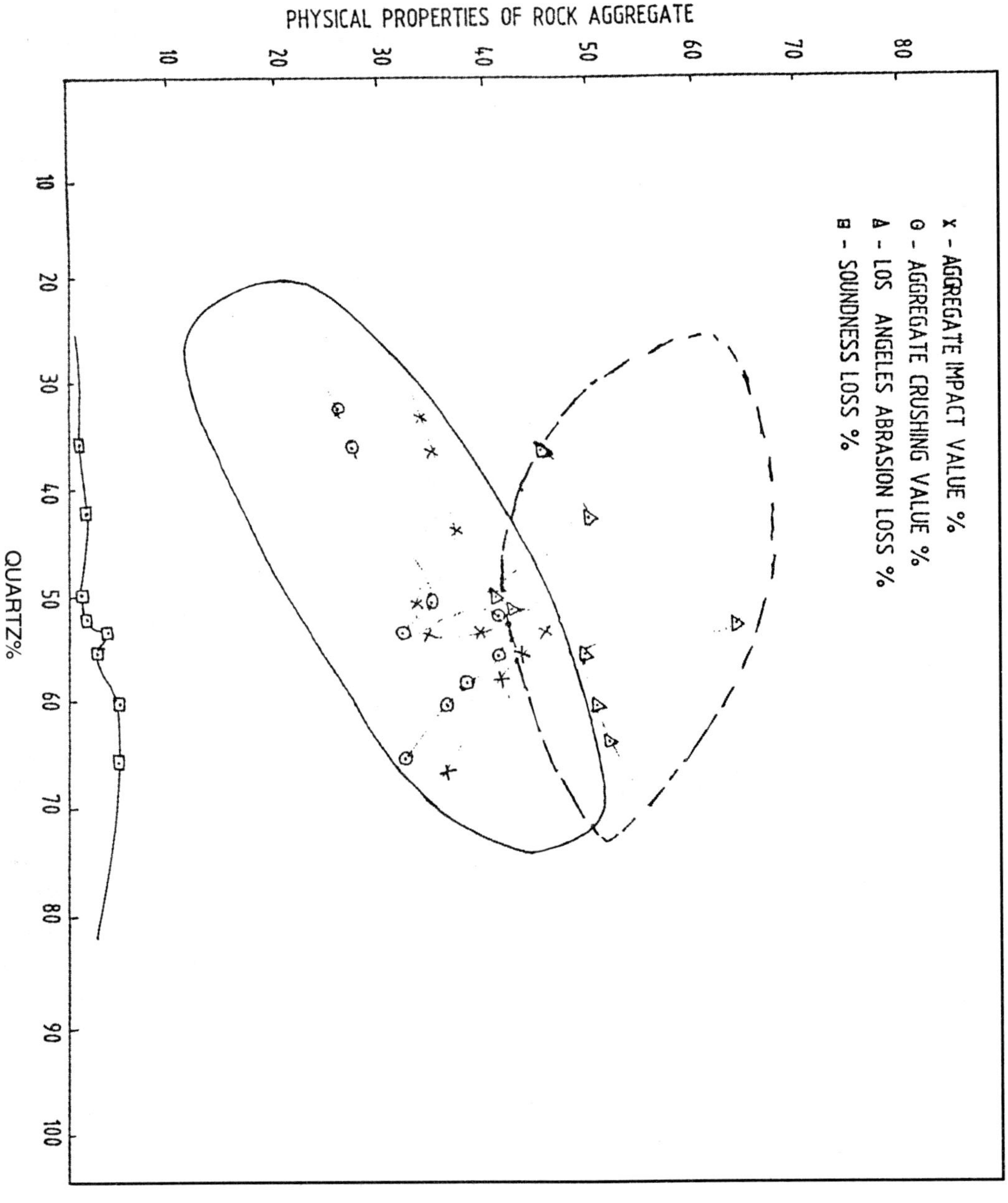

Figure 7: Plots showing correlation between percentile values of strained quartz and physical properties rock aggregates.

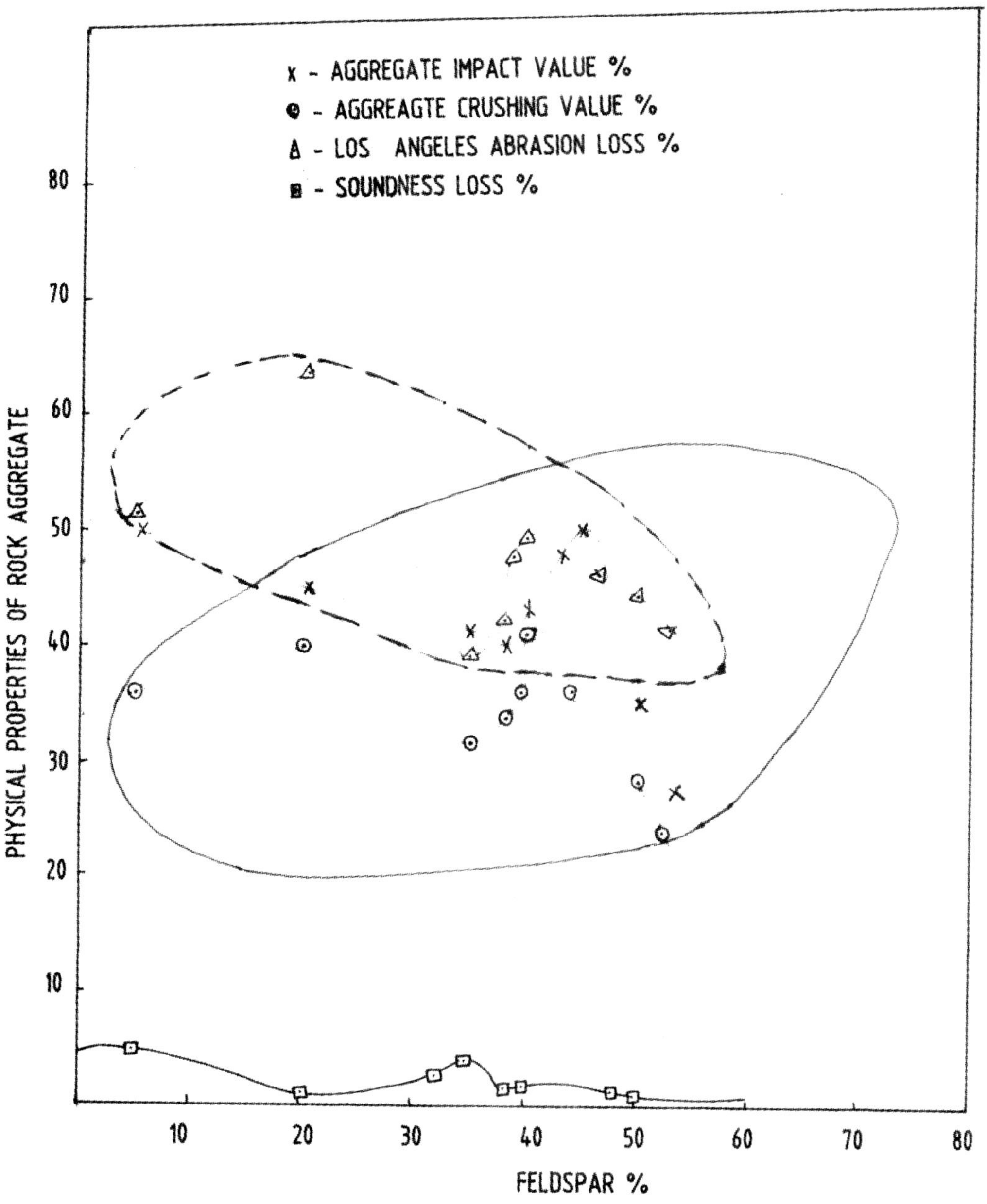

Figure 8: Graphical representation showing correlation between percentile values of feldspar present in rock aggregates.

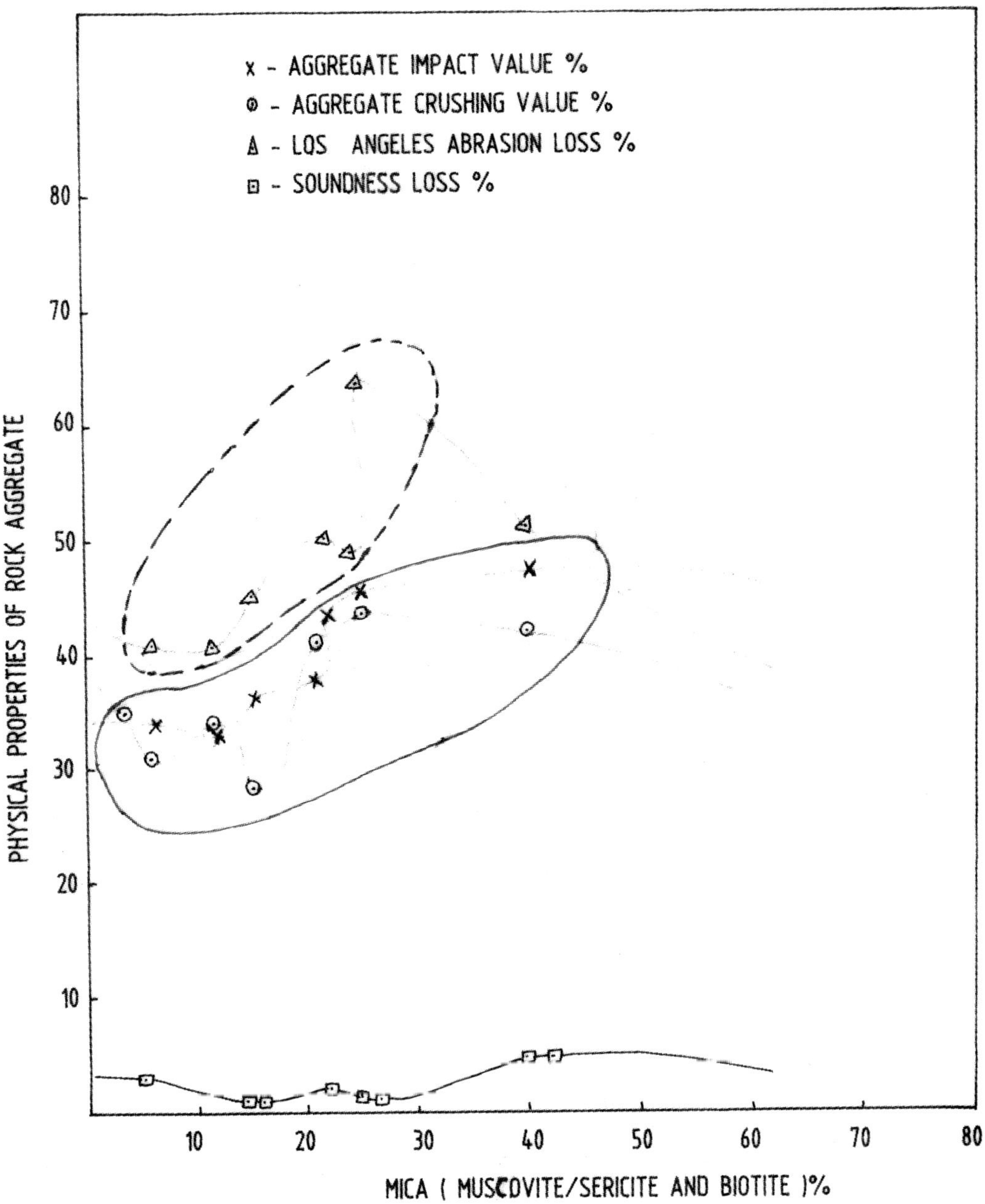

Figure 9: Graph showing correlation between physical properties and percentage of elastic minerals (muscovite/sericite and biotite).

Table 3: Gradation of fine aggregate, used in Rohtang pass highway tunnel project, Manali (H.P.)

Sieve analysis												
Sample No.	Sieve size no. IS sieve	% Retained	Cumulative Retained	% pass	Grading zone	Fineness Modulus	Specific gravity	Compressive strength	W/C Ratio	Flow	Silt and clay content	Type of Material
A-147	2.30mm 1.18mm 600micron 300micron 150micron 150micron	1.6 13.6 19.4 44.8 11.0 9.6	1.60 15.20 34.60 79.40 90.40 100.00	98.40 84.80 65.40 20.60 9.60 -	Grading zone III as Per IS: 383-1970	2.21	2.67	1 cement: 2.8 sand mortar 7 days 80.0 strength kg/cm^2 28 days 130.0 strength kg/cm^2	0.60	100 5%	2.45%	Sand Deposits from the river shoal
A-151	2.30mm 1.18mm 600micron 300micron 150micron 150micron	0.20 2.20 9.0 54.4 12.0 22.20	0.20 2.40 11.40 65.80 77.80 100.00	99.88 97.60 88.60 34.20 22.20 -	Grading zone IV as Per IS: 383-1970	1.58	2.67	1 cement: 2.75 sand mortar 7 days 85.0 strength kg/cm^2 28 days 159.0 strength kg/cm^2	0.60	100 5%	1.0%	

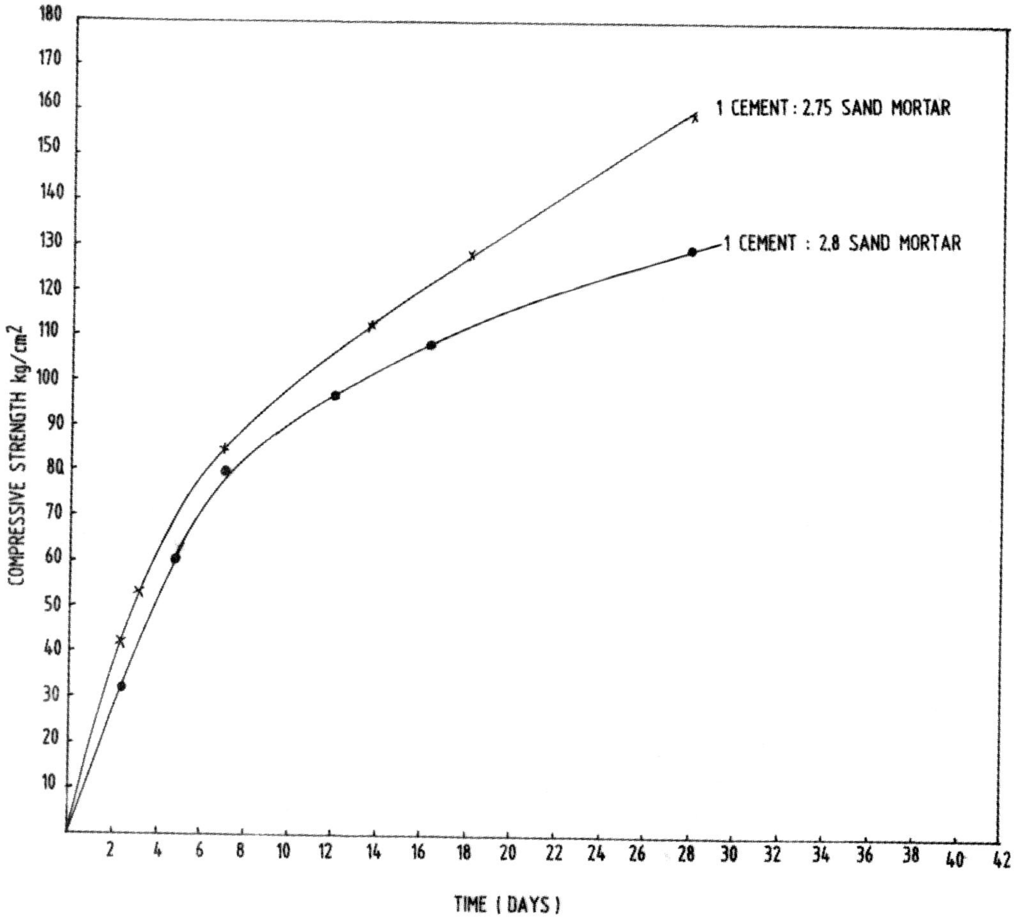

Figure 10: Graph showing variation in compressive strength of mortar with enhancement in time (days).

Generally the fineness modulus is obtained by adding the total percentage of an aggregate sample retained on each of a specified Tyler sieve series and dividing the sum by 100. It is also defined as an aggregate particle size of the grading and as such it is a fundamental parameter of particle size distribution (Popovics, 1979).

In this analysis it was also noticed that the strong mortar (A-151) was marked by low fineness modulus with low silt and clay contents than that of observed in slightly weak mortar (A-147). Hence an adverse relationship between compressive strength and finess modulus was found (fig.11).

ALKALI SILICA / SILICATE REACTIVITY TEST (CHEMICAL METHOD)

The twelve samples (Table-4) of various rock types (mainly quartzite, gneiss and schist) collected from the different quarries of Rohtang Pass Highway project, Manali were tested as per IS : 2386 (part VII)- 1963 by chemical method to asses the potential reactivity of aggregates. For this purpose about 4 kg. Crushed and sieved (passed through 300 micron IS sieve and retained on a 150 micron IS sieve) samples of rock were mixed with 1 N sodium hydroxide solution and heated at 80+1°C for 24 hours. The coarse aggregates were reduced to maximum size of about 6.3 mm by means of small jaw crusher. The material was repeatedly passed though 150 micron IS sieve and reduced up to that it can be retained on the 300 micron IS sieve. To ensure the fineness of the material the sample were washed over a 150 micron IS sieve and dried at 100 to 105°C for 20+4 hrs. Subsequently, the samples were cooled and again sieved on 150 micron IS sieve. In this way the silty and clayey coating was completely washed and removed and the retain samplesed were kept ready for test.

The test was commenced by taking three representative 25.00+ 0.05 portion of dry 300 micron to 150-micron test sample and placed each of three into a reaction container. After this 25 ml of 1.00 N caustic soda solution was added by pipette or burette. To a fourth reaction container, by means of pipette or burette, 25 ml 1 N. NaOH was added to serve as a blank. Immediately it was sealed and gently swirled to liberate traped air. The containers were then placed in a liquid bath maintained at 80+1.0 C. After 24 hours the containers were removed from the bath and boiled under flowing water for 15+2 minutes to below 30°C.

After this procedure the container were kept open and the and solution was filtered from the aggregate residue as per the prescribed method given in the IS: 2386 (Part VII) –1963; 3.6.3 to 3.6.5. The quantity of dissolved silica was determined by Gravimetric method. In this test about 100 ml. of dilute solution was transferred to an evaporating disk preferably of platinum and then 5 to 10 ml. of HCl acid was added and immediately exported to dryness on a steam bath. The procedure was repeated three times by heating each residue and then an equal amount of water was at once poured 10 to 20 ml of HCl upon the residue. After this the dish was covered and digested for 10 minutes on bath or hot plate. Then the solution was diluted by adding equal volume of water.

Immediately the solution was filtered and the separated silica was washed thoroughly with hot water. Again the filtrate was evaporated to dryness and the drying and filtering process was done for three times and finally the concentration of silica in original filterate

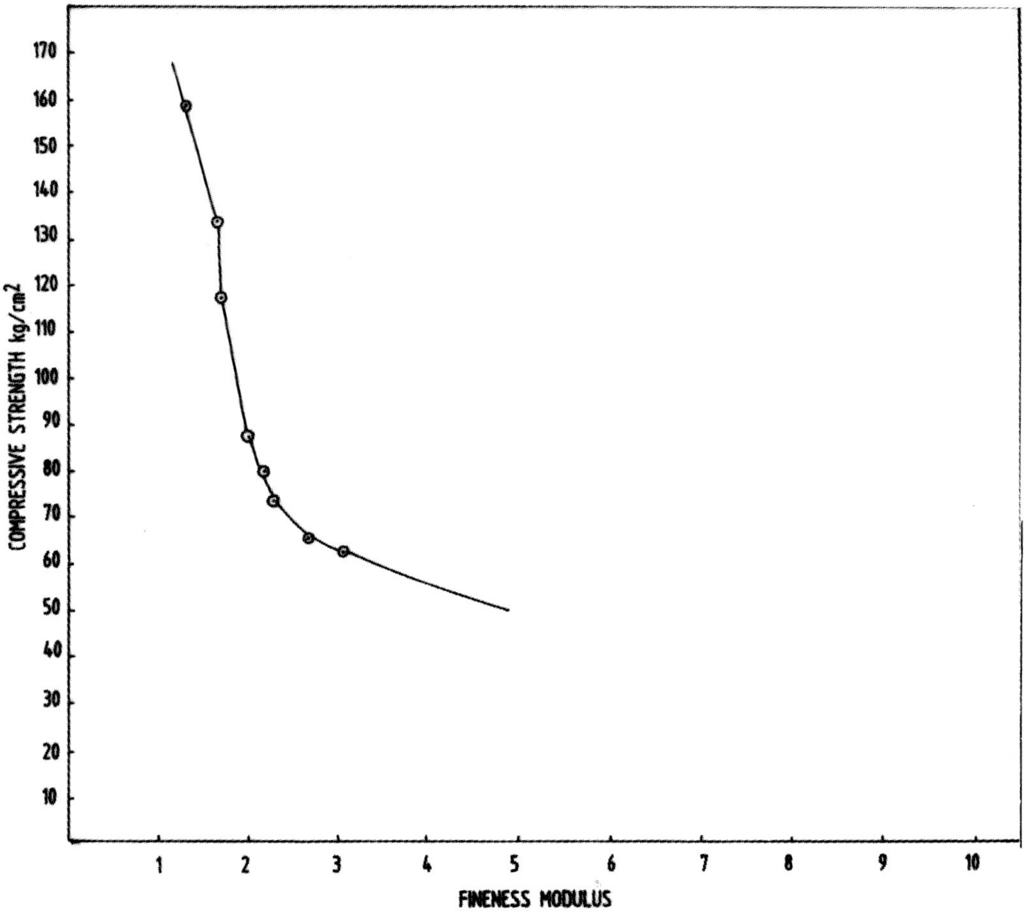

Figure 11: Relationship between compressive strength and fineness modulus of fine aggregates.

Table 4: Statement showing the test results of alkali aggregate reactivity test by chemical method (24 hrs) carried out in the samples collected from the Rohtang pass highway tunnel project (H.P.)

Sample No.	Milimoles/Liter (24 hours)		Petrographic Sample	Rock Types
	Reductionin alkalinity	Silica release		
A-117	210.00	13.32		
A-118	120.00	8.33	A-146B	Gneiss
A_119	70.00	20.55	A-149	Biotite-schist
A-120	60.00	7.99		
A-121	80.00	18.28	A-152	Gneiss
A-122	40.00	26.31	A-144A	Mica-schist
A-123	50.00	16.98	A-150	Gneiss
A-124	60.00	10.66	A-146A	Micaceous Quartzite
A-125	30.00	9.32		Quartzite
A-126	130.00	18.32		
A-127	40.00	15.32		
A-128	40.00	33.97	A-141	Mica-schist

(Sc) weight of silicia in g found 100 ml of dilute solution (w) and weight of silica in g found in blank were determined to find out the concentration of released silica (Sc).

Sc = (w – w) x 3330

The 20 ml aliquot of the dilute solution was transferred to a 125 ml Erlenmeyer flask, 2 or 3 drops of phenolphthalein solution added and titrated with 0.05 N HCl the phenolphthalein end point and the alkalinity was calculated.

$$RC = \frac{20N}{V1} \qquad (V3 - V2) \times 1000$$

Interpretation

As illustrated in the standard graph i.e. in figure 12 the most of data obtained after alkali aggregate reactivity test are falling to the left of the boundary line i.e. in the field of innocuous aggregates. The results pertaining to quartzite and mica schist composing of higher percentage of strained quartz (70-90%), altered feldspar (20-30%), mica and clayey minerals than that of other rocks may be considered more prone to alkali silica and alkali silicate reactions. Although these specimens do not fall in the deleterious zone but few of them (quartzite, mica-schist and mylonitised gneiss) releasing higher amount of silica can not be used as more qualitative and suitable aggregates for concreting.

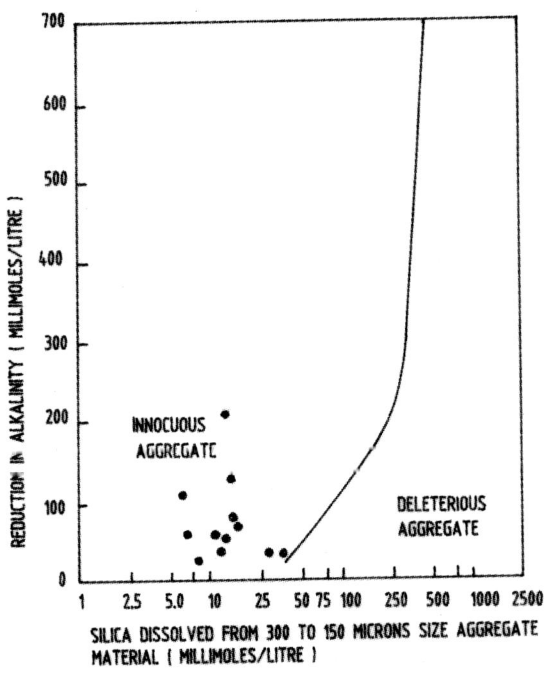

Figure 12: Results obtained from alkali-silica/silicate reactivity test (chemical method) are shown in standard curve.

Nearly 60% (SiO_2) constitutes the composition of earth crust (Sarkar and Aitcin; 1990).In the present analysis most of the dissolved silica was released from the crystalline form of silica i.e. from strained quartz and other silicates (Phyllosilicates and tectosilicates). The size of quartz grains also plays a significant role in causing expansion and deterioration in concrete structures. Except for very fine grained (smaller 0.1 mm) rocks, average grain size did not exhibit any discernible influence on the resultant expansion (Glasser, 1992). It may be summarized that the rate of expansion may be directly proportional to the percentage of strained quartz, reactive silicates and clayey minerals present in the aggregate and other admixtures of concrete.

RELATIONSHIP BETWEEN DENSITY VS. COMPRESSIVE STRENGTH AND AGGREGATE SIZE VS. COMPRESSIVE STRENGTH

The eight core samples collected at uniform distance from the main axis were tested as per IS : 2386 (Part III and Part IV) – 1963 for density and crushing strength. The density is defined as the weight of material in a given volume, measured in kilogram / per liter.

Results shown in table-5 indicate that the core specimen (A-162) of granite consisting of quartz, feldspar, mica and heavier ferro-magnesian minerals showed high value of density (2714 kg./M^3) whereas the core sample of same granitic material composed of quartz, feldspar and mica with low amount of ferro-magnesian and other heavy minerals encompassed low value of density (2556 kg./M^3). Figure 13 shows the relationship between compressive strength and density. It reveals that with an increase in density of aggregate the strength increases or in other words it may be observed the heavier rocks are more likely to be stronger than the weak rocks. A concrete constituting hard and strong rock aggregate can not be so strong because the strength of concrete not only is dependent on hardness, toughness and strength of individual mineral aggregate or rock but it depends on shape, size, density and texture of aggregate as well as firmness of cement paste and other admixture used in concreting.

Five rock core samples were tested for uniaxial compressive strength as per IS: 1979-1979 (Table – 6). The size of aggregate ranging from 2 to 6 cm. with L/D ratio varying from 1 to 2 were correlated with variable values of compressive strength (figure 13). It was observed that the aggregates size ranges from 2 to 6 cm were marked by compressive strength, varying from 200 to 500 kg/cm^3. On other hand, the aggregate of 1 to 4 cm sizes showed the compressive strength varying from 1000 to 1600 kg./cm^2. Contrary to this aggregate of smaller sizes (.5 to 3.5 cm) were marked by higher compressive strength (1000 to 1600 kg./cm^2). This study infers that with reduction in aggregate size the average compressive strength increased. Apart from this the remarkable differences were also noticed in the compressive strength of aggregate showing uniform size and in the surface area of upper and lower part of same aggregate. Assessing these investigations it can be inferred that not only differences in size but in shape, density, texture and fabric components of aggregate as well as in cement paste and other admixture affect the strength of concrete.

Table 5: Physical properties of core samples collected from the project.

Sample No.	Location of core sample	Density (average) In kg./ M^3	Crushing Strength (average) In kg.cm^2
A-162	L/B from D/S of axis Depth:	2714	898.0
	19.5m to 19.9m		
A-163A	L/B 30m U/S of axis Depth: 3.80m to 4.25m	2576	896.0
A-163B	L/B 30m U/S of axis Depth: 6.85m to 7.50m	2561	698.0
A-164A	R/B axis Depth: 14.3m to 14.5m	2568	900.0
A-164B	R/B axis Depth: 15.87m to 10.17m	2607	674.0
A-165	R/B 30m U/S of axis Depth: 12.85m to 13.20m	2556	396.0
A-166	15m U/S corner of river bed R/B side Depth: 8.40m to 9.00m	2564	782.0
A-167	30m D/S corner of river bed L/B side Depth: 4.50m to 4.80m	2580	749.0

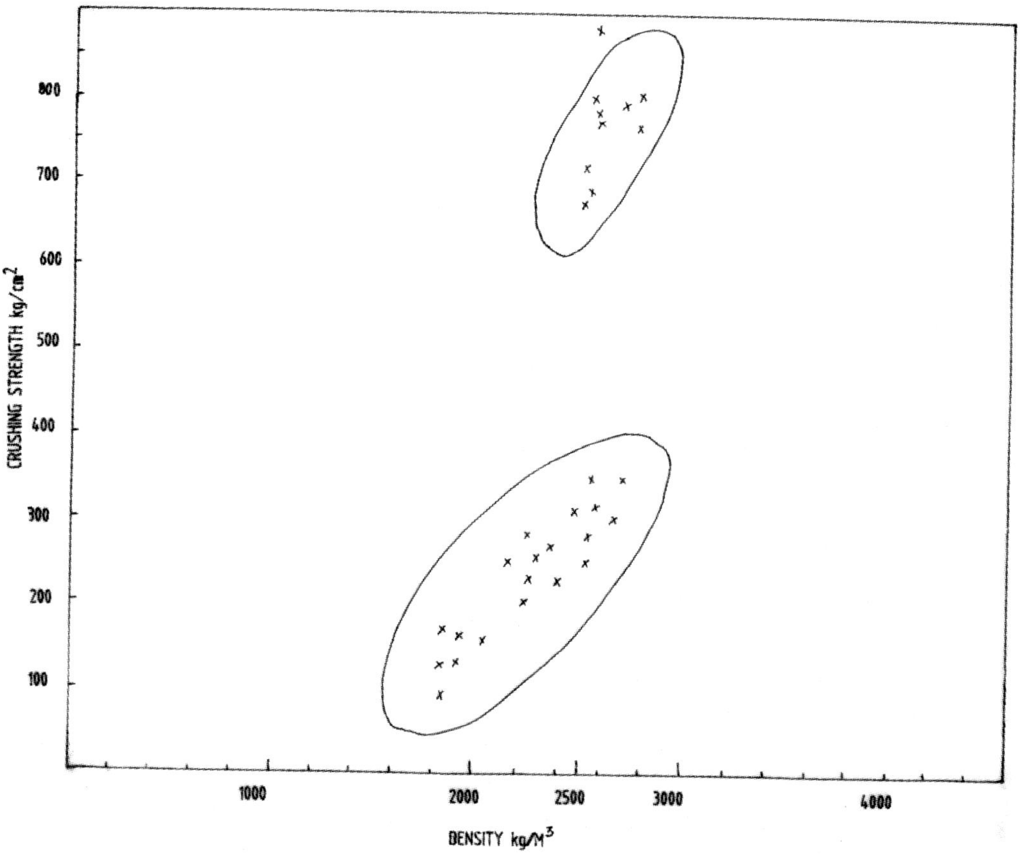

Figure 13: Graph showing relationship between crushing strength (Kg/cm^2) and density (Kg/m^3) of core samples.

DISCUSSION

Assessing all mineralogical, textural and physical/mechanical properties of coarse and fine aggregates it can be inferred that the deformational features (cleavage, foliation, fractures, kinking, deformation bands, undulose extinction recrystallization etc.) twinning, alteration effects shape and size of aggregate, type of texture and induced stresses in the porphyroblasts of quartz and feldspar and in elastic (mica) and high grade clayey minerals have adverse effects on strength of rock as well as in corresponding concrete. They can may be responsible for enhancing discernible alkali silica/silicate reaction in concrete and cause sufficient swelling, expansion and distress in concrete structures when used as aggregate along with weak cement paste and other admixtures.

Apart from these factors, the hardness, toughness and durability of aggregate i.e. resistance to weathering effects or degree of weathering (both physical and chemical) play another significant role in the determination of strength and soundness of rock and concrete. Like other unstable crystalline forms of silica (chert, chalcedony, cryptocrystalline quartz etc) the strained quartz occurred in metamorphic belts and ductile shear zones and was characterized by deformation bands, lamellar structures (Boehm lamellae; formed due to optical disorientation on closely spaced parallel planes), undulose extinction (result due to tectonic or mechanical strain caused by external stresses which produced disorientation of irregular and adjacent microscopic patches. Therefore, the high and low polymorphs of strained quartz characteristics of disordered and weak bonded structure of tetrahedral (SiO_2) shows great susceptibility to alkali silica reaction and other chemical attack. Similarly the other silicates (feldspar, mica and clayey minerals) can also cause kaolinization, oxidation and carbonic reaction and would accelerate alkali silicate reaction in high alkaline pore solution. However, in the present case all the aggregates (constituting strained quartz and other silicates) tested by chemical method did not show deleterious degree of alkali reactivity but few of them falling very close to the boundary line in the innocuous zone may be attributed to cause deleterious degree of alkali reactivity in concrete structures.

Thus it seems that the mylonitised gneiss and mica-schist composed of hard quartz and feldspar grain do not possess good strength than that of quartzite and granitic gneiss because in these rocks the quartz and feldspar grains were comparatively well cemented and interlocked. In all cases the aggregate constituting undeformed unaltered and interlocked grains of quartz and feldspar associated with well cemented matrix produced significantly higher crushing and compressive strength.

Absorption of water (porosity) and permeability are also two important physical characters of aggregate because they may strongly influence the chemical stability, hardness, strength, deformability and thermal properties of aggregate. They also control soundness and durability partly through their effect on degree of saturation. Another important aspect is size of pores which may effect the physical and chemical behaviour of an aggregate. Therefore the mylonitised mica-schist composed of higher proportion of mica, clay and other flaky minerals with lower fraction of quartz and feldspar shows substantial increase in water absorption value than that of gneiss and quartzite.

Table 6: Uniaxial Compressive Strength of samples collected from the Project (H.P.)

Sample No.	Location of Samples	Core length (L) cm	Core Diameter (D) cm	L/D Ratio	Observed uniaxial compressive strength (kg./ cm^2)	Corrected strength for L/D ratio (kg./ cm^2)	Average uniaxial compressive strength (kg./ cm^2)
A-171	No.1 (feeder Tunnel)	Rock Sample 10.00 5.00	5.00 5.00	2.00 1.00	469.00 510.00	469.00 455.00	462.00
A-172	No.2 (U/S of Adit No.1)	Rock Sample 10.00 10.00	5.00 5.00	2.00 2.00	928.00 898.00	928.00 898.00	913.00
A-173	No.3 (D/S of Adit No.1)	Rock Sample 9.80 5.00 5.00	5.00 5.00 5.00	1.96 1.00 1.00	1187.00 1643.00 1010.00	1180.00 1465.00 900.00	1181.00
A-174	No.4 (U/S of Adit No.1)	Rock Sample 5.00 6.00 6.00	5.00 3.00 3.00	1.00 2.00 2.00	510.00 920.00 595.00	455.00 920.00 595.00	656.00
A-175	No.5 (D/S of Adit No.1)	Rock Sample 10.00 10.00	5.00 5.00	2.00 2.00	536.00 980.00	536.00 980.00	758.00

Inspite of various effects of mineralogical, textural, porosity etc on physical and mechanical behaviour of rock/concrete density of aggregate i.e. the small and rounded size of aggregate associated with denser ferro-magnesian minerals with moderate compressibility of particle size in cement paste can produce a concrete of reasonable strength or in other words it may be said the density of aggregate or concrete is directly proportional to the strength of aggregate and corresponding concrete.

In contrast with relationship between aggregate size and compressive strength, the average compressive strength increased as the aggregate size decreased. Moreover, part of differences were noted in few aggregates of same size which may be probably due to differences in shape, texture, type and other mineralogical composition of aggregate. Nevertheless, a change in shape, size, density, mineralogical and textural elements of aggregate and quality of cement and other admixture adversely affect the strength of concrete.

ACKNOWLEDGEMENTS

The first author expresses his thanks to CSIR, Pool Scientist scheme New Delhi for providing financial assistance in completion of this work. We also extend our gratitude to Professor P.S. Saklani for critical review of the manuscript. For providing all laboratory facilities Director, CSMRS, New Delhi and Head, Department of Geology, Bundelkhand University are gratefully acknowledged.

REFERENCES

ACI, (1967) Cement and Concrete Terminology Publication, American Concrete Institute SP-19.

Baalbaki, W., Baalbaki, M., Benmokrane, B. and Aitcin, P.C. (1992) Influence of specimen size on Compressive Strength and Elastic Modulus Of high Performance Concrete, American Society of Testing Materials, p.113-118.

Czarnecki, B. And Gillott, .E. J (1989) Effect of different Admixtures on the Strength of Sulphur Concrete, American Society of Testing Materials, p. 109-117.

Glasser, F.P. (1992) Chemistry of the alkali aggregate reaction. In: Swamy, R.N. (ed:) The Alkali-Silica Reaction in Concrete, p. 30-53.

Kucynski, W. (1974) Law pf water absorption of Concrete aggregates in terms of Two Parameters of the Grain Size Characteristics, Materials and Structures-Research and Testing (Paris), v.7 (38), p.11-17.

Kumar, S. (1981) Metamorphism in Manali-Rohtang Pass Area, Northwest Himalaya, In: Saklani, P.S. (ed:), Metamorphic Tectonics of The Himalaya p.155-174.

Mulliок, A.K. (1992) alkali silica reaction Indian experience. In: Swamy, R.N. (ed:), The Alkali-Silica Reaction in Concrete, P, 307-333.

Popovics, S. (1979), Concrete–Making Materials, McGraw Hill Book Company, p.370.

Portland Cement Concrete Materials Manual (1989): Publication No. FHWA- ED-89-006, Federal Highway Administration, Washington, DC.

Sarkar, S.L. and Aitcin, P. (1990), The Importance of Petrological, Petrographical and Mineralogical Characteristics of Aggregates in Very High Strength Concrete in Erlin and Stark (eds:) Petrography Applied to Concrete and Concrete Aggregates, p. 129-144.

STRUCTURE AND TECTONICS OF THE YAMUNA MYLONITE ZONE, DISTRICT UTTARKASHI, UTTARANCHAL

A. K. Biyani

Department of Geology, DBS College, Dehra Dun - 248001

ABSTRACT

The Yamuna Mylonite Zone evolved along the regional Main Central Thrust and is a part of the Higher Himalayan Crystalline Unit. Homoclinal quartzite and granite of the Proterozoic age were initially mylonitised in ductile conditions during the Himalayan orogenic movement of the mid-Tertiary age and further experienced ductile-brittle deformation in subsequent periods. As a result of crystal plastic, superplastic and cataclastic deformations in conditions of middle greenschist facies, several microstructures in minerals and shear sense indicators, were produced suggesting southwestward displacement of thrust sheet. During upthrust as well as exhumation phase a part of mylonite zone experienced ductile-brittle deformation and thereby produced protocataclasites. Strain and palaeostress data reveal constrictional deformation in toe and flattening type of deformation in hinterland region at 123 to 73 MPa respectively. Quartz C-axis analysis reveals orthohombic, monoclinic and transitional symmetries in deformed tectonites.

After the emergence of the thrust sheet deformation continued further and produced several regional faults e.g. the Yamuna Sinistral Tear, Paligad and Ujargad strike-slip faults, Dihargad Reverse Fault and Hanumanganga Normal Fault. However, folds are conspicuously absent in this zone. Neotectonic upheaval in this region asymmetrically uplifted the area along and across the Yamuna Tear which reflected through several geomorphic features viz. river terraces, waterfall, gorges, etc.

INTRODUCTION

A series of intracrustal thrust faults delineate the geological boundaries of different longitudinal divisions of the Himalaya. Among these faults, the Main Central Thrust (MCT) has a unique position. Along this thrust crystallines of the Higher Himalaya were uplifted from midcrustal level to the surface and were translated for more than 100 km in south over the Lesser Himalaya. The MCT was earlier visualised as a sharp plane and is now regarded as 5-15 km wide ductile shear zone of polyphase deformational history (Gansser, 1964; Thakur, 1992). This paper discusses, for the first time, the structural and tectonic history of the Main Central Thrust Zone designated as the Yamuna Mylonite Zone in the Higher Himalayan Yamuna valley.

GEOLOGY OF THE AREA

The study area is situated between Pali and Rana villages of Barkot tehsil of Uttarkashi district of Uttaranchal, (Fig. 1) Geology of the area on regional scale as well as in parts was described by Agarwal and Kumar (1973), Pati and Rao (1981) Singh (1986), Saklani (1993) and Biyani (1991, 1996 and 1998). The Higher Himalaya is sharply separated from the Lesser Himalaya by the southern face of the Main Central Thrust Zone MCTZ. The MCTZ is a 7-8 km wide homoclinal region of intensely mylonitised rocks. The regional lithotectonic stratigraphy around the MCTZ is as follows :

Higher Himalaya	Yamuna Formation	Green schist metapelites metabasics; and quartzites and marble of Paligad Formation exposed in eyelid windows.
Main Central Thrust Zone	Yamuna Mylonite Zone	Mylonitised granite gneiss, quartzite; and metabasics, pyrophyllitic quartzite and chlorite schists
---------------------------- Main Central Thrust ------------------------------------		
Lesser Himalaya	Paligad Formation	Paligad Quartzite

The Paligad Quartzite of Paligad Formation occurs to the south of the Main Central Thrust between Wazri to Kuthnaur in a distance of 10 km. The northerly dipping milky white quartzites are, fine grained, highly jointed and feebly folded. These quartzites are equivalent to Berinag Quartzite. Radiometric ages of Uranium mineralisation by U-Pb method and Rb/Sr dates of intrusive Bandal Granite in Himachal Pradesh occurring in the quartzite yeilded 1220 Ma (Narayan Das et al., 1979) and 1800 Ma (Thakur, 1992) respectively. Due to impact of thrusting, microscopic coarsening in the quartz grains was observed near the thrust contact and the quartzite is occasionally folded into broad open folds. The 7 to 8 km wide homoclinal Yamuna Mylonite Zone (YMZ) is situated in the frontal part of the Higher Himalayan thrust slab. This zone is mainly composed of the Sayanachatti mylonitic granite gneiss, mylonitic quartzite and Wazri metabasic schist and phyllite, occupying a small area, also occur near the southern part of YMZ and are in contact with the Paligad Formation. These mylonitised rocks are overlain by the chlorite schist and pyrophyllitic quartzite between Wazri and Sayanachatti. The 6 km long transported granite pluton lies between low grade metabasic at the base and chlorite schist on the top. This Middle Proterozoic pluton is 1972 ±102 Ma in age (Singh et al., 1986). The 500 m wide Wazri Metabasic dyke was emplaced after the deformation as it is partially altered at the margins but its core consists of phenocrysts of augite and plagioclase. The Yamuna Formation comprising of biotite schist, garnet-chlorite schist and quartzite in Banas-Phulchatti area; and quartzite and fine grained bluish grey marble occurs in northeast of the YMZ. Quartzite and quartzite-marble are equivalents of Paligad Quartzite occurring as a window (Biyani, 1998).

Towards the northern end, granite gneiss is fragmented into megabreccia in a 400m wide zone designated as Rana Brittle Shear Zone (RBSZ) and beyond the RBSZ biotite schist and garnet schist are present. The northern contact of the YMZ is concealed under thick debris. Towards the southern end of the YMZ, a minor imbricate zone of nearly 0.5 km

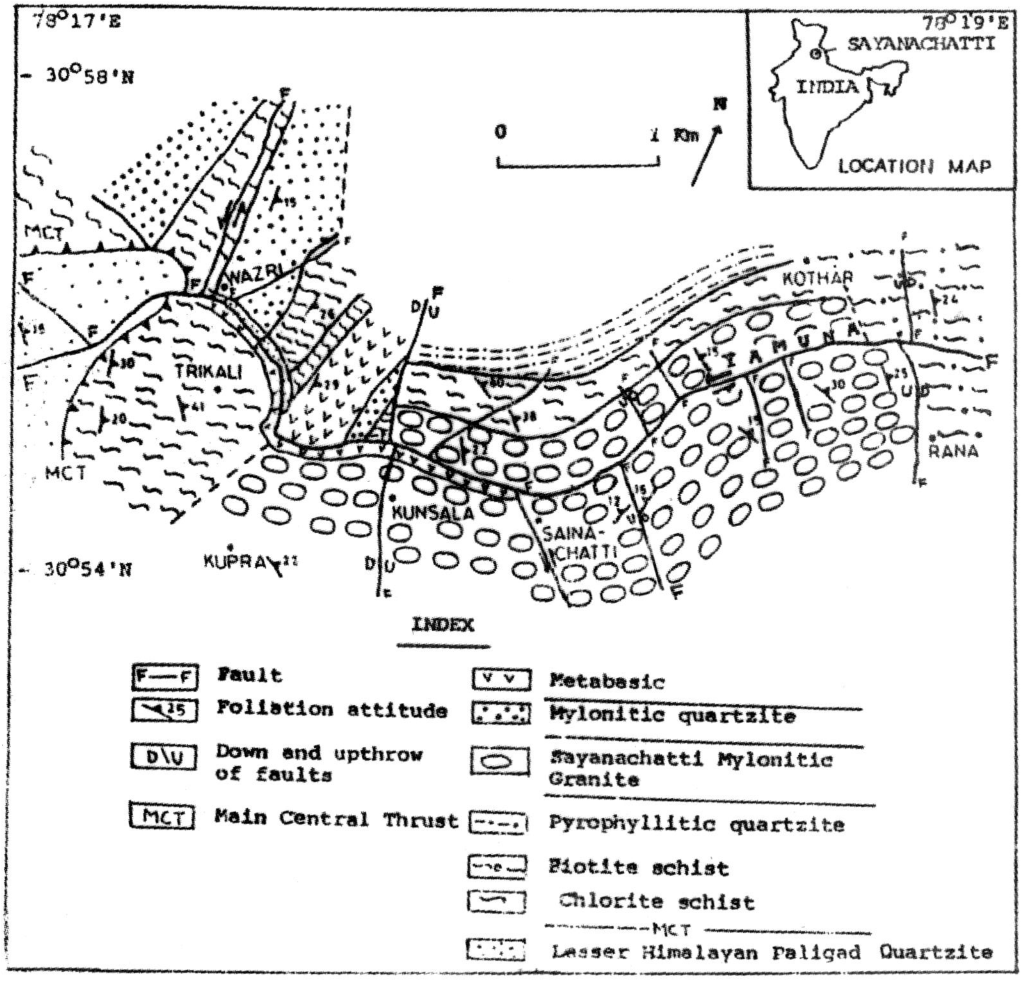

Figure 1: Geological map of Yamuna Mylonite Zone between Wazri and Rana, tehsil Barkot, district Uttarkashi.

length is observed showing repetition of Paligad Quartzite' chlorite schist and phyllite. (Biyani,1996).

BRIEF PETROGRAPHICAL CHARACTERS

Paligad Quartzite contains 95-98 percent quartz of extremly fine size. These quartz grains show thick grain boundaries (Plate 1A and 1B) and exhibit poor to moderate preferred orientation.

The YMZ consists of mylonitised quartzites, granite gneiss chlorite schist and metabasics. The strongly foliated (Sm) medium to coarse grained mylonitic quartzite (Plates 1C to 1F) is grey, brown, green, yellow and white. Quartz grains occur as porphyroclasts and are also very fine grained recrystallised grains. The clasts are globular (Plates 1C and 1F), elongated (Plate 1D) to irregular in shapes having a rim of extremely fine grained quartz showing mortar texture (Plates 1C, 1E and 1F). These clasts show undulose exitinction, cracks perpendicular to foliation (Plate 1E), microfaulting (Plate 1C), Boehm lamellae (Plate 1D) and secondary twinnings. The size of the clasts decreases away from the thrust contact. The fine grained matrix quartz (Plate 1E) shows straight extinction and sharp grain boundaries with triple junction (Plate 1F) phenomenon. The quantity of matrix gradually decreases in the inner side of the YMZ.

The greyish white fine to coarse grained, well foliated (Sm), S-type, mylonitic granite gneiss (Plate 2A to 2D) shows augen texture (Plates 2E and 2F). and contains 10-25 percent of quartz grains occuring both as finely crystallized matrix (Plates 1H and 2A-D) and as porphyroclasts; (size ratio 1:2500). Orthoclase also occurs in porphyroclasts and in matrix and occupies 50-70 percent of rock volume. Clasts in many cases show sharply truncated grain boundaries and also show core mantle texture (Plate 2A) on the grain boundaries. Orthoclase is frequently present with microcline (Plate 2C) Grains show both inclined and undulose extinctions. Orthoclase grains are also elongated due to strong shearing effect (Plate 2D). Microcline (3-5 percent) displays bent cleavage and Tartan, albite and pericline types of twining (Plate 2B and 2C). Plagioclase shows bent cleavage, kink structure and shows alteration into epidote, zoisite and sericite. Muscovite (4 percent) and biotite (2.6 percent) constituting the foliation (Sm) occurs in laths, fractured and shreaded grains (Plate 2B) Granite gneiss also contains layers of chlorite schist.

The chlorite schist occurring in the frontal part of thrust sheet (Plate 2G) is low grade metamorphic product of basic rock containing porphyroblasts of albite and chlorite and actinolite, zoisite, quartz, white mica and calcite. Well foliated pyrophyllitic quartzite contains quartz as major constituent (60-75 percent) of and 20-30 percent bluish green pyrophyllite.

The poorly foliated fine to medium grained, dark blackish green Wazri Metabasics consist of augite and plagioclase phenocrysts partly altered into uralite, chlorite, epidote while its margins are composed of actinolite, chlorite, quartz and sericite.

DEFORMATION IN THE YAMUNA MYLONITE ZONE

Structural analysis of YMZ reveals polyphase nature of deformation. The YMZ is a part of Higher Himalaya nappe or thrust sheet which was mylonitised in ductile condition all along its length during Himalayan orogenic movement of Tertiary age. These rocks were

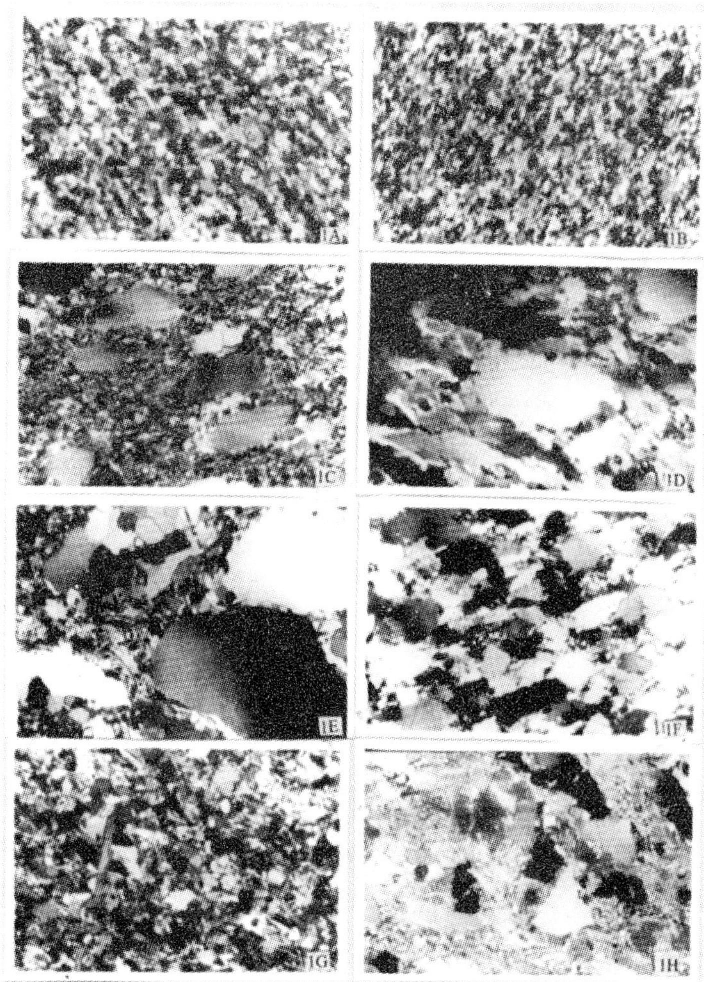

Plate 1: Photomicrographic traverse across Mylonitic zone. 1A - Paligad Quartzite between Wazri and Paligad. Extreamly fine grained nearly pure quartzite showing indistinct grain boundaries and polygonisation. 1B - Quartzite, 0.1 km SW of MCT, showing enlargement in grain size. Most of the grains are showing straight extinction due to thermal event surpassed deformational episode. 1C - Mylonitic quartzite, just on the MCT, showing large percentage of recrystallised matrix of fine size and clasts are showing deformational and tensional cracks. 1D - Ribbon shaped porphyroclast showing serrated high angle grain boundaries and strong plastic deformation. 1E - Megasize porphyroclasts displaying core and mantle structure and sigmoidal foliation. 1F - Globular grains in mylonitic quartzite showing mortar texture on periphery and various shear sense structures such as sigma and delta shapes of grains, microfaulting, elongated grains, globular grains and tails of porphyroclasts. 1G - Quartzite exposed in eyelid window near Banas, development of texture in normal metamorphic condition. 1H - Protomylonitic granite, from SW end of YMZ, exhibiting swirlling of mylonitic foliation, recrystallisation of quartz grains, perthite in orthoclase, tectonic twinning and kinking in plagioclase. 79X and XPL condition for all photographs. the inner side of the YMZ. Ribbon shaped quartz grains were formed under ductile conditions (Plate 1D). Pyrophyllitic quartzite shows 60 to 70 quartz and 20 to 30 percent pyrophyllite and remaining minerals are muscovite, chlorite and garnet with clayey matrix.

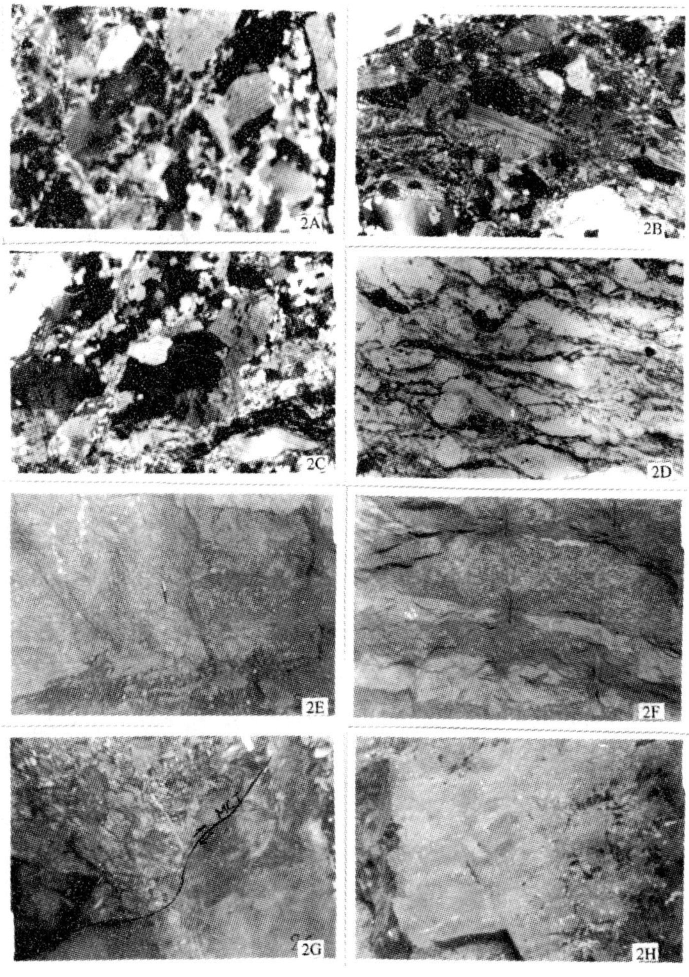

Plate 2: Photomicrographs of mylonitic granite and field photographs 2A. Typical mylonitic granite showing several shear sense structures. Strong mylonitisation leading through crushing of ground mass 2B. Another view of mylonitisation however, percentage of matrix decreased but recrystallised quartz grains are observable at several places. Note pull apart structure in orthoclase, in the centre. 2C - Protomylonitic granite showing incipient conversion of orthoclase into microcline, down left and mica fish down right. 2D - S-C structure in mylonitic granite showing rolled and globular shaped feldspar grains and excellent development of mylonitic foliation. 79X and XPL condition for all photographs. 2E - Field photograph of mylonitic granite showing two sets of foliation. Location - 1.5 km SW of Sayanachatti near adit. 2F - Extensional tectonics in mylonite zone. Pinching and swelling of quartz veins resulting into formation of boudins, location - same as 2E. 2G - Main Central Thrust near Wazri. Dark tone, towards right, representing chlorite schist and light tone, in left part, showing quartzarenite, brittle crushing is also visible. 2H - Mylonitic granite showing jointing and brecciation. Location near Sayanachatti.

brought on the surface by the Main Central Thrust from 15-20 km depth and subsequently pushed in south direction (Valdiya,, 1984). During upliftment and southward migration the YMZ further suffered brittle - ductile and brittle deformation of localised nature. The fabric gives evidence of several phases of episodic deformations.

A. *Deformation in Ductile Regime and Micro-Structures*

Continuous and cohesive deformation caused by plastic or viscous flow is the main attribute of ductile deformation. It is usually takes place at more than 15 km depth where temperature is above 250^0C (Ramsay, 1980). Ductile deformation is classified into low (<20 Mpa), medium (20-200 Mpa) and high stress (>200 Mpa) regimes. Deformation in ductile regime occurs through a complex set of crystal plastic processes, diffusional flow and cataclastic flow (Schmid, 1982 and Ghosh, 1993).

These mode of deformations occur in different P-T conditions through various creep processes operating on grain scale such as super plastic creep controlling the diffusion flow with in the grains (Nabbro-herring creep) or on the grain boundary (Cobble creep). Crystal plastic deformation is controlled by dislocation creep, glide controlled creep, power-law creep and solution creep. Dislocation creep operates through various mechanisms like dislocation glides, recovery, dynamic crystallisation, annealing and Harper-Dorn Creep (Poirier, 1985). These processes produce variety of micro structures in minerals of mylonite zone.

A.1 Microstructures in Minerals Of YMZ

Crystal plastic deformation which affected the granite gneiss and quartziteis marked by twins in quartz and feldspar (Plates 1H and 2A-D), extinction in quartz, feldspar and in mica (Plates 1C-2C), deformation lamellae in quartz (Plate 1D), Boehm lamellae (Plate 1E), presence of equiaxial grains (Plate 1F) and subgrains with low angle (Plate 1D) and high angle (Plate 1E) grain boundaries. Many of these subgrains show elongation (Plate 1D).

In superplastic deformation elongated grains are converted into equiaxial grains with weak preferred orientation. Paligad Quartzite on the both sides of the MCT shows superplastic deformation (plates 1A-1E). Cataclastic flow was manifested by decrease in grain size (Plate 2A) sharp angular boundaries of clasts and grains (Plate 1H), grains size fault (Plate 2B), off-set twin lamellea (Plate 2A) microcrushed zone (Plate 2D), pervasive cracks (Plate 2A) in quartzite and granite gneiss.

Brittle fracturing on grain scale is quite common in granite gneiss.

A.2 Shear Sense Indicators

The YMZ is profusely rich in shear sense indicators. Among important shear indicators observed in the mylonite zone are sigma (ó) structure (Plates 1C, 2C, 2D), delta (s) structure (Plates 1E and 2A), sigmoidal foliation (Plate 2D) porphyroclast tail (Plate 1C), S-C structure (Plate 2D), deformation Lamellae (Plate 1C), ribbon quartz grains (Plate 1D), pull-apart structure (Plate 2B), rotation of veins (Plate 3A), fractured and displaced grains (Plates 1H, 2A and 2B) and book shelf gliding (plate 2A). These indicators were developed in different strain and temperature conditions associated with non-coaxial deformation (Lister and William,1979, Simpson and Schmidt, 1983; Ramsay and Huber, 1987 and others). These indicators suggest sinistral sense of displacement along the walls of the YMZ and these indicators point out southward direction of transport of thrust sheet.

A.3 Strain Measurement

In Yamuna Valley, strain was estimated by Singh (1991) and Saklani (1993) independently. Singh (1991) estimated three dimensional finite strain at MCT 2.3:1.8:1, distortional strain, Es, 1.11 and Lode's ratio, v, 0.72 and near Paligad calculated finite strain is 1.8:1.5:1, distortional strain, Es, -0.72 and Lode's ratio (v) 1.7. Saklani (1993) reported decrease in strain from MCT I, near Yamunotri, to MCT III near Wazri from 0.829 to 0.563 and also reported that during thrust transport finite strain intensity was high and varying between 0.676 to 0.937. He also reported constrictional type of deformation at edge of thrust sheet and flattening type of deformation away from edge. Singh (1991) reported progressive coaxial deformation during the development of MCT involving layer parallel shear and believed it to be cause of ductile shear zone and during second episode of deformation strain pattern changed to plain strain because of non-coaxial deformation in simple shear condition. Jain and Anand (1988) carried out strain study in adjoining Pabar, Rupin and Supin valleys and reported flattening type of deformation. Strain study reveals that MCT is a shear thrust and a continuous strain gradient existed from higher structural level to lower structural level in the valley.

A.4 Quartz C-Axis Fabric Analysis

The C-axis were measured under universal stage and data were plotted on equal area net as procedure described by Turner and Weiss (1963). The fabric plot of C-axis of eight oriented samples (XZ section) covering the entire YMZ and adjoining area is shown in fig. 2 and the results are summarized in the Table 1. The oriented thin sections used in petrofabric analysis also partly used for paleostress calculation (1B of Lesser Himalaya Paligad Quartzite; 1C, 1D, 1E and 1F of mylonitic quartzite of Higher Himalaya and 1H of mylonitic granite).Three samples were selected from outside of the YMZ namely from Lesser Himalayan Paligad Quartzite (115) located at 1 km south of MCT, pyrophyllitic quartzite (108A) located 20m above the roof of mylonitic granite gneiss and biotite schist (229B) and these sections show orthorhombic symmetry. Three samples of the YMZ namely, mylonitic quartzite (118 and 47A/2) and mylonitic granite (227C) exhibit monoclinic symmetry. Two samples of pyrophyllitic quartzite (108A) and schistose quartzite (20/2) from left bank of the Yamuna river, display orthorhombic to monoclinic symmetry.

The near symmetrical orthorhombic pattern of c-axis plot with type 1 of cross girdle pattern indicates flattening type of deformation under plain strain or coaxial condition of deformation in which C-axis tends to lie on small circle girdle near periphery showing maximum shortening in Sm direction. Monoclinic symmetry can be described to noncoaxial deformation within mylonitic zone, the orthorhombic symmetry was due to simple flattening K = 0 and plane strain (K=1) condition.

There is a dominant tendency for the C-axis to be oriented parallel to the direction of slip. The monoclinic symmetry further shows that sinistral sense of slip or shear in

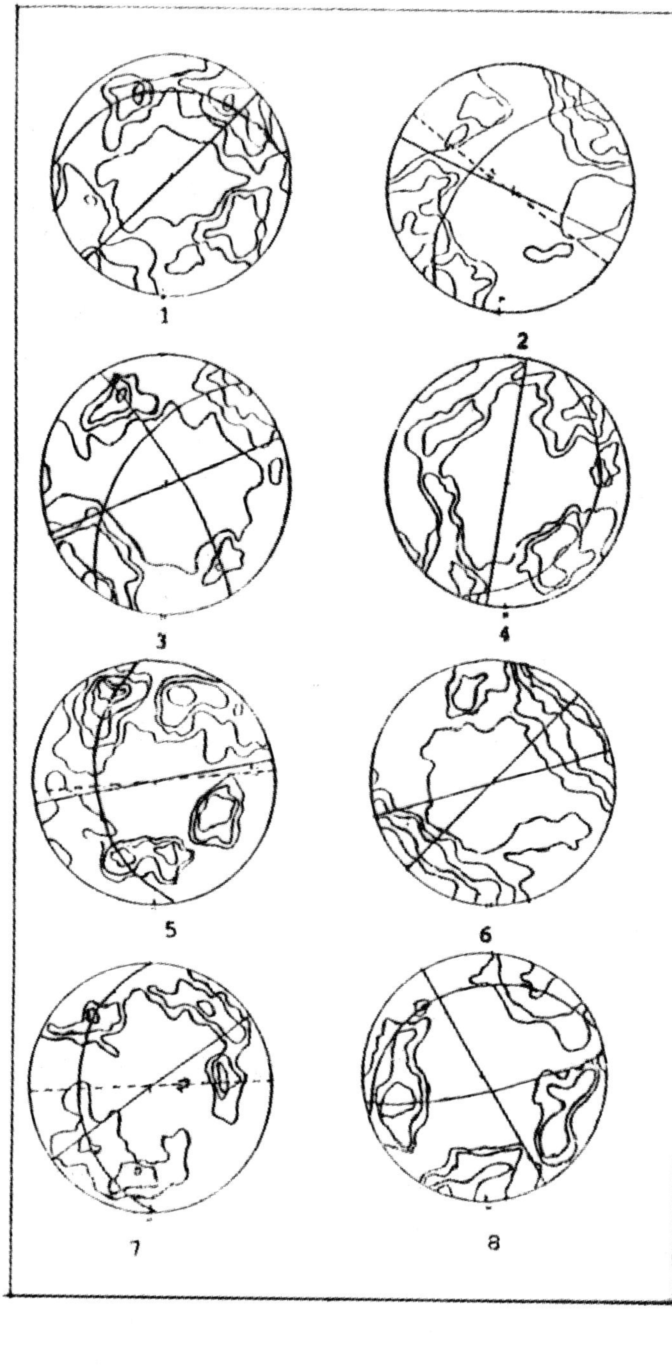

Figure 2: Quartz C-axis plot of oriented thin section. 1 - Lesser Himalayan quartzite, 2 - Mylonitic quartzite near Wazri, 3 - Quartzite at the roof of granite near Wazri, 4 - Pyrophyllitic quartzite, 5 - Right Bank mylonitic quartzite, 6 - Left Bank mylonitic quartzite near Trikhali, 7 - Mylonitic granite gneiss near Rana, 8 - Biotite schist between Rana and Pindki. Contours 1-2-4-7-77% per 1% area. Number of measured grains and other details are given in the Table No.1.

Tabel 1: Sumary of Quartz Fabric Analysis of Yamuna Mylonite Zone

S.No.	Sample No.	Location	Rock Type	Mineral Lineation Direction	N	Symmetry	GirdlePatern	Thrust Transport with direction of shear
1.	115	Wazri	Quartzite (LH)	N 40°E	224	Orthorhombic	Type-I Crpss Girdle	-
2.	118	Wazri	Cataclastic Quartzite (CH)	W22°S	203	Monoclinic		WNW-ENE Sinistral Slip
3.	118A	Wazri	Quartzite on top of granite	N68°E	185	Ortho-Mono	Tpe-I Cross Girdle	-
4.	121A	Opposite Kunsala	Pyorophyllitic Qzt. On ranite		187	Orthorombic	Type-1 Girdle	-
5.	47A/2	Opposite Trikhli	Cataclastic Quartzite	N75°E	215	Monoclinic		WW NW - EE SE Sinistral Slip
6.	20/2	Tr.khli	Schistose Quartzite	N56°E	222	Ortho-Mono	Type-1 Girdle	
7.	227C	Near Rana	Granite	N56°E	75	Monoclinic	W-E Dextral	
8.	229B	Pindki	Biotite-Quartz Schist	N30°E	175	Orthorhombic	Type 1 Cross Girdle	

quartzites while dextral sense in mylonitic granite near Rana. This reversal in sense of slip in granite can be due to brittle deformation zone (plate 2E).

A.5 Piezometry

Several piezometers such as deformation induced dislocation density, recrystallised grain size and subgrain size were helpful for palaeostress estimate. Limitations of various peizometers have been discussed by Schmid (1982) and Ghosh, (1993). Recrystallised quartz grains of mylonitic quartzite and of granite gneiss were used for palaeostress calculation. The recrystallised grain size near MCT was less than 12μm while it increased systematically in hinterland of MCTZ.

The palaeostress was calculated by using the formula

$$ó = AD\text{-}m$$

where 'is differential stress, D is grain size and A and m are constants. The value A and m are provided by Mercier et al., Twiss and Ord and Christie (in Ghosh ,1993). For stress determination mica and other impurities free portion were selected (Plates 1C, 1D, 1E and 1F) and the grain size D was calculated by the following equation (Saha, 1989).

$$D = \frac{1}{n}\sum_{i=1}^{n}(D_i D_i)^{1/2}$$

where n is the number of grains measured, between 120-160 for each thin sections, and Di and Di are the length and breadth measured in microscope using micrometer. The computed palaeostresses value using different constants are given in Table 2.

Table 2: Computed palaeostress value from mylonitic quartzites (Plates 1C-1F). The flow stress is calculated by using constant provided by : I. Mercier et. al. (1977); II. Twiss (1977); III. Ord and Christie (1984) in Ghosh (1993).

Plate No.	Palaeostress (Mpa)			
	I	II	III	Average
1C	68	115	185	123
1D	57	94	151	101
1E	50	83	136	90
1F	41	71	108	73

On the basis of field observations and laboratory study of mesoscopic structural data it is inferred that two events of deformation were present in the YMZ. The D1 event is prominent and widespread and was responsible for S1 and Sm while later event, D2, was considerably weak and produced S2 foliation (Biyani, 1996).

MACROSCOPIC STRUCTURES

Macroscopic structures are dominated by faults of different origin in the YMZ. The important faults present in the area are as follows :

A. Main Central Thrust

The sinuous Main Central Thrust defines the tectonic contact between Lesser Himalaya (foot wall) and the Higher Himalayan thrust sheet (hanging wall). The sharp contact is exposed at Trikhli, Wazri and in Paligad valley where it dips at 28^0, 35^0 and 67^0 respectively from north to northeast (Plates 2G and 3A). The MCT was dislocated by nearly 200 m by the Yamuna Tear and 105 m by Paligad strike-slip fault. In Paligad valley, a dilatant (opening or cavity caused by extension) is present just near the MCT in a crushed zone (Plate 3B). The MCT was visualised as a very low angle listric thrust which flattens in depth (Thakur, 1992) but its moderate to very steep dip in the YMZ is probably due subsequent locking of thrust movement and transfer of movement to other faults (Yamuna Tear and Paligad and Ujargad strike-slip faults). Saklani (1993) proposed duplex structure in the this area and noted the presence of MCT I, MCT II and MCT III,

C. OTHER FAULTS

Main Central Thrust and imbricate zones evolved simalteneously on the surface. Parallel to trace of MCT some fractures were developed and one such fracture is the, Dihargad Reverse Fault, (Plate 3C). A magnificant exposure of this fault is present on the right bank of the Yamuna river showing 12 m dip separation with 74^{0}dip in NE. Mylonitised granite gneiss was also severely crushed along the fault plane (Plate 3D and 3E) producing crushed fault breceia. Further 3 km in the NE direction a normal fault was developed near the village Hanumanchatti in the Hanumanchatti metabasic. This fault is named as Hamnumanchatti Normal Fault (HNF). Dipping in NE with 60^0 showing dip separation of 20mg the Yamuna Sinistral Tear was recognised by Oldham (in Valdiya, 1980). This Yamuna Sinistral Tear shows 100-180 m displacement of Higher Himalayan lithounits including of the YMZ and also displacement of same order in tectonic elements like the MCT, DRF and HNF and other faults. Other supporting evidences of tear are straight course of the Yamuna, presence of cold and hot water springs and strike slip type fault solution of recent micro seismicity along the Yamuna river (Khattri et al., 1989). The Yamuna Tear appears to be a creeping fault (Rubin et al., 1999). Infact in Himalaya several transverse pairs comprising of left lateral and right lateral faults are present (Virdi, 1979).

Two other sinistral strike-slip faults namely the Paligad and Ujargad are present in the SW part of the YMZ. Their location near the edge of thrust sheet support the model given by Mckinstry (1953) and Chinnery (1966), Diplacement in Paligad fault Fig.4, is 80 m while in the Ujargad fault it is 45 m.

At Trikhli (Plate 3F), Kunsala and Sayanachatti steeply dipping faults running across the Yamuna river and showing strike displacement are also observed.

Plate-3 : Field photographs of mylonite zone. 3A - MCT is making sharp contact between dark toned chlorite schist of Higher Himalaya top right, with light toned Paligad Quartzite. Also sinistrally displaced quartz veins across MCT is faintly visible. Location - Pali in Paligad Valley. 3B - Crushed zone mylonitic forming a dilatant, 5m from MCT in Paligad Valley, indicating initial ductile and subsequent brittle deformation. 3C - High angle Dehargad reverse fault in mylonite granite between Sayanachatti and Rana. 3D - Cataclasite in mylonitic granite due to ductile brittle deformation resulting into disorientation of granite fragments, and on the surface two sets of slickensides not clearly visible, are present. 3E - Brittle deformation of mylonitic granite near surface condition cause formation of megafragments. Location - Near Rana, Left bank of Yamuna river. 3F - Very steep angle fault near Trikhli, very close to MCT, causing abrupt change in height - Left Bank of the Yamuna river.

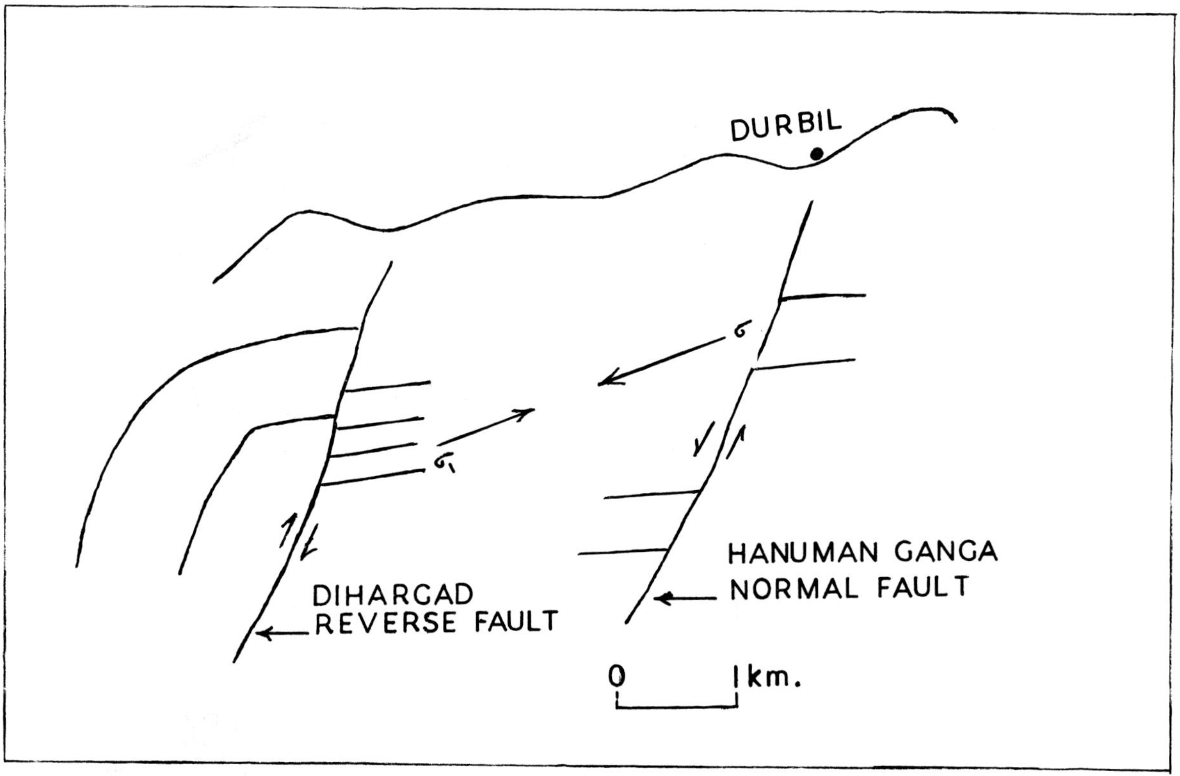

Figure 3: Field relationship between Dihargad reverse fault and Hanumanganga normal fault; a conjugate pair produced by inclined ό axis.

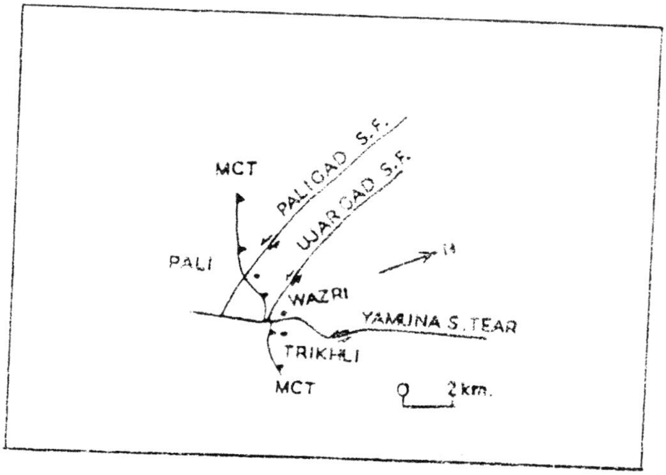

Figure 4: Field relationship between MCT and Yamuna Tear (Ist order fault) and second order Paligad and Ujargad sinistral strike-slip faults.

NEOTECTONICS

Ample evidences of neotectonic activities are present in many parts of the YMZ. Evidences of neotectonic deformations in the YMZ are :

A. ***Rana Brittle Zone:*** This 400 m wide brittle zone is developed at NE end of the YMZ, near Rana village. Granite gneiss and biotite schist are thoroughly crushed into fragments of few cms to many tens of meters length (Plates 2H and 3D-E). Supporting evidence of deformation are presence of minor imbricate zone e.g., profuse development of quartz veins, fault plane striations, joints of random orientations, occurrence of springs and seepage zones and waterfalls, abrupt change in topographic elevation, sudden asymmetric widening of cross profile of the Yamuna valley on the left side and abrupt change in height of river terraces.

B. ***River Terraces:*** Two to three sets of asymmetrical river terraces are present along the Yamuna river. These terraces show variable upliftment in different sectors. The first terrace is 0.5 to 2.5 m above water level of river and second terraces is 3 to 50 m in height and making vertical to steeply sloping scarp. Third terrace of 1 to 15 m height is not uniformly developed in the region. The terraces are 2-3 m to 200 m wide and face morphology of terraces indicate that area was rapidly uplifted and at several places across the terraces waterfall and rapids are formed. A six step water fall is present in Trikhli (Plate 3F). Terraces are absent in the tributaries of the Yamuna river.

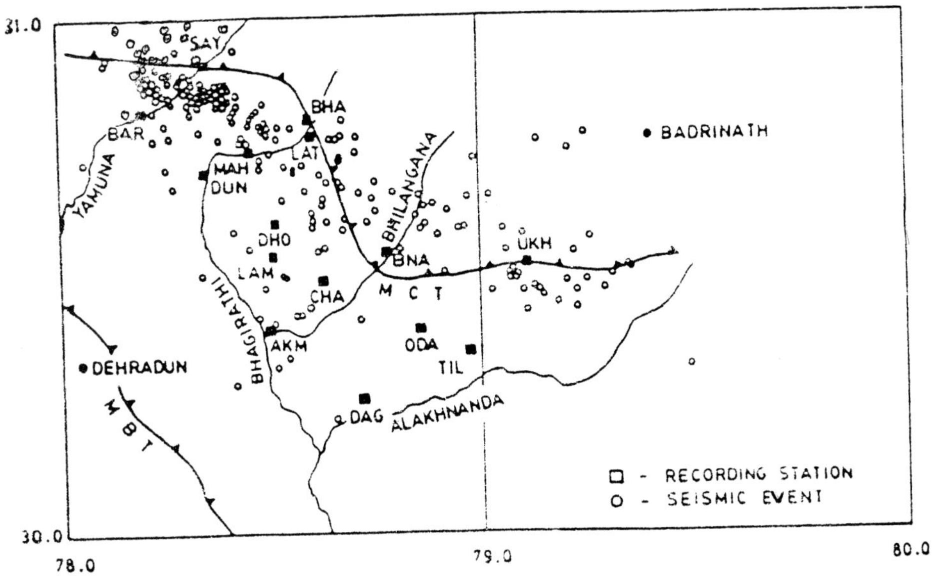

Figure 5: Micro seismicity in and around Yamuna Mylonite Zone between 1979-1980 and 1984-1986 (Khattri et al., 1989). Note SAY, BAR local names of villages in abbreviated form i.e. SAY means Sayanachatti and BAR means Barkot.

C. **Dilatant:** A dilatant is extensional feature formed by expansion of compressed rock when it gets exposed to surface (Twiss and Moores, 1992). A 30 m wide 2.5 m deep and several meter height dilatant is present near the MCT in Paligad valley. Highly crushed but coherent mylonitic quartzite is exposed in this dilatant (Plate 3B).

DISCUSSION AND CONCLUSION

In Higher Himalayan Yamuna Valley, the mylonite zone was formed in ductile conditions of lower to middle greenschist facies. The conspicuous effects of mylonitisation are discernible in quartzite and quartzo-feldspathic schists. However, pelitic rocks responded poorly to creep deformation process due to high ductility. During exhumation stage, still at considerable depth, from 5-10 km, the granite gneiss was affected by localised ductile-brittle deformation with formation of protocatalasite showing disoriented angular clasts of millimeters to less than decacentimeters length. Finally, nappe sheet emerged on the surface along a listric plane due to underthrusting of the Lesser Himalaya. In the Yamuna Valley, southward movement of the crystalline thrust slab was checked by a northern bulge in the Lesser Himalaya. This bulge is result of resistance created by the subsurface northern Himalaya in underthrusting of the Lesser Himalaya. This arrested the movement of southward translating leading edge of Chail thrust sheet and as a result of this the trailing edge and middle portion of the thrust sheet started gaining elevation thereby, 'ó1 and ó3' interchanged their orientation. Hence a series of normal faults Hanumanganga and Rishiganga in the Yamuna Formation and, strike-slip faults (Yamuna Sinisital Tear and Paligad and Ujargad strike slip faults) were formed. Due to Further continuation of tectonic activity in Quaternary period terraces along the Yamuna river were ratified with development of gorges, dilatants, waterfalls and cold and hot water springs.

ACKNOWLEDGEMENTS

Author is grateful to Dr. V.C. Thakur, Director, Wadia Institute of Himalayan Geology, DehraDun and Dr. B.K. Pandey, ex-Head, Department of Geology, D.B.S. College, DehraDun for guiding this work for D.Phil. Degree.

REFERENCES

Agarwal, N.C. and Kumar, G. (1973). Geology of Upper Bhagirathi and Yamuna Valleys. Uttarkashi district, Kumaun Himalaya. Him. Geol., Vol.3, pp.1-23.

Biyani, A.K. (1991). Hydrogeothermometry of hot springs waters of Central Himalayan Yamuna Valley, Uttarkashi, U.P., In : Mountain Resource Management and Remote Sensing, Ed. Gupta, P.N. and Ray, A.K., Surya Publication, DehraDun, pp.116-121.

Biyani, A.K. (1996). Structural and metamorphic history of Higher Himalayan zone of Phulchatti to Yamunotri area with special reference to Main Central Thrust in Garhwal Himalaya. Unpublished D.Phil. thesis, HNB Garhwal University, Srinagar, pp.250.

Biyani, A.K. (1998). A study of inverted metamorphism in the Higher Himalayan Yamuna Valley. Uttarkashi district, Uttar Pradesh, Jour. Geol. Soc., India, V.51, pp.609-617.

Chinnery, M.A. (1966). Secondary Faulting. Can. J. Earth Sci., V.3(2), pp.163-190.

Gansser, A. (1964). Geology of the Himalayas. Wiley Interscience, London, pp.289.

Ghosh, S.K. (1993). Structural Geology. Pergamon Press, pp.598.

Jain, A.K. and Anand, A. (1988). Deformational and strain patterns of an intracontinental collision ductile shear zone an example from the Higher Garhwal Himalaya. Jour. Struct. Geol., V.10(7), pp.717-739.

Khattri, K.N. and Ramesh Chander, Gaur, V.K., Sarkar, I and Sushil Kumar (1989). New Seismological results on the tectonics of the Garhwal Himalaya. Proc. Indian Acad Sci. (Earth Planet, Sci.), Voi. 98(1), pp.91-109.

Lister, G.S. and Williams, P.F. (1979). Fabric development in shear zones : Theoritical controls and observed phenomena. Jour. Struct. Geol. V.1, pp.283-297.

McKinstry, H.E. (1953). Shears of second order. Am. J. Sci., V.251, pp.401-444.

Narayan Das, et al. (1981). Uranium in metamorphites around Brijrani gad, Raunsal, Tehri Garhwal (U.P.); Himalaya In : Metamorphic Tectonites of the Himalaya, Saklani P.S. ed. Today and Tomorrow's Publishers. pp.111-122.

Pati, U.C. and Rao, P.N. (1981). The Main Central Thrust in U.P. Himalaya. In Contemp. Geol. Res. Himalaya, Ed. A.K. Sinha, BSMPS, Publ. DehraDun, pp.125-129.

Poirier, J.P. (1985). Creep of Crystals. Cambridge University Press, Cambridge, 260 pp.

Ramsay, J.G. (1980). Shear zone geometry. a review. Jour. Struct. Geol., V.2, pp.83-99.

Ramsay, J.G. & Huber, M.I. (1987). The Technique of Modern Structural Geology, V. 2, Folds and Fractures. Academic Press, London, pp.309-700.

Rubin, A.M., Gillard, D. and Got J.L. (1999). Streaks of microearthquakes along creeping faults. Nature, V.400 (6745), pp.635-641.

Saha, D. (1989). The Calidonian Skerrols Thrust, SW Scotland microstructure and strain. Jour. of Struct. Geol., V. 11, No.5, pp.553-568.

Saklani, P.S. (1993). Geology of the Lower Himalaya (Garhwal), International Books and Periodical Supply Service, Delhi, pp.246.

Schmid, S.M. (1982). Microfabric studies as indicators of deformation mechanisms and flow laws operative in mountain building. In : Mountain Building Processes, Ed. Hsu K.J., Academic Press, London, pp.95-110.

Singh, K. (1991). Strain variation in the Berinag Thrust sheet and its relationships with the Main Central Thrust in the Yamuna and Tons Valleys, Garhwal Himalaya. Jour. Him. Geology, Vol. 2(1) : 71-88.

Singh, R.P., Singh, T.P., Bhanot, V.B. and Mehta, P.K. (1986). Rb-Sr. ages of gneissic rocks Riheegangi, Bhatwari, Hanumachatti and Naitwar areas of the Central Crystalline Zone of Kumaun Himalaya, U.P. Ind. Jour. Earth Sci., V.13, pp.197-208.

Simpson, C. and Schmidt, S.M. (1983). An evaluation of critiria to deduce the sense of inovement in sheared rocks. Bull. Geol. Soc. Am., 94, pp.1281-1288.

Turner, F.J. and Weiss, L.E. (1963). Structural Analysis of Metamorphic Tectonites. McGraw - Hill Book Co., pp.1-545.

Valdiya, K.S. (1980). Geology of Kumaun Lesser Himalaya. Wadia Institute of Himalayan Geology, DehraDun, pp.1-291.

Valdiya, K.S. (1984). Aspects of Tectonics. Tata McGraw-Hill Publishing Company, New Delhi, pp.1-319.

Virdi, N.S. (1979). On geodynamic significance of megalineaments in the Outer and Lesser region of Western Himalaya. Him. Geol., V. 9(1), pp.79-99.

GEOLOGICAL AND MORPHOTECTONIC STUDIES USING SATELLITEIMAGES OF A PART OF TEHRI-UTTARKASHI DISTRICT, LESSER GARHWAL HIMALAYA

Pandey, A.C*, Saklani, P.S and Dubey, C.S****

Haryana State Council for Science & Technology (HSCST), Chandigarh.

** Department of Geology, University of Delhi, Delhi-110007

ABSTRACT

Geological mapping in Garhwal Himalayas is always challenging as major part of the terrain is inaccessible and rock exposures are exposed mainly along the road and river sections. The regional coverage of the terrain acquired by satellite images provide pictorial representation of lithological units, structures (in the form of lineaments), drainage, geomorphic landforms, landuse and erosional characteristics. The satellite-based information coupled with ground information and ancillary data helped in preparation of accurate geological and lineament maps. In the present study the geological-lineament mapping of 2700 sq km area was carried out on 1:50,000 scale in Tehri-Uttarkashi districts of Lesser Garhwal Himalaya using IRS-1C satellite images. The various lithological units mapped in the area were grouped into Simla, Krol, Jaunsar and Crystalline groups of rocks. The regional thrusts, viz; Garhwal, North Almora, Pratapnagar, Budhakedar and Jutogh were delineated in the area. A total number of 1096 lineaments were marked which exhibit four prominent trends, viz; NE-SW, NW-SE, N-S and E-W. The Bhagirathi valley exhibits multicyclic evolution of its landscape. The morphotectonic evaluation of river profiles of the Bhagirathi was done based on the analysis of transverse and longitudinal profiles. The cross valley profiles of the Bhagirathi river shows prominent breaks in the valley slopes reflecting periods of dynamic rejuvenation due to epirogenic uplift. The longitudinal profile of the Bhagirathi and its major tributaries shows prominent nick points, which further confirm rejuvenation of the valley. The five levels of river terraces in the Bhagirathi valley were delineated which suggest atleast five episodes of regional uplift of valley. Based on the geological and morphotectonic studies it can be remarked that major drainage channels of the Bhagirathi basin were developed along the major structural planes formed during different phases of regional deformation and tectonics.

INTRODUCTION

The investigated area represents a part of the Lesser Garhwal Himalaya and is situated in Tehri and Uttarkashi districts of Uttaranchal State. It is located between 78°15' - 78o45' E longitudes and 30°15' - 30°50' N latitudes and measures for about 2700 sq km in the Survey of India topographical sheets 53 J/5, J/6, J/7, J/10, J/11, J/9.

The thrust tectonics greatly influenced the structural setup. The area under investigation is constituted by a number of nappes, namely, the Krol, Chail (I, II, III) and Crystalline nappes (I, II). The geological and lineament mapping was carried out on a regional scale using False Colour Composite (FCC) images(Fig.1) on 1:50,000 scale of Indian

Remote Sensing (IRS) 1C satellite. The river profiles (transverse and longitudinal) of the Bhagirathi and its major tributaries and the fluvial river terraces present in the Bhagirathi and Bhilangna rivers were analyzed with reference to lithology and structures to comment on the morphotectonic evolution of Bhagirathi valley.

GEOLOGY OF THE AREA

The investigated area consists of Simla Group, Krol Group, Jaunsar Group and Central Crystalline Group of rocks (Fig.2). These rocks are confined to nappes delineated by well-defined thrusts. Due to erosion, these rocks occur in the form of windows and klippes. The rocks generally exhibit NW-SE strike trend with NE and SW dips at 30^0-55^0.

The Simla group in the study area is represented by grey, green, purple and black slates. On the satellite image these slates are characterised by light to medium tone, fine drainage texture, subdued topographic expression, rounded ridges and intensive cultivation.

The Krol group is constituted by Blaini, Infrakrol, Krol and Tal formations. Blaini formation consists of cherry red and purple slates with subordinate quartzites. It shows gradational contact with Infrakrol formation but is marked by unconformable contacts with Nagthat formation. Satellite image characteristics of the Blaini formation include a light tone, fine drainage texture with subdued topography whereas the Nagthat formation has medium tone, medium to coarse drainage texture and bold topographic expression with sharp ridge lines. The Infrakrol formation is represented by greyish green, black and buff slate. On the satellite image the Infrakrol formation is characterised by a light tone, fine drainage texture and cultivation. The area is mainly covered by Krol C member which is characterized by dark grey and bluish-grey massive crystalline and gypsiferous limestone. It shows medium tone, coarse drainage texture with deep rill and gully erosion, scanty vegetation, undulating topography and broken ridgelines. The Tal formation occurs in the core of the Krol syncline and is characterised by medium to coarse grained, yellowish brown ferruginous quartzite (Upper Tal) and dark grey to black carbonaceous and pyritic shales associated with subordinate quartzites (Lower Tal). The Lower Tal quartzites exhibit tectonic contact with Chandpur phyllites (occurring as klippe) at Banali. This contact can be demarcated on the satellite image by a bold topography with sharp strike ridges whereas the Chandpur phyllite is characterised by a smooth topographical expression with intensive cultivation.

In the Garhwal Himalaya the Chandpur phyllite, Pratapnagar quartzite and Kotga-Banali Calc Zone constitute the Jaunsar group/Chail group (Saklani, 1993). The older Chail formations, namely, the Chandpur phyllite and Pratapnagar quartzite exhibit tectonic contacts with the calcareous zone and hence the Paturi/Nagthat quartzite, Bhelunta limestone and Bhainga slate of the Calc Zone represent the youngest members of the Chail/Jaunsar Group.

The Chandpur phyllites, which are thrust over the Kotga Banali Calc Zone comprise of green sericite-chlorite-quartz rich slaty member and greyish green sericite-chlorite rich schistose member. The Pratapnagar quartzite is fine to coarse grained, and tectonically overlies the Kotga Banali Calc Zone along the Pratapnagar Thrust. The Pratapnagar quartzite is generally massive but becomes schistose at Sainj and Siguni along the Budhakedar thrust (MCT-III) and Dharkot thrust (NAT) zones respectively. On the satellite image the Pratapnagar quartzite is characterized by a medium to dark tone, joint controlled drainage, bold topographic expression with prominent strike ridges, scanty cultivation with a thick forest cover. The Kotga-Banali-Calc Zone occurs in the form of NW-SE trending windows (Pratapnagar, Dunda and Uttarkashi windows) and exhibits tectonic contacts with Chandpur and Pratapnagar formations. On the satellite image the extent of Kotga-Banali-Calc Zone was demarcated. The Bhelunta limestone member of the Calc Zone is characterized by a coarse drainage texture, deep gullies, more rill erosion and scanty vegetation while Bhainga slate exhibits a subdued topography with intensive cultivation. Metabasics are well exposed within the rocks of Pratapnagar formation and Kotga Banali Calc Zone. Due to their smaller extent, the demarcation of these posed a problems in their identification on the satellite images.

The rocks of the Crystalline group are represented by the Budhakedar formation and Jutogh formation. Both the formations are characterised by distinct mineral assemblages and are in tectonic contacts with one another. The Budhakedar formation is represented by granite gneisses (mylonitic gneiss, streaky gneiss and porphyroblastic gneiss) associated with quartzofeldspathic schists and migmatites (Pandey 1993). The rocks of the Budhakedar formation are thrust over the Pratapnagar formation along the Budhakedar thrust. On the satellite image the rocks of the Budhakedar formation exhibit medium to dark tone, coarse drainage texture and less dissected topography. The Jutogh formation overlies the Budhakedar formation along the Jutogh thrust.

LINEAMENTS IN THE AREA

The area under investigation is traversed by a number of thrusts of regional extent which show NW-SE trend with northeasterly dips varying 25°-45°, except for the North Almora thrust which dips in south (Fig.2). These tectonic units were also described by Saklani (1993), Valdiya (1976) and Jain (1971). The Garhwal thrust exists between the Nagthat quartzites and Chandpur phyllites. In the study area, the Garhwal thrust is traced over a distance of about 40 km from Pokhri (southwest of Chamba) to Kyari (west of Chinyalisaur). It generally trends NW-SE with a northeasterly dip of 30°. The Chandpur phyllite is in tectonic contact with Pratapnagar quartzite along the North Almora Thrust (NAT). It can be traced for a distance of 70 km from Dang Chaunra (southeast of Chandrabadni) to northwest of Nagal with a NW-SE trend and southwesterly dips. A highly subdued topographical expression with distinct erosional characteristics developed of Chandpur phyllites as against the bold topographical expression of Nagthat / Pratapnagar quartzites and were found helpful to delineate the Garhwal / North Almora thrust on satellite imagery. A branch of North Almora thrust, designated as Pratapnagar thrust by Saklani (1993) was also delineated on satellite images. It is demarcated over a distance of about 35 km in the area from Kandi Khal (east of Deol) to Pujargaon (east of Chinyalisaur). A relative absence of surface drainage in the limestone member and subdued topographical expression of the slaty member of the Kotga-Banali Calc Zone helps in distinguishing it from the Pratapnagar quartzites on satellite images. The **Budhakedar Thrust (MCT-III)** is a major tectonic break, which delimits the vast stretch of Pratapnagar quartzite from the

Budhakedar formation of gneisses and migmatites. On the satellite images Budhakedar thrust is delineated for a distance of about 48 km across the Bhagirathi and Bhilangna valleys from east of Ghansyali to northwest of Sainj. The image characteristics like medium tone, joint controlled drainage, forest cover and relative absence of cultivation in the Pratapnagar formation as compared to dark tone, coarse drainage texture and less dissected topography in the Budhakedar formation were utilised in delineating the thrust contact between the two formations. The **Jutogh Thrust (MCT-II)** is a tectono-metamorhic break between the low grade and medium grade rocks of Budhakedar and Jutogh formations respectively. On the satellite images it is delineated for a distance of about 34 km across the Bhagirathi and Bhilangna valleys. Due to pronounced denudation along the thrust zone with luxuriant vegetation cover in the region, the recognition of this lithological change becomes difficult even on satellite images.

In the Bhagirathi valley, a number of transverse and oblique faults were delineated. A number of N-S trending transverse faults like Ganwan fault, Rano fault and Shyalam fault were delineated within the Pratapnagar quartzites towards north of Uttarkashi. A major E-W trending fault is located at Uttarkashi, along which the Bhagirathi river flows in an E-W direction for about 7 km (Fig. 3a). A transverse fault along the Jalkur gad trending NE-SW passes through Bhainga displacing the Pratapnagar thrust for about 350m. (Fig. 3b). In the Bhilangna valley near village Deol, a NE-SW trending fault having a continuation of about 4 km was observed which displaced the North Almora thrust for a distance of about 400m.The Bhilangna and Balganga rivers flow along the major tear faults. Such tear faults trending NW-SE and NE-SW are present at Chamyala and Raunsal respectively (Fig. 3c).

Based on the satellite image interpretation the four groups of lineaments were established in Tehri-Uttarkashi region, which trend NW-SE, NE-SW, N-S and E-W (Fig.3). It is to state that these groups were closely linked with thrusts, faults, joints and anticlinal/ synclinal axes. Many NW-SE trending lineaments were developed along the Budhakedar Thrust, North Almora Thrust, Garhwal Thrusts etc. The NE-SW and NW-SE trending lineaments are expressions of the regional joint patterns. The N-S lineaments show development parallel to faults and faulted anticlinal fold axes controlling a number of drainage channels e.g., Jalkur, Rano and Syalam gad. The E-W lineaments show preferred orientation but exhibit less control on drainage (Pandey, 2001).

There are approximately 1096 lineaments present in the Krol, Chandpur, Pratapnagar and Crystalline blocks (Table 1-4 and Fig.4).

Table 1: Direction and lengthwise occurrence of lineaments in Krol Block.

Length	NE-SW	NW-SE	N-S	E-W	Total
Upto 2 km.	38	38	9	19	104
2-4 km.	17	8	8	7	40
> 4 km.	5	2	1	7	8
Total	60	48	18	26	152

Table 2: Direction and lengthwise occurrence of lineaments in Chandpur block

Length	NE-SW	NW-SE	N-S	E-W	Total
Upto 2 km.	63	36	30	45	174
2-4 km.	36	24	15	17	92
> 4 km.	7	8	5	5	25
Total	106	68	50	67	291

Table 3: Direction and lengthwise occurrence of number of lineaments in Pratapnagar block.

Length	NE-SW	NW-SE	N-S	E-W	Total
Upto 2 km.	93	72	49	49	263
2-4 km.	57	50	38	27	172
> 4 km.	12	9	7	3	31
Total	162	131	94	79	466

Table 4: Direction and lengthwise occurrence of number of lineaments in Crystalline block.

Length	NE-SW	NW-SE	N-S	E-W	Total
Upto 2 km.	38	32	20	28	118
2-4 km.	30	8	9	18	65
> 4 km.	2	0	0	2	4
Total	70	40	29	48	187

The origin of these lineaments can be ascribed to tectonic forces, which caused various phases of folding. In sequence to these phases of folding, the thrusting of rocks of the Central Crystalline, Pratapnagar, Chandpur and Krol groups took place, which was accompanied by transverse faulting. The joints also developed in conjugate sets during the folding episodes.

The lineament rose diagram (Fig.5) indicates presence of four prominent trends in the area. These are N55^0 E-S55^0 W, N50^0 W-S50^0 E, N90^0 E-S90^0 W and N10^0 E-S10^0 W. The rocks of Tehri-Uttarkashi area underwent four phases of deformation during the Himalayan orogeny (Saklani, 1993). The various structural features viz., axial plane foliation, three to four sets of joints (strike joints, dip joints and oblique joints), faults and folds were developed due to these episodes of tectonics. During the development of E-W trending major lineaments (e.g., Uttarkashi fault) the maximum Principal compressive stress (ó) was directed from N-S and the least Principal compressive stress (ó) was along E-W direction. The intermediate Principal compressive stress (ó) was vertically directed (Fig.6, 7). The minor lineaments were developed around N30^0 W- S30^0 E and N30^0 E-S30^0 W. The major lineaments can be correlated with E-W plunging folds (Phase I) of Saklani (1971) in Garhwal and Naha and Ray (1970) in Simla.

Figure 1: False Colour Composite Image of Tehri-Uttarkashi Area.

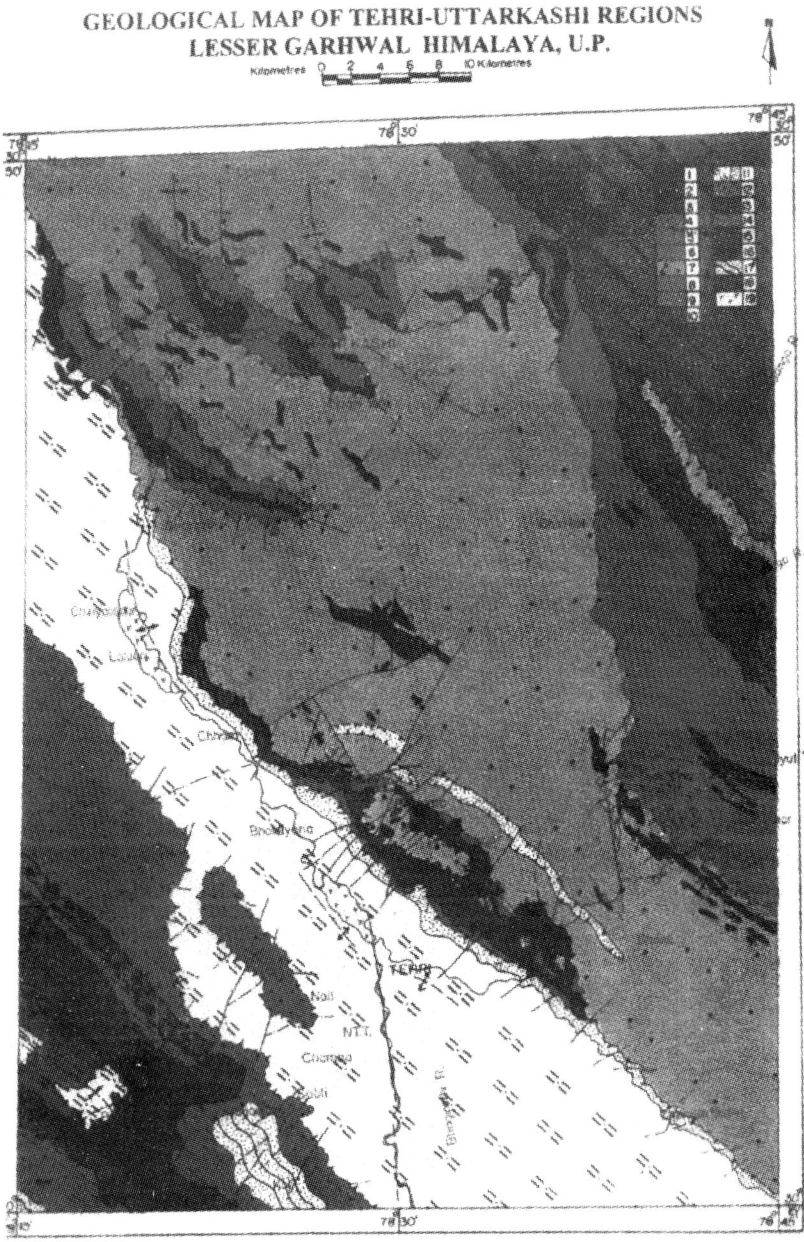

Figure 2: Legend of Geological map as per number shown in the map-
1. Kyanite-Sillimanite Schist and Biotite Gneiss 2. Porphyroblastic gneiss and Migmatite 3. Garnet Biotite Schist 4. Cataclastic Gneiss 5. Augen Gneiss 6. Streaky Gneiss 7. Pratapnagar Quartzite 8. Paturi Quartzite 9. Bhainga Slate 10. Bhelunta Limstone 11a. Slaty Phyllite 11b. Schistose phyllite 12. Nagthat Quartzite 13. Blaini Boulder Bed 14. Infrakrol slate 15. Krol Limestone 16a. Upper Tal Shale and Quartzite 16b. Lower Tal Quartzite 17. Simla Slate 18. Metabasic 19. Alluvium.

Figure 3: (a-c). FCCs showing delineation of thrust and faults in different parts of the area.

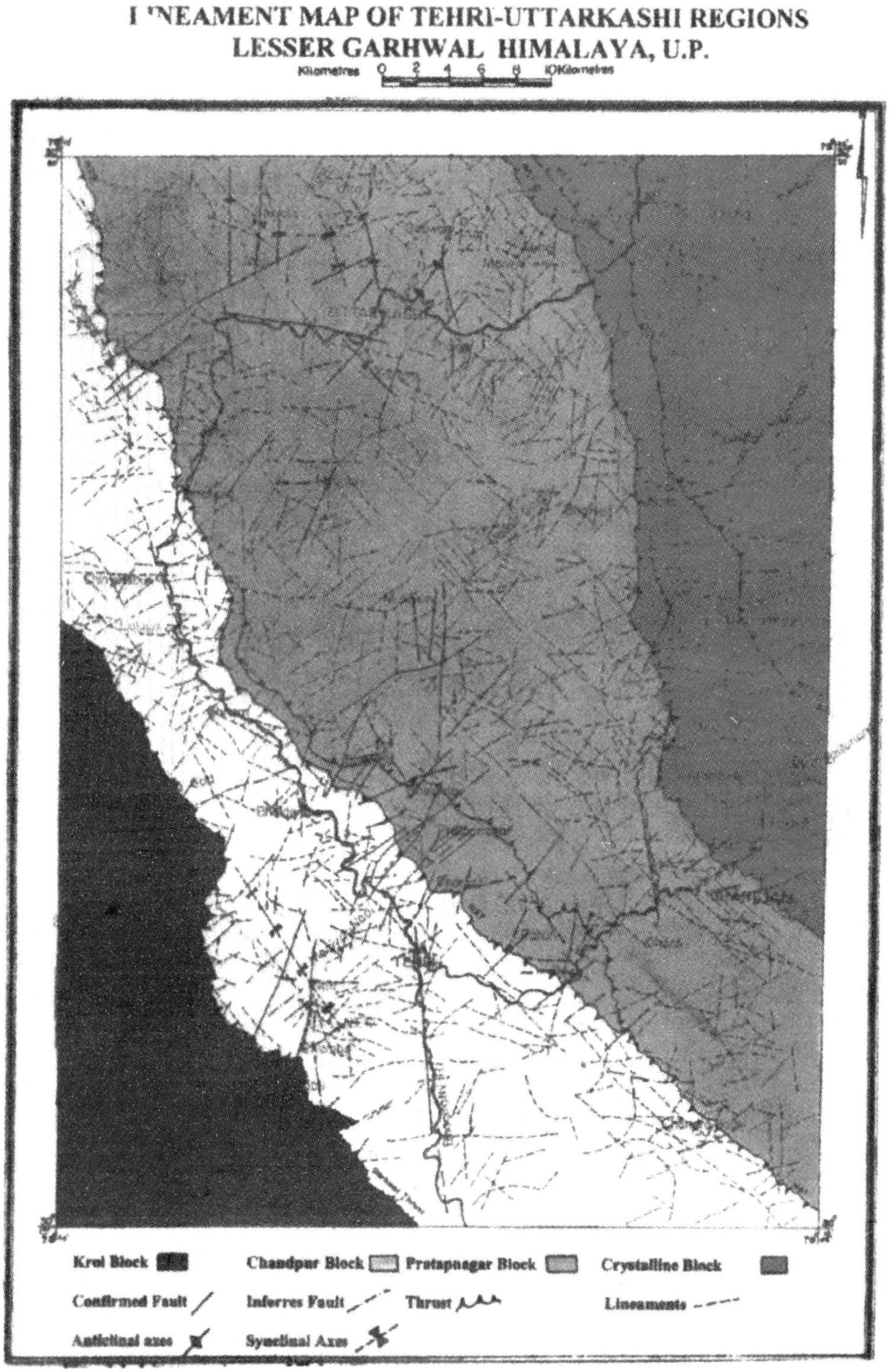

Figure 4: Lineament Map of Tehri Uttarkashi Area.

The NW-SE trending major lineaments along N50° W-S50° E together with minor lineaments along N20° E- S20° W and N80° E-S80° W were developed due to stress transfer from N50° E-S50° W (the maximum Principal compressive stress direction ó). These lineaments can be correlated with NW-SE plunging regional folds (Phase II) of Saklani (1971) in Garhwal and Bhattacharya and Niyogi (1971) in Simla and thrusts (NAT, MCT-III etc.) and faults (i.e., Gadolia tear fault) were produced during this phase of deformation.

The maxima of N-S lineaments (like Syalam fault, Rano fault, Jalkur fault and Tehri fault) along N10°E-S10° W indicates the direction of least compressive stress (õ) The minor lineaments were formed along N70° E-S70° W and S50° E-N50° W. The major N-S lineaments can be correlated with N-S and N-NE plunging folds (Phase III) of Merh (1966 in Kumaon) and Saklani (1970 in Garhwal).

The major lineaments produced along N55° E-S55° W (e.g., Dewal fault, Raunsal fault, Bhaldiyana fault, Sainj fault) indicate that the maximum Principal compressive stress (D) acted along N55°W-S55°E. Minor lineaments formed around N85° W-S85° E and N25° W- S25° E and can be related with N-S and NE-SW folds(Phase IV of Saklani, 1971).

Bharktya and Gupta (1982) based on lineament tectonic interpretation in the Garhwal-Kumaon Himalaya inferred that lineaments were produced by maximum Principal compressive stress directed along N30°E-S30°W. The lineaments in the Tehri-Uttarkashi region were produced during different phases of regional deformation and tectonics.

MORPHOTECTONICS

The transverse and longitudinal river profiles of Bhagirathi and its major tributaries and the development of fluvial river terraces along the Bhagirathi and Bhilangna rivers were analyzed with reference to lithology and structures for morphotectonic study.

The Bhagirathi valley displays a variety of transverse river profiles indicating action of geomorphic processes (glacial, fluvio-glacial and fluvial) in the development of valley. Based on the superposed cross-valley profiles of the Bhagirathi river (Fig. 8) it can be remarked that fluvial processes at Gaumukh resulted into valley deepening by 900m. Downstream in the crystalline terrain (between Gaumukh and Sainj) the average rate of valley deepening was 25:1000 (ie. 25m drop in 1000m.) whereas it was 12:1000 in the quartzitic terrain (between Sainj and Dharasu) and 5:1000 in the phyllitic terrain (between Dharasu and Deoprayag). The higher value of valley deepening in the crystalline terrain was due to fluvio-glacial activity whereas fluvial activity was prominent in the quartzitic and phyllitic terrains. The glacial activity in the area was reported by Sharma and Owen (1996) at Dharali.

The longitudinal profiles of a river are sensitive to ongoing process of uplift and can be used to recognize active structures (Seeber and Gornitz, 1983). The longitudinal profiles of the Bhagirathi river and its major tributaries were prepared to understand the nature of channel development during the morphotectonic evolution of the Bhagirathi basin (Fig. 9). The longitudinal profile of the Bhagirathi valley between Bhatwari and Tehri indicates break in channel gradient (nick points) at a number of places indicating adjustment of the channel

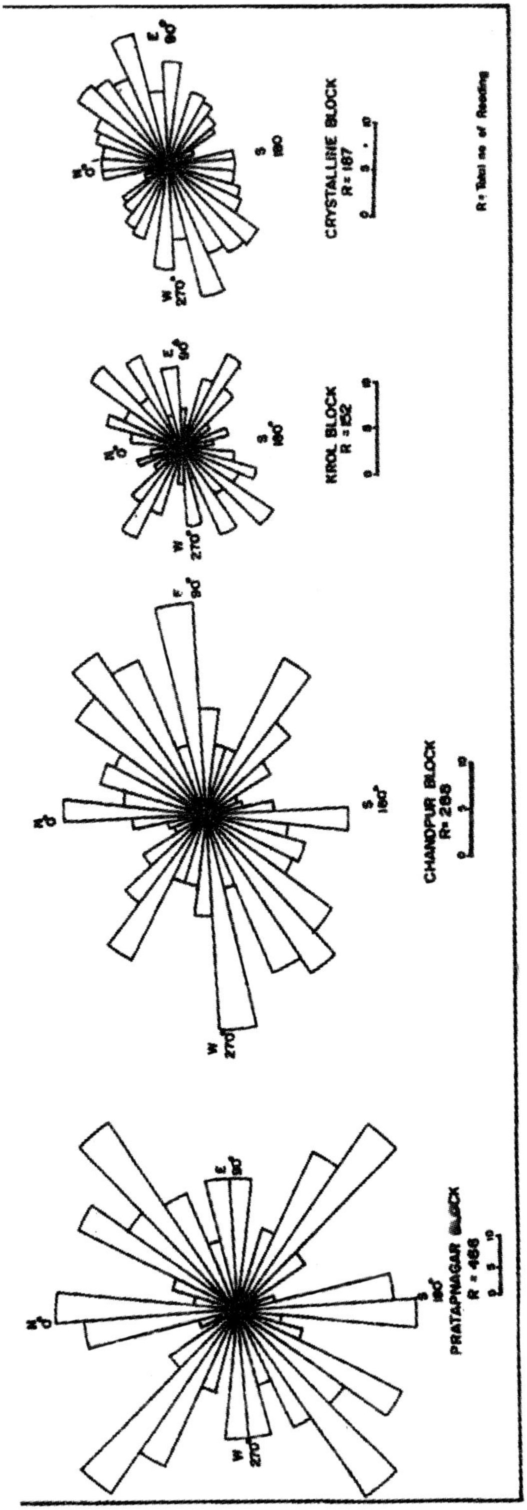

Figure 5: Rose diagram showing directional distribution of lineaments in different tectonic blocks.

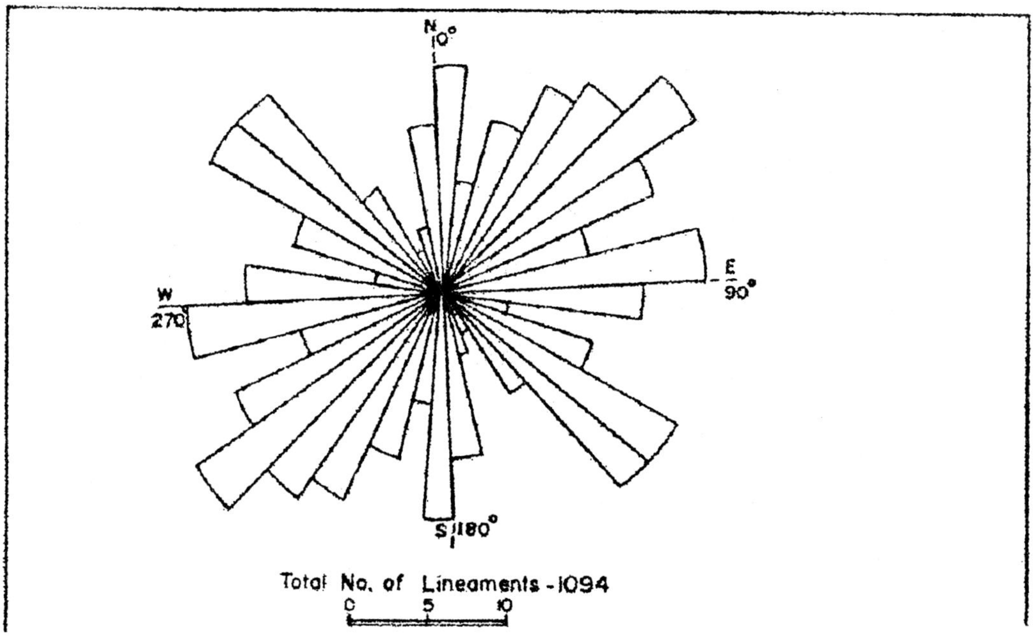

Figure 6: Rose diagram showing directional distribution of lineaments in Tehri-Uttarkashi area.

with geodynamic processes and lithological variations. In the Crystalline terrain, the channel gradient is steep. A change in the valley thalwag is observed at the contacts of crystallines with quartzites along the Budhakedar thrust. The development of nick point can be due to lithological changes and Budhakedar thrust.A nick point (800m) coincides with the surface trace of North Almora thrust (NAT) at Dharasu. Due to the thrust the quartzites overlie the phyllites. The marked change in the river profile across the NAT can be attributed to neotectonic activity. The Bhilangna river constitutes a major tributary of Bhagirathi river, where a major nick point (700m) along the valley thalwag is noticed at Deol along the North Almora Thrust. The river gradient across the NAT is reduced significantly in the phyllitic terrain. In the Balganaga river (tributary of the Bhilangna river) a prominent nick point (1100m) at Chattera is present near the Bhudakedar thrust. A significant decrease in the river gradient has been noticed across the Budhakedar thrust. The lowering of valley gradient from crystalline terrain to quartzitic terrain across the thrust can be ascribed to the development of river terraces. The Dharamganga river (tributary of the Balganaga river) shows minor nick points in the valley profile reflecting minor lithological variations. The Jalkur gad, a tributary of the Bhagirathi, shows marked variation in its valley profile. It shows major nick points in valley at 2000m, 1600m, 1400m, 1200m, 1000m, 960m, 920m, 880m, 760m and 720m indicating several stages of adjustments. The Jalkur gad flows along a faulted course. Its valley profile between 1000m to 720m exhibits 6 nick points. The development of nick points at 920m and 880m is attributed to the Pratapnagar Thrust and North Almora Thrust. The Seansu gad, a tributary of Bhagirathi river, shows a major nick point at 1000m where it is traversed by the Garhwal Thrust. The Syalam gad and Asi Ganga flow along the axes of faulted anticlines. Both show nick points at 1400m and 1200m indicating their development due to faulting.

The fluvial terraces along the Bhagirathi and Bhilangna rivers are mostly depositional and are paired as well as unpaired. There are five levels of fluvial terraces distinctly seen at Uttarkashi, Dunda, Jugyara, Chham and Tehri, (i.e.,T1 to T5.) The T1 represents the oldest terrace, which occupies the highest elevation in the valley section whereas the T5 represents the youngest terrace lying above the present flood plain (T0) of the Bhagirathi river. There are only four levels of terraces (T1 to T4) in the Bhilangna valley, which occur at Tehri and Simalsu.

Khan et al. (1982) described five, four and one of the levels of fluvial terraces in the Bhagirathi, Bhilangna and Jalkur valleys. Singh and Saklani (1978) described the unpaired depositional terraces of local extent along the Bhilangna river at Raunsal and Saruna. The semi-circular to elliptical terraces occurs at 1200-1350 m and 900m.

In order to understand the relative disposition of terraces in the Bhagirathi and Bhilangna valleys, the height of the river terraces above mean sea level were plotted (along Y axis) against their corresponding positions along the valley thalwag (Fig.10). The T5 terrace at Uttarkashi is located at a height of about 4m from the valley floor whereas at Tehri it is located at a height of about 40m from valley floor. This difference can be is due to tectonic uplift. The gap between the T1 and T5 is maximum (224.86m) at Jugyara. It decreases on both upstream and downstream of this location. It is 64.74m at Dunda and

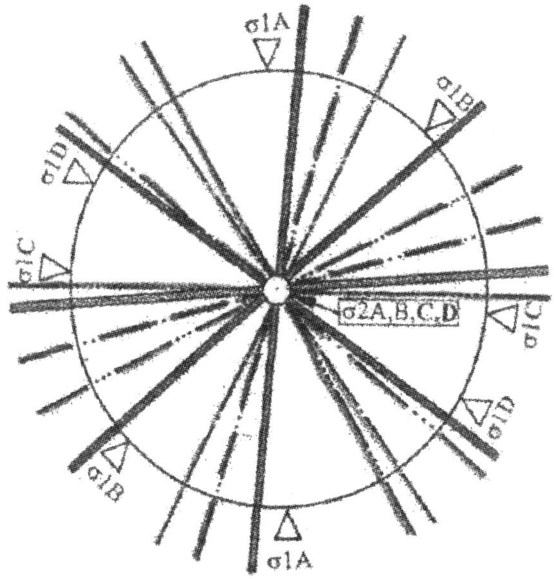

Deformation Phase	Principal Compressive Stress Dir. (σ1)	Prominent Lineaments Dir.	Minor Lineaments Dir.
I	N-S	E-W	N30°W & N30°E
II	N50°E-S50°W	N50°W-S50°E	N20°E & N80°E
III	N80°W-S80°E	N10°E-S10°W	N70°E & S50°E
IV	N55°W-S55°E	N55°E-S55°W	N85°W & N25°W

Figure 7: Schematic diagram showing directional distribution of tectonic forces and prominent and minor lineament's direction in Tehri-Uttarkashi area.

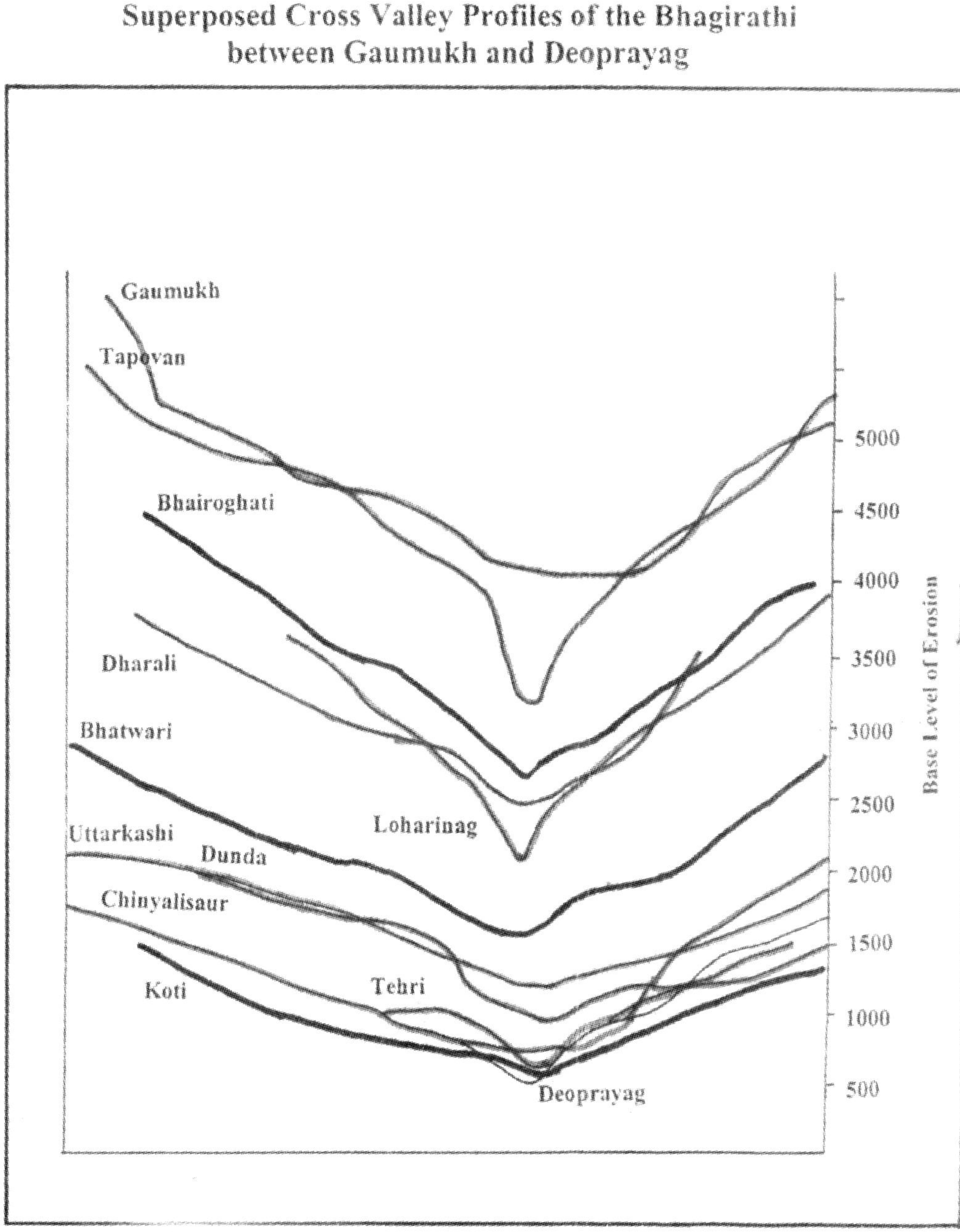

Figure 8: Superposed Transverse river profiles along the Bhagirathi river.

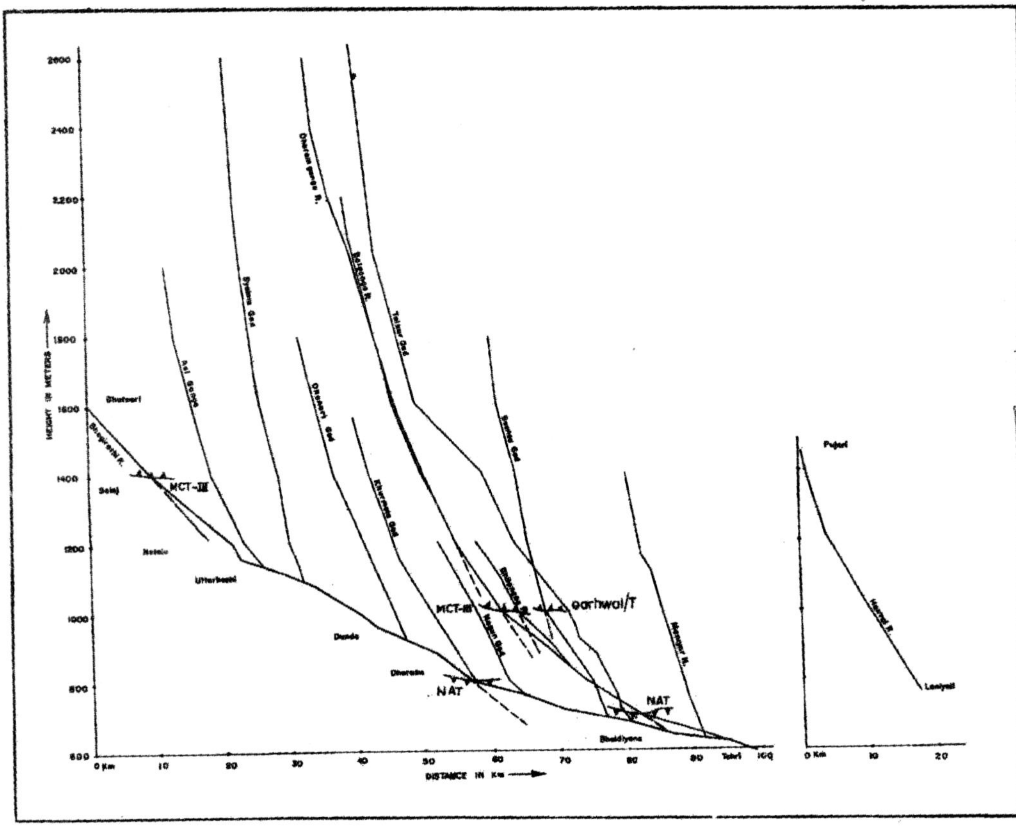

Figure 9: Longitudinal river profiles of the Bhagirathi and its major tributaries.

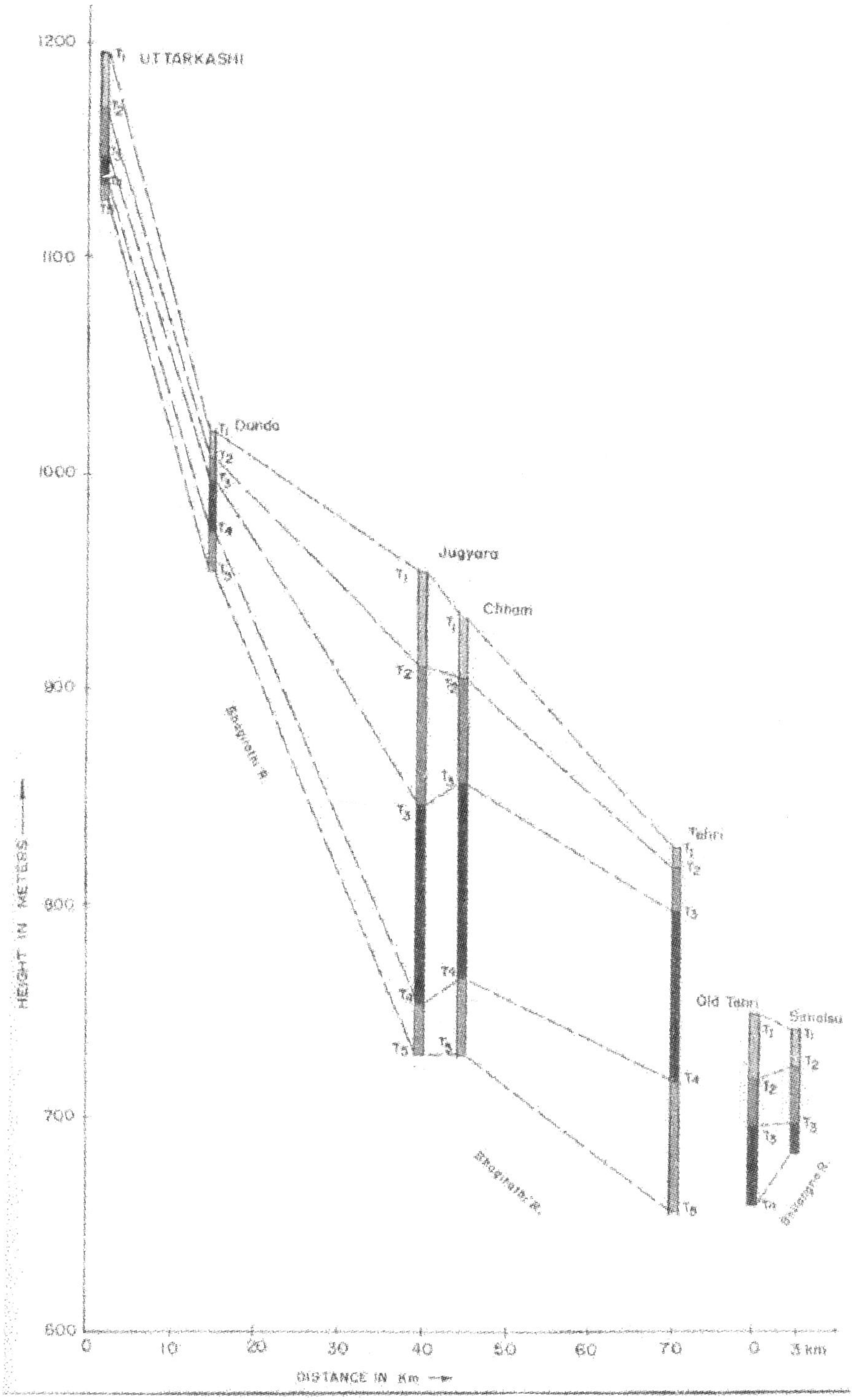

Figure 10: Schematic diagram showing disposition of river terraces along the Bhagirathi and Bhilangna rivers.

65.48m at Uttarkashi, while towards the downstream it is 203.13m at Chham and 169.21m at Tehri. Possibly the intensity of uplift was maximum in the Jugyara-Chham sector, which gradually decreased towards Tehri and Dunda-Uttarkashi regions. The variation in the tectonic uplift in these areas can be ascribed to the North Almora Thrust. It has also been noticed that the along the Bhagirathi valley, the gap between T4 and T5 terraces is maximum in Tehri area. It indicates that T5 terrace was formed due to uplift in Tehri region. Valdiya (1992) inferred 6 levels of river terraces in the Chham-Bhaldiyana regions and attributed then to three episodes of uplift along the NAT during Late Pleistocene period. In the Bhilangna valley, the four levels of river terraces present at Simalsu and Tehri show terrace gap between T1 and T4 as 57.70m and 89.07m respectively. This is also due to a relatively higher rate of tectonic uplift towards the down stream direction.

The diverging and converging terraces can be ascribed to tectonic tilting (Nakata, 1972). It has been observed that along the Bhagirathi and Bhilangna rivers the terraces generally show an increase in the height towards downstream direction. This diverging trend of the terraces towards downstream direction indicates the NE/N tilting of the area due to tectonic uplift from south to north. The five levels of river terraces in the Bhagirathi watershed indicate successive regional uplift of the area. It was about 10-44m for the formation of the oldest terrace (T1), 18-66m for T2, 13-94m for T3, 8-58m for T4 and 6-16m for T5 terraces.

CONCLUSION

The geological realm in the Tehri-Uttarkashi districts is constituted by the Krol, Chail (I, II, III) and Crystalline nappes (I, II). The presence of different drainage pattern, joints, erosional characteristics and tonal contrast evident on the satellite image were utilsed in demarcation of major lithological units viz; Chandpur phyllites, Nagthat-Pratapnagar quartzites, Kotga Banali Calc zone and gneissic and schistose rocks of Crystalline terrain. Satellite image interpretation greatly helped in demarcation of regional extent of thrusts and faults which otherwise can not be mapped throughout their extent due to inaccessible terrain. There are 1096 lineaments delineated in the area, which were distributed into four groups, trending NW-SE, NE-SW, N-S and E-W. Many NW-SE trending lineaments developed along the Budhakedar Thrust, North Almora Thrust, Garhwal Thrusts etc. The majority NE-SW and NW-SE trending lineaments are expressions of the regional joint patterns. The N-S lineaments show development parallel to faults and faulted anticlinal fold axes controlling the drainage channels. The Bhagirathi Basin exhibits multicyclic evolution of its landscape. The cross valley profiles of the Bhagirathi river show prominent breaks in the valley slopes reflecting periods of dynamic rejuvenation due to epirogenic uplift of the land. The longitudinal profile of the Bhagirathi shows prominent nick points confirming its rejuvenation. The development of five levels of river terraces in the Bhagirathi valley suggests atleast five episodes of regional uplift of valley.

BIBLIOGRAPHY

Bharktya, D.K. and Gupta, R.P. (1982). Lineament tectonic interpretation from LANDSAT images in Garhwal Himalaya. Him. Geol., 12, 1-11.

Bhattacharya, S.C. and Niyogi, D. (1971). Geologic evolution of the Krol Belt in Simla Hills, H.P., Him. Geol. 1, 178-212.

Drury, S.A., (1987). Image interpretation in geology, Allen and Unwin, London.

Jain, A.K. (1971). Stratigraphy and tectonics of Lesser Himalayan region of Uttarkashi, Garhwal Himalaya, Him. Geol., 1, 25-58

Khan, A.A., Dubey, V.S., Sahgal, M.N. and Awasthi, S.C. (1982). Terraces in the Himalayan tributaries of the Ganga. Jour. Geol. Soc. India, 23, 392-401.

Merh, S.S. (1966). A preliminary note on the structural history of Central Kumaon Himalayas, Bull. Geol. Soc. Ind., 5, 1-7.

Naha, K. and Ray. S. (1970). Metamorphic history of the Jutogh Series in the Simla klippe, Lower Himalaya. Contr. Mineral and Petrol. 20, 117-184, Berlin.

Nakata, T. (1972). Geomorphic History and Crustal Movements of the foot hills of the Himalaya, Inst. Geog. Tohoku University Sendai. 177 p.

Pandey, A.C., (1993). Geology of a part of Northern Uttarkashi region with special reference to Boudinage structures. M.Sc dissertation submitted to University of Delhi, Delhi.

Pandey, A.C., (2001). Geology and Morphotectonics of a part of Lesser Garhwal Himalaya (Tehri-Uttarkashi districts) using Remote sensing and GIS techniques. Unpublished Ph.D. thesis submitted to University of Delhi, Delhi.

Saklani, P.S. (1970). Metamorphism in rocks of Garhwal Himalaya. Jour. Geol. Soc. Ind., 23, 263-266.

Saklani, P.S. (1971). Structure and tectonics of Pratapnagar area, Garhwal Himalaya. Him. Geol., 1, 75-91.

Saklani, P.S. (1993). Geology of the Lower Himalaya (Garhwal). International Books and Periodicals Supply Services, New Delhi, 246 p.

Singh, S. and Saklani, P.S. (1978). Some geomorphologic observations in the Ghansyali area, Garhwal Himalaya. Him. Geol. 8, 813-821.

Valdiya, K.S. (1976). Himalayan transverse faults and folds and their parallelism with subsurface structures of North Indian plains, Tectonophysics, 32, 353-386.

Valdiya, K.S. (1992). Must we have high dam in the geodynamically active Himalayan domain? Current Science, 63, (6) 289-296.

Virdi, N.S. (1979). On the geodynamic significance of mega-lineaments in the Outer and Lesser regions of Himalaya. Him. Geol. 9 (1), 79-99.

TECTONIC GEOLOGY OF THE MAIN CENTRAL THRUST, GARHWAL - UTTARANCHAL

P.S. Saklani

Geology Department, University of Delhi, Delhi-110 007

ABSTRACT

Tectonic aspects of lithologies exposed in the zone of Main Central Thrust (MCT) are described and discussed. The crystallines and metamorphics occur as nappes. The MCT-I lies at the base of the Vaikrita nappe and is separated by the kyanite-sillimanite-staurolite schists and gneisses from metamorphics of the Jutogh nappe. The gannetiferous mica schists and gneisses are medium grade rocks of the Jutogh nappe, which tectonically overlie the low grade rocks of Budhakedar formation along the MCT-II. The low grade Budhakedar formation consists of the chlorite- chloritoid – quartz – sericite schists and gneisses (Budhakedar nappe) and these are thrust over the massive white Pratapnagar Quartzite by the MCT-III. The lithotectonic units were examined in the Bhaigrathi and the Bhilangana Valleys with reference to composite MCT. The MCT-I, MCT-II and III form a large duplex of which MCT-I is the roof thrust and MCT-III as the floor thrust. The intermediate unit is an imbricate and is characterized by the MCT-II (=Branch line thrust). In the Yamuna Valley, the MCT-II consists of the MCT-IIa and b. The ramps and flats were the carriers of the composite MCT. The area of the Bhagirathi Duplex (270 sq. km.) was shortened for about 17 km. The shortening percentage in the upper and Lower Duplexes was about – 55%. The Yamunotri Duplex consists of upper (Phulchatti) and the lower Wazri horses. The area of the Upper Phulchatti horse is 8 sq. km. and underwent shortening for about –110%. The lower Duplex i.e. the Wazri horse (18 sq. km. in area) was shortened by 1 km. (i.e. –14%) by the MCT-IIb. It moved along a flat from 5 to 10 km while its shortening was 1-4 km. The propagation of the MCT-III along a flat was for about 7 km. with a ramping from 3 to 4 km. The shortening percentage of both the duplexes was about –60%. The high grade Vaikrita rocks were formed at about 610^{0}C ± 30^{0}C. The retrograde conditions existed along the MCT-II (i.e., 418-429^{0}C). The rocks of the Budhakedar nappe were formed at 500^{0}C. These metamorphics occurring at a depth of about 16 to 18 km ramped for about 8 km during the Main Central Thrusting.

INTRODUCTION

The investigated area is located between 70° 15' to 79° 10' E longitudes and 30° 30' to 31° 0' N latitudes (about 2000 square km) in the Higher and Lesser Himalaya. Although field work was carried out in the valley regions of the Bhagirathi, Bhilangana, Mandakini and the Yamuna yet the cross sections discussed in the present paper are along the Bhagirathi and the Yamuna valleys (Fig.1). All these valley- regions lithologically consist of Central Crystallines and Metamorphics of low, medium and high grades delineated by the composite Main Central Thrust (MCT) i.e. MCT-I, MCT-II and MCT-III (Table 1).

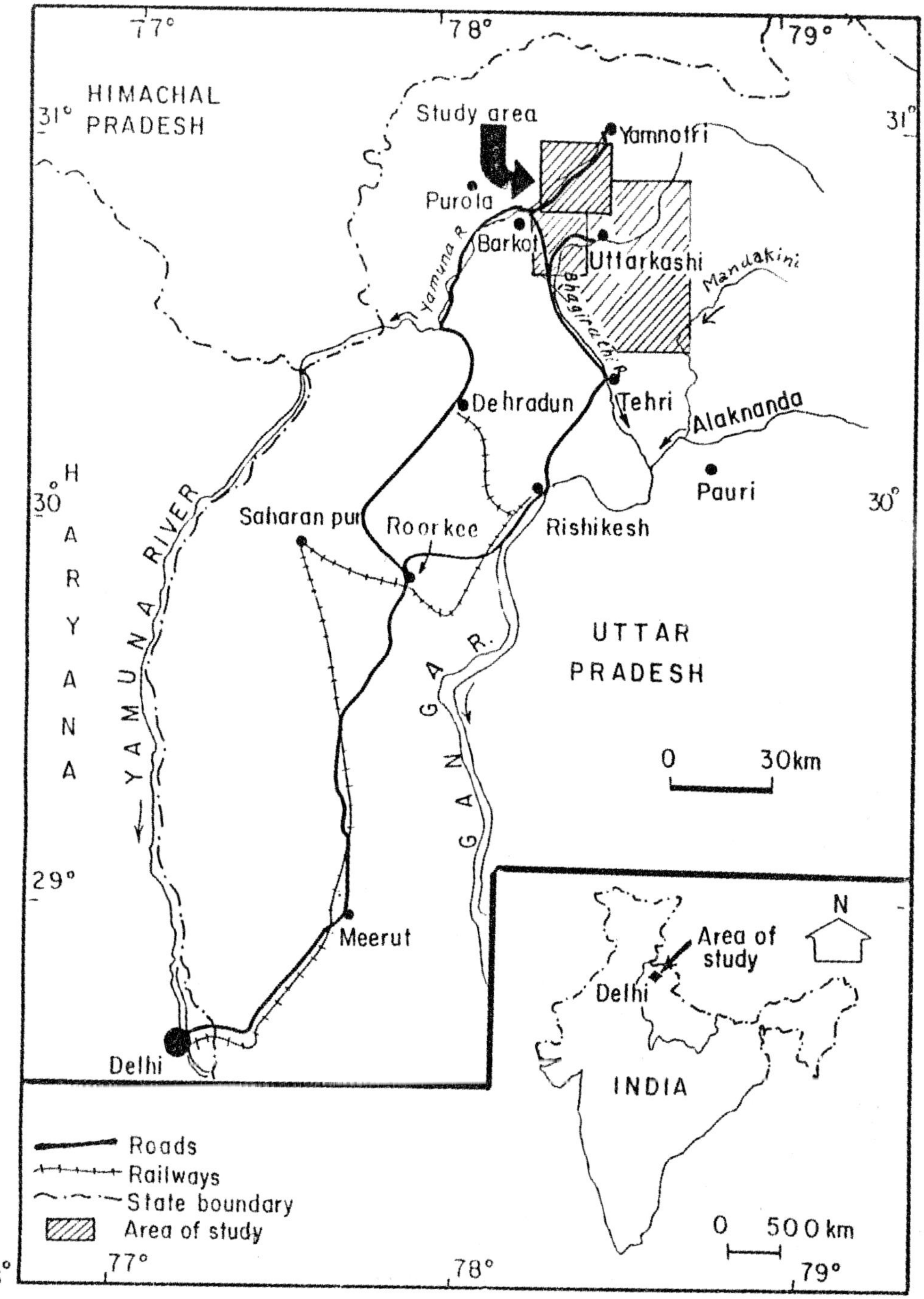

Figure 1: Location Map of the area.

Table 1. Tectonostratigraphic Units of the Bhagirathi and Yamuna Valleys

Bhagirathi Valley		Yamuna Valley	
Joti Fm. (High grade)	Kyanite-sillimanite-staurolite-chloritoid bearing mica schist and porphyroblastic gneiss	Yamunotri Fm. (High grade)	Kyanite-staurolite schist and gneiss
	MCT-I (Vaikrita)	MCT-I (Yamnotri)	
Bhatwari Fm. (Medium grade)	Biotite schist, porphyroblastic Gneiss and migmatites	Biph Fm. (Medium grade)	Porphyroblastic gneiss with marble
			—MCT-IIa
	MCT-II (Jutogh)	MCT-II	Porphyroblastic gneiss and biotite schist with marble
			—MCT-IIb
Budhakedar	Quartz feldspathic schist with graite gneisses	Syanachatti	Migmatized granite gneiss

TECTONIC GEOLOGY

The tectonostratigraphic setting of the Bhagirathi and Yamuna valleys is given in Table 1. The gneisses, schists and associated migmatites exposed in both the valley regions with important lithological characters are given in Fig.2 and Table 2 and in Fig.3 and Table 3.

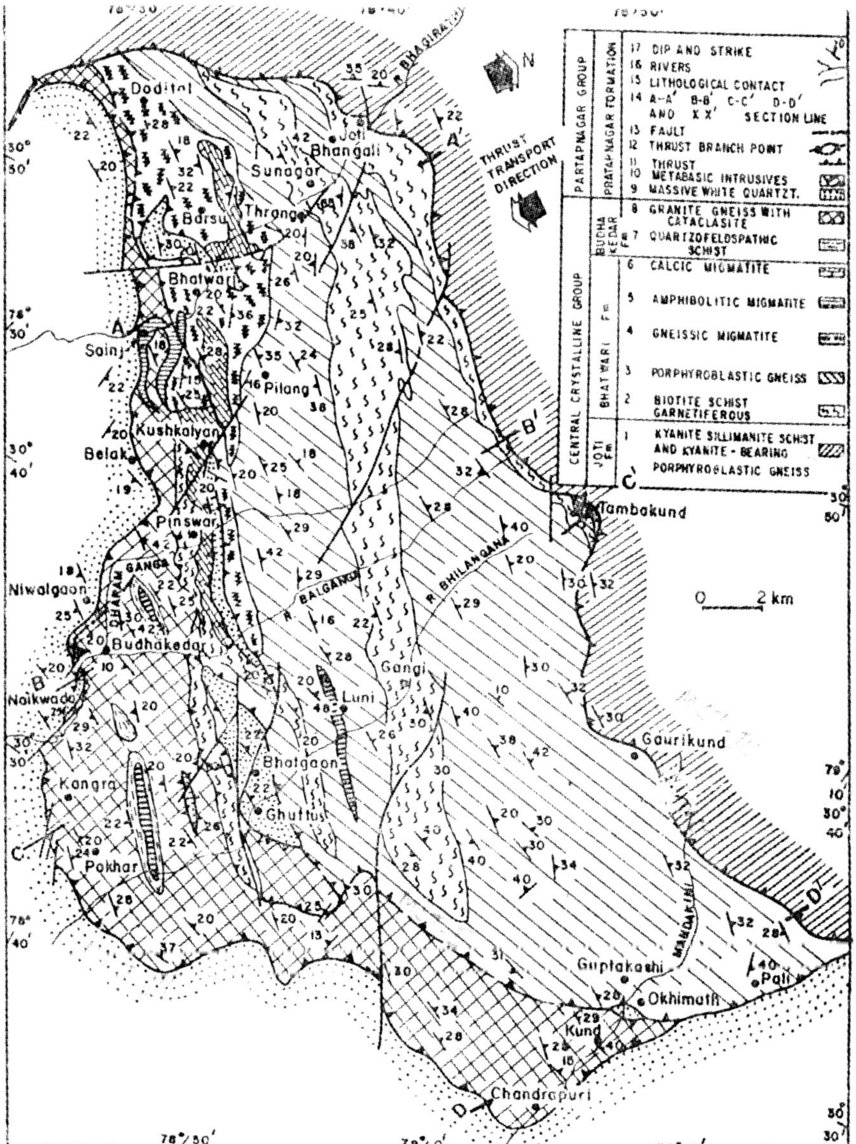

Figure 2: Geological map of imbricates and the Main Central Thrust in the Bhagirathi, Balganga, Bhilangana and Mandakini Valley regions

Table 2. Central Crystallines and Metamorphics of the Bhagirathi Valley

Group	Formation	Lithounits
CENTRAL CRYSTALLINE GROUP	MCT - II (Jutogh Thrust)	
	Bhatwari/ Jutogh Formation	Garnet-biotite zone (Bio-Grt-Chl-Qtz, Chl-Bio-Mus-Qtz, Mus-Grt-Kfs-Plag-Qtz, Grt-Ep-Chl-Mus-Qtz, A Ep-Chl-Mus-Hbl-Ctd-Qtz)
		Kyanite-garnet zone (Ky-Grt-Bio-Mus-Kfs-Plag-Qtz, Ky-Mus-Kfs-Qtz-Bio)
		Sillimanite-muscovite zone (Sill-Ky-Mus-Kfs-Qtz+ Grt, Sill-Mus-Bio-Kfs-Qtz, Grt-Hbl-Bio-Plag-Qtz, Ky-Sill-And-Mus-Qtz-Dumortierite)
		Sillimanite - K-feldspar zone (Sill-Bio-Kfs-Qtz, Bio-Kfs-Plag-Qtz), Quartzofeldspathic schist (biotite-quartz schist,
	Budhakedar Formation	Quartz-chlorite-sericite schist (Biotite-quartz schist, Quartz-chlorite-sericite schist)
		Granite-gneisses and Migmatites (Augen-gneiss, prophyroblastic Gneiss,mylonitised augen-gneiss, Kfs-Bio-Chl-Tour-Qtz-Sericite, Kfs-Plag-Qtz-Mus-Bio, Hbl-Qtz, Hbl-Bio-Plag-Qtz)
	MCT - III (Chail ? Thrust)	
Pratapnagar Group	Pratapnagar Quartzite	Massive white quartzite which is, at places, pebbly, ferruginous and mylonitised (Quartz-sericite schist)

The granite gneissic group has three varieties viz. Granite gneiss, streaky gneiss and mylonitic/cataclastic gneiss and was designated by the author as the Budhakedar formation. The mineral assemblages of the Budhakedar formation are:

- Quartz-k-feldspar-chlorite-sericite-tourmaline ± epidote
- Quartz-muscovite-chlorite-plagioclase.

These low grade gneissic rocks are exposed in the Bhagirathi valley near Sainj and Syanachatti in the Yamuna valley, respectively. The Budhakedar formation associated with quartzofeldspathic schist is characterised at its base by the MCT-III and is tectonically overlain by the Bhatwari formation in the Bhagirathi valley and Biph formation in the Yamuna valley along the MCT-II (= Jutogh thrust).

The characteristic feature of this medium grade formation (Bhatwari/Biph) is its association with the migmatites. The Bhatwari formation is less exposed in the Bhilangana vallley but is predominantly exposed in the Bhagirathi and the Yamuna valleys.

The mineral assemblages of the Bhatwari/Biph formation are :

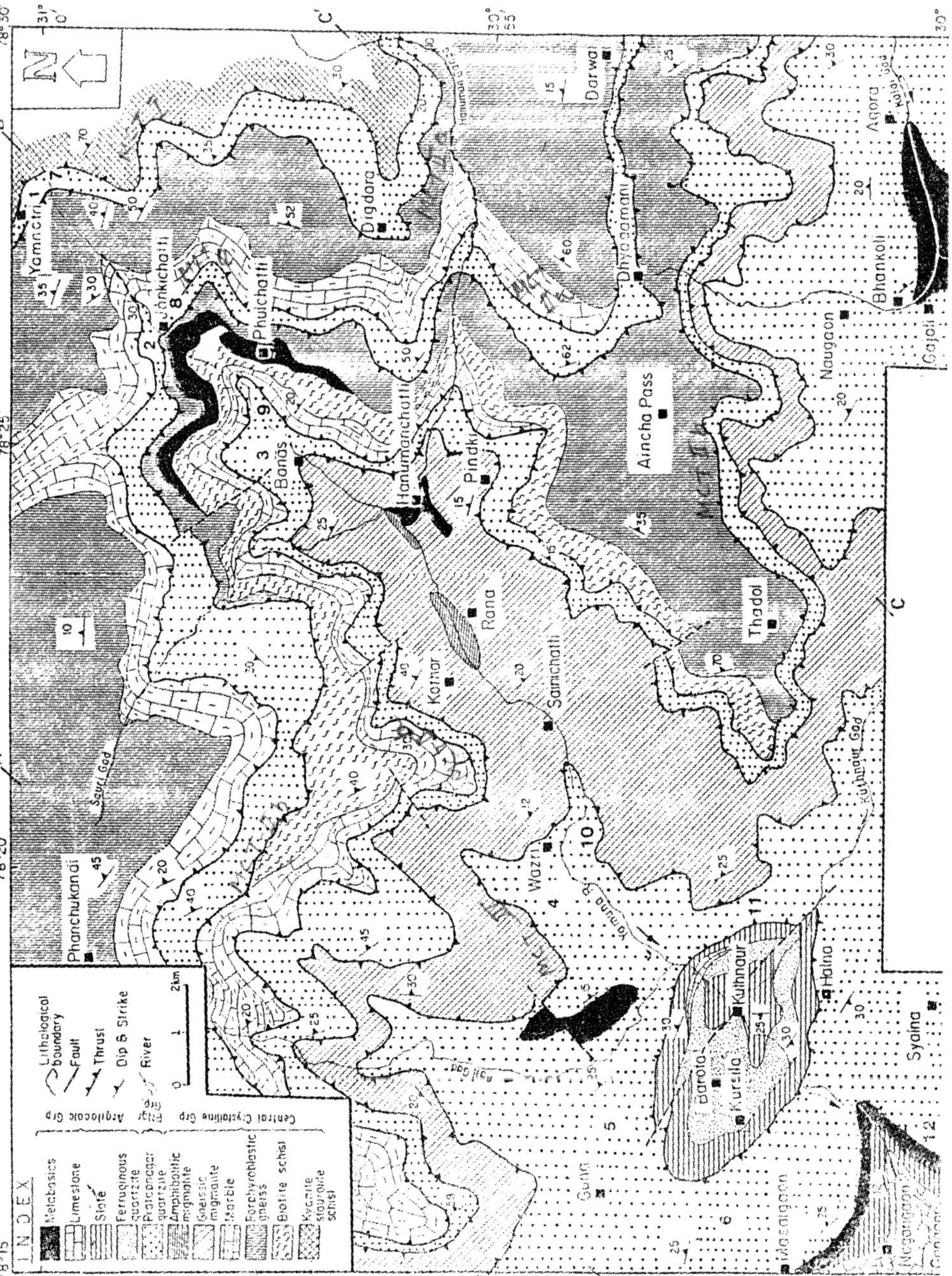

Figure 3: Geological map of the Yamuna Valley (Uttarkashi-Garhwal Himalaya)

Table 3. Central Cyrstallines and Metamorphics of the Yamunotri Valley.

Group		Formation	Lithounits
CENTRAL CRYSTALLINE GROUP	Yamunotri	Kyanite-Staurolite schist	Grey, brown, buff coloured, medium-to coarse-grained kyanite, staurolite, garnet, quartz, plagioclase, mica schist with feldspathic streaks. Kyanite crystals tend to develop strong preferred orientation.
	MCT - II (Jutogh Thrust)		
	Biph	Porphyro-blastic Gneiss	Greyish green, fine-to coarse-grained, folded rock. Augen of feldspar and quartz occur as porphyroblast defining the main gneissic foliation. Bands of marble (200 m in thickness) are also present which contain tremolite, diopside and quartz. Near thrust, it is sheared and mylonitised.
		Biotite-Schist	Green to dark green, medium-to coarse-Grained folded schist. Quartz veins parallel to schistosity are present. Marble bands consist of calcite, quartz and tremolite.
	Syanachatti	Gneissic Migmatite	Grey, fine-to coarse-grained, quartz-feldspar muscovite-biotite-chlorite gneissic migmatite.Enclaves of metabasics occur in the rock.
		Amphibolite Migmatite	Brown, dark green, grey, fine-to coarse-grained banded rock consisting of hornblende, plagioclase, biotite and quartz. Enclaves of metabasics are common.
	MCT - III (Budhakedar Thrust)		
Pratapnagar	Pratapnagar	Quartzite	Creamy white, reddish pink to green, fine-to coarse-grained rock. Intercalations of chlorite-sericite schist are common. Near the thrust the rock is characterized by effects of crushing and mylonitization.

- Quartz-muscovite-chlorite-biotite-garnet ± chloritoid
- Quartz-biotite-garnet-plagioclase ± k-feldspar
- Garnet-hornblende- biotite- plagioclase.

The main rock of the Bhatwari/Biph formation is the porphyroblastic biotite gneiss which is associated with chlorite – quartz-garnet schist occurring in lensoidal bands.

The migmatites of the Bhagirathi and Yamunotri valley regions are gneissic, amphibolitic and calcic. The migmatites exposed at Bhatwari in the Bhagirathi valley treasure various types of folds and minor structures.

The Bhatwari/Biph formation is tectonically overlain by the Joti/Yamunotri formation along the MCT-I. It consists of sillimanite-staurolite schist and porphyroblastic gneiss. It is grey to brown or buff, medium to coarse grained and is kyanite-staurolite-garnet bearing. The mineral assemblages of this unit are:

- Quartz-biotite-garnet – staurolite plagioclase
- Quartz-muscovite-biotite-garnet-k feldspar ± plagioclase
- Quartz-biotite-kyanite–sillimanite – plagioclase– k-feldspar.

It is amply clear that the MCT-I, II and III indicating the metamorphic breaks also represent the thrust planes and this led to a great amount of confusion also. On re-examination of the rocks in the field, it was established that MCT-I, II and III do not necessarily represent the metamorphic breaks. In fact, the author has found kyanite- sillimanite in the so called low to medium grade gneisses and schists of the Budhakedar and Bhatwari/Biph formations.

It was also observed that relatively there is dominance of low grade minerals in the Budhakedar; medium grade minerals in the Jutoghs and high grade minerals in the Vaikrita formations. This led to the conclusion that the status of the composite MCT is controversial. Taking into account the investigations of the previous workers it became problematic to regard whether composite MCT is formed by two or three planes. Japanese geologists (Hashimoto et al., 1973) proposed existence of two Main Central Thrusts, namely, the lower MCT-I and upper MCT-II. The lower MCT-I was regarded by them to be equivalent of the classical MCT of Heim and Gansser (1939) and the upper MCT or MCT-II corresponded to the real Main Central Thrust of French geologists (Le Fort, 1975; Pecher, 1989). Fuchs and Frank (1970) established the existence of two thrusts at the base of the Tibetan slab (upper and lower Main Central Thrust). Hodges et al. (1988), Hubbard and Harrison (1989), MacFerlane et al. (1992), Schelling (1992), Harrison et al. (1993) also favoured the existence of the Main Central Thrust at the base of the Tibetan slab. Valdiya (1980) supported the existence of two thrusts i.e., the Munsiari and the Vaikrita. He opined that the Munsiari thrust is the real Main Central Thrust but regarded the classical Main Central Thrust of Heim and Gansser (1939) as the Chail Thrust. The author is in agreement with Upreti (1999) that the origin, timing of movement and associated metamorphism along the Main Central Thrust holds the key to many questions about the evolution of the Himalaya.

Gansser (1993) revised his earlier view of 1939 and opined that in the Himalayan region Main Central Thrust I and Main Central Thrust II are present. Most of the geologists are in agreement that thrusting in the Himalaya started from north and culminated towards the south. In the opinion of the author, the oldest thrust in Higher Himalayan region should be MCT-I and should not be MCT-II as designated by Japanese and French geologists and so should be with regard to the view of Gansser (1993). Therefore, the tectonic contact between the high grade and medium grade metamorphics i.e. the Vaikrita Thrust represents the MCT-I which is also the upper MCT of Japanese and French geologists.

Likewise the contact between the medium grade and low grade metamorphics is the MCT-II which is equivalent to the Jutogh Thrust or the Munsiari Thrust of Valdiya (1980). The classical MCT of Heim and Gansser (1939) and Gansser (1964, 1993) is, infact, the MCT-III which was designated by the author (1993) as the Budhakedar Thrust. This MCT-III is an equivalent of the lower MCT of Hashimoto et al. (1973), Le Fort (1975) and Chail Thrust of Valdiya (1980).

Saklani and Bahuguna (1983, 1986, 1989), Saklani et al. (1991), Bahuguna and Saklani (1988), Saklani (1993), Saklani (1998) propounded the concept of composite Main Central Thrust in the Garhwal Himalaya occurring as a duplex system of thrust. Accordingly, the Vaikrita thrust is the roof thrust i.e. MCT-I and Jutogh thrust is the branch line thrust i.e. MCT-II while the Budhakedar thrust is the MCT-III and also represents the floor thrust. The duplex geometry of the thrusting has considerably resolved the status of the MCT. Percentage of metamorphic minerals of various grades, ductile shear zones, folding and fracturing patterns should be taken into consideration to establish the thrusts. Previously, the metamorphic grade criteria was mainly adopted by some geologists to delineate the thrusts (Valdiya, 1980; Hashimoto et al., 1973 and Le Fort, 1975).

The age of the granite gneisses is different in time. Krummenacher (1978) reported the radiometric dates of the biotite in gneisses and migmatites as 47 m.y. and also 720 m.y. Ashgirei et al. (1977) remarked that the Jutogh metamorphics range from 13 to 85 m.y. and also 730 m.y. The age of the biotite in Mandi and Chor granites ranges from 20 to 50 m.y. and also 500 m.y. (Jager et al., 1971). Mehta (1977) described the age of mica of Manali metamorphics as 10 to 28 m.y. and 368 m.y. It is interesting to note that the age of Almora-Askot gneiss is 1620 m.y. by Powell et al. (1979); 1983 m.y by Kwatra et al., (1989) and 1810 m.y. by Trivedi et al. (1982). Trivedi et al. (1984) opined that the Ramgarh granite in Kumaun is 1800 m.y. while the Almora granite is 550 m.y. in age. Frank et al. (1973) hold the view that the Bandel granite in Himachal Pradesh is 1840 m.y. in age.

Such a type of variation in the geochronology of the Himalayan Crystallines and Matamorphics have made the metamorphic history of these rocks problematic and complex. Verma (1989) holds the view that the metamorphism of the Himalayan rocks was pre-Himalayan (300 to 1800 m.y.) and Himalayan (30 m.y.).

It is now an established fact that the composite Main Central Thrusting in the Himalaya took place in Tertiary times. The variation of age in one group of gneisses is from 1620 to 1983 m.y. indicating there by their the formation in Precambrian times. Another group is of the view that Himalayan Crystallines and Metamorphics ranged in age from 368 to 730 m.y. It broadly corresponds to the impact of Caledonain/Hercynian orogeny. The third group of granite –gneisses ranged in age from 10 m.y. and is mainly Tertiary. It is during this period that large scale thrusting, shearing and imbrication of rocks came into existence in the Himalaya. Remobilisation, variation in temperature – pressure conditions and deformation were associated with these tectonic processes. The thrusting occurred along the flats and ramps with formation of schuppen and duplex structures.

The two broad tectonic units of thrust zones are hangingwall and footwall which are changeable with reference to number of duplexes. According to Harrison et al. (1999) hanging wall deformation of the MCT took place at ~ 22 m.y. (Hodges et al., 1996; Coleman, 1998) and there existed a broad shear zone below the Great Himalayan Crystallines which remained active between 8-4 m.y. Harrison et al. (1999) opined that the brittle fracturing within the hanging wall of the MCT occurred at *ca* 3 m.y. Due to up section cutting by the MCT the rocks were imbricated and caused the variation of thickness of the Great Himalayan Crystallines.

It is, therefore, very essential to re-examine the Himalayan metamorphics in the light of duplex mechanism of thrusting and imbrication which can possibly also explain the variation in the age of the crystallines and metamorphics of the Himalaya.

Keeping in view the duplex geometry, it may be remarked that the southern migration of the stresses which were responsible for the formation of thrust took place due to footwall collapse. The geological cross sections of the Bhagirathi and Yamunotri valleys reveal that the major duplex constituted by the Central Crystallines is divisible into upper and lower duplexes.

The upper duplex in all the valleys is bound upsection by the MCT-I (Vaikrita - roof) and southwards by the MCT-II (Jutogh). The MCT-II (Jutogh) further represents the roof of the lower duplex which is marked by the MCT-III (Budhakedar - floor). The upper and lower duplexes in different valleys are characterised by tectonic horses.

BHAGIRATHI DUPLEX

The Bhagirathi duplex (Fig. 4) roofed by the MCT-I (Vaikrita) and floored by the MCT-III has a horizontal distance (L') of 11.9 km. The deformed structural thickness of the rocks (H') between the MCT-I (roof) and MCT-III (floor) measures about 6.8. km. Since the rocks are folded only an average stratigraphic thickness (t') was inferred i.e., 6.8 k.m.

The area (A) of the Bhagirathi duplex is for about 270 sq. km. On dividing the area of geological cross-section by its average stratigraphic thickness, the initial length (L_0) was inferred as about 39.7 km. and the shortening distance (L_0 39.7 km- L' 22.7 km) which is 17 km. Keeping in view the shortening distance of upper and lower duplexes the shortening percentage is inferred about -55% with a contraction ratio of 0.57.

In an attempt to prepare the restored section (Dahlstrom, 1970; Boyer and Elliott, 1982) along the line A-A' of the cross section in place of structural thickness the average stratigraphic thickness was considered. The Bhagirathi duplex prior to deformation was 39.7 km in length. Initially the MCT-I (Vaikrita) ramped from the sole for about 39 km in NE at the base of Joti Formation. The total ramping of MCT-I from the sole was 8.2 km which brought the Joti Formation to rest over the porphyroblastic gneiss of Bhatwari Formation and then the MCT-I travelled as a flat for 12.1 km at the top of porphyroblastic gneiss. After this, MCT-I ramped again from 3 to 5 km to bring the Joti Formation to rest over the migmatites. In the second stage of thrusting the MCT-II (Barsu Branch line) ramped for about 2.5 km to merge with the MCT-I and the Barsu horse was formed. Another group of rocks along the branch line of MCT-II travelled for a distance of about 17 km as a flat from Barsu branch line towards SW and ramped for about 3.2 km to merge with MCT-I. This resulted into the formation of Bhatwari horse. In the final stage, the MCT-III (Budhakedar-floor) at the base of granite gneiss ramped for about 1.7 km to merge with the MCT-I with the formation of the Sainj tectonic horse (Lower duplex). This differential ramping was due to ductility contrasts. The tectonic elements of the Bhagirathi duplex are given in Table 4.

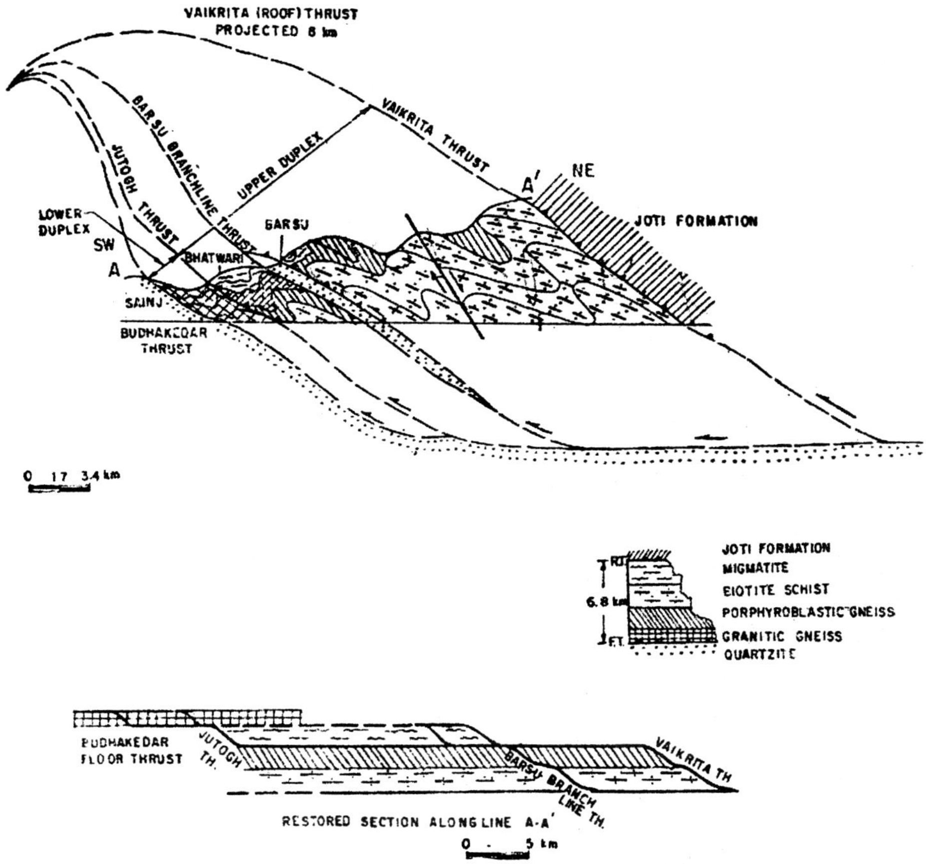

Figure 4: Geological Cross-Section and Restored Section along line A-A' in Figure 2.
1. Kyanite-sillimanite gneisses and schists, 2. Biotite schist, 3. Porphyroblastic gneiss 4. Amphibolitic migmatites, 5. Calcic migmatite, 6.Granite gneiss, 7.Quartzite

YAMUNOTRI DUPLEX

This Central crystalline duplex is constituted by upper and lower duplexes (Saklani et al., 1991). The upper duplex is represented by Phulchatti horse while the lower duplex consists of Wazri horse along the Phachukandi-Guna section (A-A'). The upper duplex (i.e. Phulchatti horse) consists of medium grade porphyroblastic gneiss and marble. Its roof is delineated by the MCT-I while the floor is defined by the MCT-II. The current length (L') of Phulchatti horse is 3.5 km. The deformed thickness (H') of the rocks is about 2.2 km which is three times more than the average stratigraphic thickness (t': 0.7 km). The cross sectional area (A) measures 8 sq.km. The initial length (L_0) is for about 10.5 km. This horse shortened by 7 km (i.e. –110% of shortening) with a contraction ratio of 0.3. The lower duplex (i.e. Wazri horse) shows current length (L') of 6 km. The structural thickness (H') is for 3 km and the stratigraphic thickness (t') measures 2.6 km. The cross sectional area of the horse is for about 18 sq km with an initial length of 7 km. This horse was shortened by 1 km with shortening of -14%. The contraction ratio remained 0.8.

Table 4. Tectonic Elements of the Bhagirathi Duplex

Current length L'	Structural thickness H'	Stratigraphic thickness t'	Cross-sectional area A	Initial length Lo	Shortening distance Lo-L'	Contraction ratio L'/Lo	Shortening percentage	In [l+(L'-Lo/Lo)]x100 where ln is the natural log
22.7 km	11.9 km	6.8 km	270 sq km	39.7 km	17.0 km	0.57	-55.8%	Bhagirathi Duplex
21.8 km	10.4 km	5.7 km	227.2 sq km	39.8 km	18.0 km	0.54	-60%	Upper Duplex
14.5 km	7.4 km	3.7 km	107.3 sq km	29.0 km	14.5 km	0.5	-69%	Barsu horse
5.0 km	3 km	1.9 km	15 sq km	7.8 km	2.8 km	0.63	-45%	Bhatwari horse
2.0 km	1 km	0.9 km	2 sq km	2.2 km	0.2 km	0.90	-9.5%	Lower Duplex Sainj

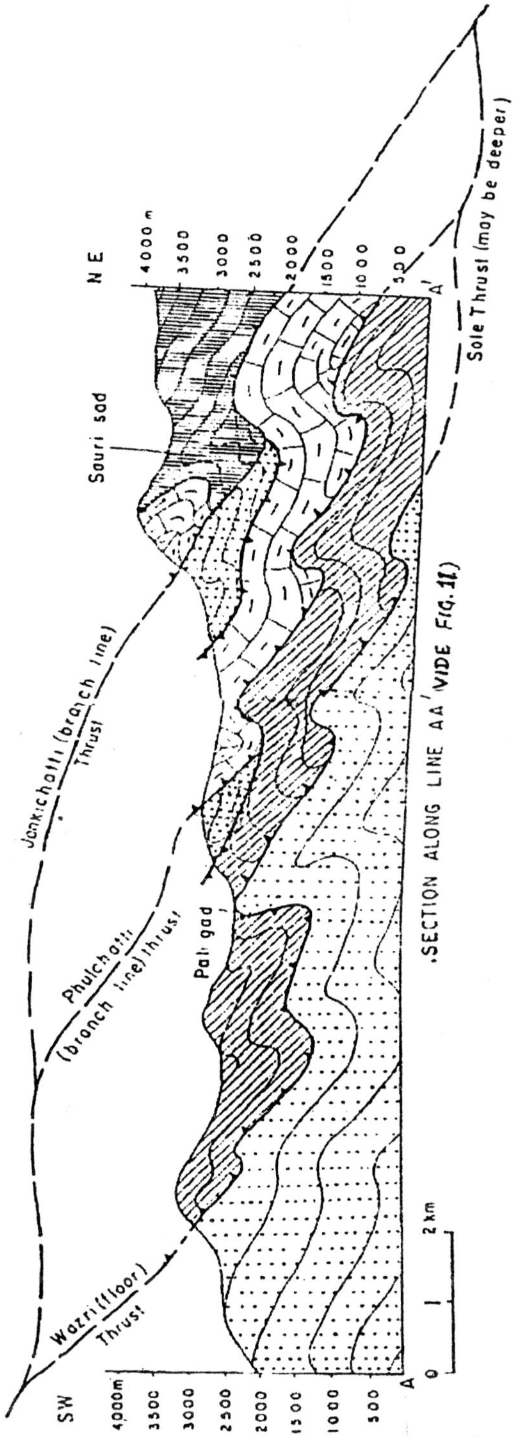

Figure 5: Geological cross section and restored section along line A-A in figure 3.

Table 5. Tectonic Elements of the the Yamunotri Duplex (A-A' Section)

Structural Units	Current length L' (Km)	Structural thickness H' (Km)	Stratigraphic thickness t' (Km)	Cross sectional area A (sq km) A (Sq km)	Initial length Lo (Km)	Shortening distance Lo-L'(Km)	Shortenning* percentage	Contraction ratio L'/Lo	Structural thickening H'/t'
Phachukandi Guna subduplex	13.5	5.2	3.3	70.8	21.1	7.6	-44.9	0.6	1.5
Wazri Horse	6.0	3.0	2.6	18.0	6.9	0.9	-14.2	0.8	1.1
Phulchatti Horse	3.5	2.5	0.7	7.9	10.6	7.1	-109.868	0.3	3.4

* ln [l+(L'-Lo/Lo)] x100 where ln is the natural log.

The restored section revealed that the Yamunotri Duplex had an initial length (L_0) for about 21 km. The MCT-IIb, moved along a flat for 5 to 11 km while it ramped for about 1-4 km and merged with the MCT-IIa. The horizontal translation of MCT-III (floor thrust) along a flat was for about 7 km with reference to MCT-IIb and at this stage it ramped up for about 3 to 4 km.

Based on the above analysis it can be said that the medium grade Jutogh Nappe along the Yamuna valley (Phulchatti and Jankichatti horses) underwent 8.5 km of shortening during formation of MCT-II while the Budhakedar Nappe was shortened for about 8 km during the MCT-III. The tectonic elements of the Yamunotri Duplex are given in Table 5.

The lithologies of the medium grade and low grade during Main Central Thrusting were shortened for about - 60% with a shortening distance varying from 1 to 17 km. More or less similar conditions prevailed in the Bhilangana and Mandakini valleys.

According to Harrison (1991) the ramping of the MCT was reactivated earlier than *ca* 8 m.y. or possibly it remained active with a reduced rate throughout the Middle Miocene. The presently exposed rocks of the hangingwall ramp of the MCT were affected by deformation in Early Miocene.

CONCLUDING REMARKS

It is observed that the Vaikrita Nappe is made up of high grade minerals in the Higher Himalayan zone. This unit is separated from the medium grade Jutoghs by the MCT-I.

As remarked earlier, the author (1993) reported the occurrence of kyanite-sillimanite-staurolite schists within the Lower/Lesser Himalayan zone, but majority of the minerals are of medium grade. Possibly a hot spot developed in the zone of Jutogh Nappe culminating into progressive regional metamorphism or the Himalayan metamorphism. The metamorphics of the Budhakedar Nappe largely represented a retrograde unit.

According to Harrison et al. (1999) the rocks near the MCT reached upto kyanite grade at 8 kbar while away from the MCT at higher levels the sillimanite grade were formed at 3-4 kbar. Th.-Pb. Monazite dating in rocks of upper Garhwal Himalaya yields late Miocene age. K-Ar dates indicate 2-16 m.y. for upper MCT hanging wall zone and 6-3 m.y. for the metamorphics within and adjacent to the MCT (Hubbard and Harrison 1989, Copeland et al., 1991; Metcalfe, 1993).

Saklani (1993) based on microprobe analysis of garnet-biotite pairs described that the temperatures ranged from 461° – 570 °C in medium grade Jutogh Nappe (Singh et al., 1996). Harrison et al. (1999) opine that the growth of garnet took place at a P-T of 6-8 K-bar and 550 ± 30 °C and that this growth occurred at a depth of ~ 25 km at ~ 6 m.y. In contrast to this, the kyanite-sillimanite schists were formed upto 610°C ± 30° in Vaikrita zone. Specially at the contact of Jutogh thrust the garnetiferous schists were affected by lowering of temperatures (i.e. 418-429°C) due to retrogression. The Budhakedar formation (granite gneiss and quartzofeldspathic schists) were formed at 500°C/4K bar while medium to high grade varities were formed at 5.5 ± 0.5 K-bar. Keeping these into consideration it can be remarked that the metamorphics occurring at a depth of about 16-18 km ramped vertically associated with the Main Central Thrusting for about 8 km. The restored sections of the Bhagirathi and Yamunotri valleys also confirm this feature.

REFERENCES

Ashgirei, G.D., Sinha, A.K., Sinha, A.K., Pande, I.C. and Mullik, B.C., 1977. Contribution to the geology, geochronology and history of regional metamorphics of Himachal Himalaya, Him. Geol., 7, 102-117.

Bahuguna, V.K. and Saklani, P.S., 1988. Tectonics of the Main Central Thrust in Garhwal Himalaya, Geol. Soc. Ind., 31(2), 197-209.

Copeland, P., Harrison, T.M., Hodges, K.V., Mareujol, Le Fort, P., Pocher, A., 1991. An early Pliocene thermal perturbation of the Main Central Thrust, Central Nepal : Implications for Himalayan tectonics. Jour. Geoph. Res., 96, 8475-8500.

Coleman, M.E., 1998. U-Pb constraints on Oligocene-Miocene deformation and anatexis within the Central Himalaya, Marsyandi Valley, Nepal, Am. Jour. Sci., 298, 553-571.

Frank, W., Toni, M. and Purtscheller, F., 1977. Geology and petrography of Kulu-Spiti Lahul area. In : Himalayan Sciences de la terre, CNRS, 26B, 147-172.

Fuchs, G. and Frank, W., 1970. Geological investigations in Western Nepal and their significance for the geology of the Himalaya. Geol. Rundschau, 59, 552-580.

Gansser, A., 1964. Geology of the Himalayas, Interscience Publishers, London, 289p.

Gansser, A., 1993. Facts and Theories of the Himalayas, Eclogae Geol. Helv., 84(1), 33-59, 1991.

Harrison, T.M., Copeland, P., Hall, S.A., Quade, J., Burner, S., Ojha, T.P., Kidd, W.S.F., 1993. Isotopic preservation of Himalayan/Tibet Uplift, denudation and climatic histories of two molasse deposits. Jour. Geol., 101, 159-177.

Harrison, T.M., Grove, M., Lovera, O.M., Catlos, E.J., Andrea, J.D., 1999. The origin of the Himalayan anatexis and inverted metamorphism : Models and constraints. Jour. Asian Earth Sci., 17, 755-772

Hashimoto, S., Ohata, Y. and Akiba, C., 1973. Geology of the Nepal Himalayas : Tokyo, Saikon Publishing Co., 286p.

Heim, A. and Gansser, A., 1939. Central Himalayas : Geological observations of the Swiss Expedition 1936, Mem. Soc. Helv. Sci. Nat., 73(1), 245p.

Hodges, K.V., Parrish, R.R. and Searle, M.P., 1996. Tectonic evolution of the Central Annapurna range, Nepalese Himalaya, Tectonics, 15, 1264-1291.

Hodges, K.V., Hubbard, M.S., Silverberg, D.S., 1988. Metamorphic constraints on the thermal evolution of the Central Himalayan Orogen. Phil. Trans. Royal Soc. London, A326, 257-280

Hubard, M.S. and Harrison, T.M., 1989. 40 Ar – 39 Ar age constraints on deformation and metamorphism in the Main Central Thrust Zone and Tibetan Slab, Eastern Nepal Himalaya, Tectonics, 8, 865-880.

Jager, E., Bhandari, A.K., Bhanot, V.B., 1971. Rb-Sr. age determination on biotites and whole rock sample from Mandi and Chor granites, H.P., Eclog. Geol. Helv., 64, 521-527.

Kwatra, S.K., Bhanot, V.B., Kakar, R., Kansal, A.K., Hedge, C., 1989. New Rb. Sr. Radiometric data for granitic rocks, Chor area and review of Isotopic ages on these rocks from Western Himalaya and geological interpretation. In : Saklani, P.S. (Ed.) Metamorphism, Ophiolites and Orogenic belts, 277-289, Today and Tomarrow's Publishers, New Delhi.

Le Fort, P., 1975. Himalayas, the collided range. Present knowledge of the continental arc, American Jour. Science, 275A, 1-44.

Mac Ferlane, A., Hodges, K.V., Lux, D., 1992. A structural analysis of the Main Central Thrust Zone Langtang National Park, Central Nepal Himalaya, Geol. Soc. Am. Bull., 104, 1389-1402.

Mehta, P.K., 1977. Rb-Sr. geochronology of the Kulu-Mandi belt, its implications for the Himalayan tectogenesis. Geol. Rundschau, 66, 156-175.

Metcalfe, R.P., 1993. Pressure, temperature and time constraints on metamorphism across the Main Central Thrust Zone and High Himalayan Slab in the Garhwal Himalaya, In : Searle, M.P. and Treolar, P.I. (Eds.), Himalayan Tectonics, Sp. Pub., Geol. Soc. London., 74, 485-509.

Pecher, A., 1989. The metamorphism in Central Himalaya, Journal of Metamorphic Geology, 7, 31-41.

Powell, C.Mc A., Crawford, A.K., Aramstrong, R.L., Prakash, R., Wynnedwards, H.R., 1979. Reconnaissance Rb-Sr. dates for the Himalayan Central gneiss, North Western Himalaya, Ind. Jour. Earth Sci., 6, 139-151.

Schelling, D., 1992. The tectonics, stratigraphy and structure of the Eastern Nepal Himalaya, Tectonics, 11, 925-943.

Saklani, P.S. and Bahuguna, V.K., 1983. Main Central Thrust Zone and associated imrbicate structures in Chhatera area, Garhwal Himalaya, In : Saklani, P.S. (Ed.) Himalayan Shears, Himalaya books, N. Delhi, 1-9.

Saklani, P.S. and Bahuguna, V.K., 1986. Thrust tectonics of Garhwal Himalaya. In : Saklani P.S. (Ed.), Himalayan Thrusts and Associated Rocks, 1-25, Today and Tomorrow's Publishers, N. Delhi.

Saklani, P.S. and Bahuguna, V.K., 1989. Geothermometry of Central Crystallines of the Bhagirathi Valley, Garhwal Himalaya, Metamorphism, Ophiolites and Orogenic Belts, In : Saklani, P.S. (Ed.), 1-16, Today and Tomorrow's Publishers, N. Delhi.

Saklani, P.S., Nainwal, D.C. and Singh, V.K., 1991, Geometry of the composite Main Central Thrust (MCT) in the Yamuna Valley, Garhwal Himalaya, India, N. Jb. Geol. Palaeont. Abh., 364-380.

Saklani, P.S., 1993. Geology of the Lower Himalaya, Garhwal, International Books and Periodicals, Delhi, 246p.

Saklani, P.S., 1998. Geology of the Garhwal Himalaya : Retrospect and Prospect, Presidential Address, Earth System Sciences, Ind. Sci. Cong., Calcutta, 32p.

Singh, V.K., Singh, S.P. and Saklani, P.S., 1996. Metamorphic and geodynamic evolution of the Central Crystallines of the Garhwal Himalaya, India, N.Jb. Geol. Palaeon. Abh., 199(1), 89-109.

Trivedi, J.R., Gopalan, K., Sharma, K.K., Gupta, K.R. and Choubey, K.C.M., 1982. Rb-Sr age of the Gaik granite, Ladakh batholith, NW Himalaya, Proc. Ind. Acad. Sc., 91, 65-73.

Trivedi, J.R., Gopalan, K., Valdiya, K.S., 1984. Rb-Sr. ages of granitic rocks within Lesser Himalaya nappes, Kumaun India, Jour. Geol. Soc. Ind., 25(10), 641-654.

Upreti, B.N., 1999. An overview of the stratigraphy and tectonics of the Nepal Himalayas, Jour. Asian Earth Sciences, 17, 577-606.

Valdiya, K.S., 1980. Geology of Kumaun Lesser Himalaya, Wadia Inst. Him. Geol., Dehradun, 291p.

Verma, P.K., 1982. The Himalayan metamorphism : Evolution of Metamorphic Belts. Sp. Pub. Geol. Soc. London, 43, 377-383.

TECTONIC EVOLUTION OF THE ALMORA NAPPE ALONG THE CHHARA - SOMESHWAR TRANSECT, KUMAUN HIMALAYA.

Mallickarjun Joshi and A. N. Tiwari

Department of Geology, Banaras Hindu University, Varanasi-221005.

ABSTRACT

The Almora Nappe in Kumaun Lesser Himalaya is flanked by the North Almora Shear Zone (NASZ) and the South Almora Shear Zone (SASZ) which represent the northern and southern exposures of the Basal Shear Zone. The Ramgarh thrust separating metamorphics of the Almora Nappe from the tectonically underlying metasedimentaries has also been traced in the northern flank of the Almora Nappe. The mylonites exposed in the vicinity of the Almora Thrust are reworked Saryu Formation rocks which with decreasing degree of mylonitization grade into the unmylonitized Saryu Formation schists and gneisses toward the central part of the nappe. The structural history of the area is preserved in three distinguishable structural regimes, viz. the pre - shear zone structures, the syn - shear zone structures and the post - shear zone structures. The pre - shear zone structures, viz. F_1- and F_2- tight to isoclinal folds developed in the unmylonitized quartzites of the Saryu Formation with subhorizontal to about 25^0 NNE plunging fold axes and the F_3- open folds with NNW plunge. The syn - shear zone structures are represented by a range of shear sense indicators, viz. shear bands, asymmetric porphyroclast systems (s_a - , s_b - and d type), mica fish, pressure shadows and the characteristic Lattice Preferred Orientation (LPO) of quartz. The late - to - post shear zone structures are preserved as folded mylonites with southerly dipping axial planes in the northern flank and northerly dipping axial planes in the southern flank of the nappe. The fold axes of these folds roughly trend parallel to the strike of the shear zone. The steep brittle faults affecting the mylonites of the shear zone represent the post - shear structures. These brittle faults are responsible for the neotectonic rejuvenation of the Almora Nappe due to a still continuing northward movement of the Indian plate.

INTRODUCTION

During the Lower-Middle Eocene period the convergence of Indian and Sino - Tibetan plates (Molnar and Tapponier,1975) resulted in the crustal shortening that was largely accommodated by south directed deformation, tectonic transport and thrusting all along the arcuate length of the Himalayan range. The Himalayan rocks are affected by two major thrusts, viz. the Main Boundary Thrust (MBT) – which separates the Siwaliks from the Lesser Himalaya in the south and the Main Central Thrust (MCT) in the north – delimiting the Higher Himalayan Crystallines (HHC) from the Lesser Himalayan sedimentaries. Further north, the Higher Himalayan metamorphics are separated from the Tethys-sedimentary sequence by the South Tibetan Detachment System (STDS). The MCT runs parallel to the Himalayan strike. The Higher Himalayan Metamorphic Belt rocks were tectonically transported southward to the Lesser Himalaya, which now occur as large and small nappes and klippen, tectonically resting over the Lesser Himalayan sedimentaries throughout the Himalayan strike. Almora Nappe is one of the largest nappes in Himalaya which is located in the Kumaun Himalaya (Uttranchal). Heim and Gansser (1939) and Gansser (1964) identified the Almora Nappe. Heim and Gansser (1939) separated the Almora Nappe from the Garhwal Nappe by a thrust, viz. the South Almora Thrust. Valdiya (1962,1976a,b and 1978a,b) suggested that the Almora Nappe is the southeastern extension of the Jutogh Nappe of Himachal Pradesh. The rocks of Almora Nappe were thrust over the rocks of Ramgarh Nappe. The latter has been correlated with the Chail Nappe of Pilgrim and West (1928) in Himachal Pradesh by Valdiya (1980) and Joshi (1999). Most of the workers (Gansser,1964; Merh and Vashi,1965; Merh,1968; Powar, Gairola and Dixit,1969; Valdiya,1980; Gairola and Joshi,1980 and Joshi 1999) believe that the Almora Nappe rests over a thrust outcropping along the northern and southern margins of the nappe and is named as the North Almora. Thrust (NAT) and the South Almora Thrust (SAT) respectively. While some workers do not agree with the existence of Almora Nappe (Mishra and Sharma, 1972; Mehdi et al., 1972; Saxena and Rao, 1975 and Bhattacharya, 1987) and believe it as a steep reverse fault, viz. the North Almora Fault (NAF) instead of a steep south dipping NAT.

Not agreeing with Valdiya (1980), i.e., with the possibility of joining the North Almora Thrust with the South Almora Thrust around Paithani Parabolic trend; Jain (1971) and Kumar and Agrawal (1975) believe that the North Almora Thrust (NAT) extends further westwards where it is referred to by various local names, viz. the Dharasu Thrust (Jain, 1971), the Dharkot Thrust (Saklani,1993) and the Srinagar Shear (Bhargava,1972). Further, Joshi (1999) designated the NAT of Heim and Gansser (1939) and Valdiya (1980) as the North Ramgarh Thrust (NRT). A detailed evolutionary history of basal shear zone of Almora Nappe has been established on the basis of three clearly distinguishable structural regimes, viz. (a) pre - shear zone structures, (b) syn - shear zone structures and (c) post - shear zone structures observed across the Chhara - Someshwar transect of the Almora Nappe.

GEOLOGICAL SETTING OF THE AREA

The rocks exposed in the area constitute two distinct lithostratigraphic units, viz. the Ramgarh Group and the Almora Group. In the southern part of the nappe SAT separates the rocks of Ramgarh Group from those of the Almora Group and in the northern parts of the nappe. The NAT separates the rocks of Almora Group from rocks of the Ramgarh Group. Rocks of the Ramgarh Group are separated from the underlying sedimentaries by the NRT in the north and by the SRT in south (Joshi,1999). Rocks of the lower part of Ramgarh Group consist of augen granitic gneisses, while micaceous quartzites, interlayered phyllitic schists, quartzites and marbles comprise the upper part of the Ramgarh Group. All the rock types of Ramgarh Group are largely mylonitized on account of their involvement in large scale thrusting during emplacement of the nappe.

Almora Group of rocks consist of mica schists, garnet - mica schists and micaceous quartzites of Saryu Formation. The upper part of Saryu Formation is intruded by a granite batholith called as the Cambrian Champawat Granitoids (CCG) (Singh et. al., 1993, Joshi et al., 1994). The Gumalikhet Formation of Almora Group comprises of black carbonaceous phyllites alternating with black metagreywackes.

Table.1: Lithotectonic Succession of the Almora Nappe (After Joshi, 1999)

Tectonic Unit		Someshwar-Binsar Sector		Khairna-Suyalbari Sector
		Champawat Granodiorite (Not mappable in the area)		Champawat Granodiorite (Not exposed in the area)
		Gumalikhet Formation Graphite Schist, biotite-rich, metagreywacke		Gumalikhet Formation (Not exposed in the area)
Almora Nappe (Almora Group)		Saryu Formation Mica schist, garnet-mica schist, micaceous quartzite & gneiss		Saryu Formation Mica schist, garnet-mica schist, & micaceous, quartzite, gneiss
	NORTH ALMORA SHEAR ZONE (BASAL SHEAR ZONE)	Saryu Formation (Mylonitized) Protomylonite, mylonite & ultramylonite from mica schist, garnet-mica schist, mica-quartzite & gneiss	SOUTH ALMORA SHEAR ZONE (BASAL SHEAR ZONE)	SaryuFormation (Mylonitized) Protomylonite, mylonite & ultramylonite from mica schist, garnet-mica schist & micaceous Quartzite
---Almora Thrust-		-----North Almora Thrust ----- (Not mentioned by earlier workers)		-----South Almora Thrust----- (South Almora Thrust; Valdiya 1980)
		Nathuakhan Formation Mylonite & ultramylonite from phyllite, schist, quartzite		Nathuakhan Formation Mylonite & ultramylonite from phyllite, schist, quartzite
		Betalghat Member: Mylonitized phyllite & marble lenses		Betalghat Member: Mylonitized phyllite & marble
Ramgarh Nappe (Ramgarh Group)		Debguru Porphyroids Protomylonite, mylonite & ultramylonite (Protolith-granite & quartz-rich granitoid)		Debguru Porphyroids and ultramylonite (Protolith-granite & quartz-rich granitoid)
-Ramgarh Thrust-		-----North Ramgarh Thrust ----- (North Almora Thrust; Valdiya, 1980)		-----South Ramgarh Thrust ---- (Ramgarh Thrust; Valdiya, 1980)
Krol Nappe (Jaunsar Group)		Berinag Formation Coarse quartz arenite		Nagthat Formation Quartz arenite/orthoquartzite
(Damtha Group)		Rautgara formation : quartzite (subgreywacke/sublitharenite & slate)		Not exposed

The Almora Nappe comprises mylonitized rocks of the Basal Shear Zone, exposed at both the northern and southern flanks of the nappe, over which the unmylonitized rocks of the Almora Group were tectonically transported (table 1). The Basal Shear Zone of the

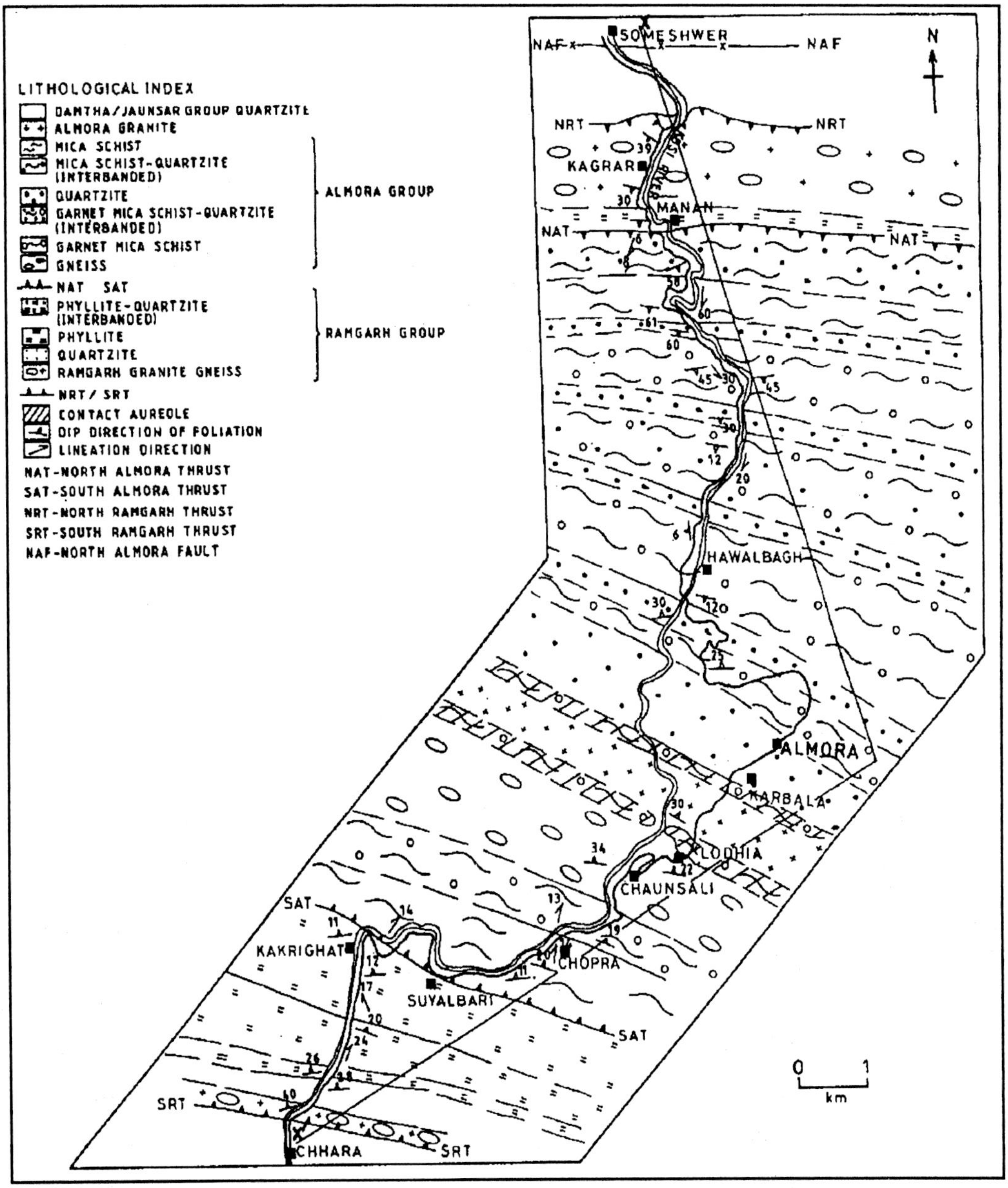

Figure 1: Geological map of Chhara - Someshwar Transect Kumaun Himalaya.

Almora Nappe is constituted of the protomylonites, mylonites and ultramylonites of the Ramgarh Group and partly by the protomylonites, mylonites and ultramylonites of the Saryu Formation of the Almora Group (Fig 1). Ramgarh Thrust is the tectonic discontinuity at the sole of the Basal Shear Zone.

STRUCTURAL HISTORY

Most of the mesoscopic structural features observed in field are substantiated by the development of microstructures. The area comprises three clearly distinguishable structural regimes, viz. the pre - shear zone structures, the syn - shear zone structures and the post - shear zone structures.

Schists and gneisses of the Saryu Formation exposed in the central part of the area extending from Chaunsali to Hawalbagh (Fig 2) are unaffected by mylonitization and preserve the structures developed prior to mylonitization. These are considered as pre - shear zone structures. The low dipping S_0 planes are defined by lithological banding striking E - W to WNW - ESE which are observed in the central parts of the nappe. The quartzites of Saryu

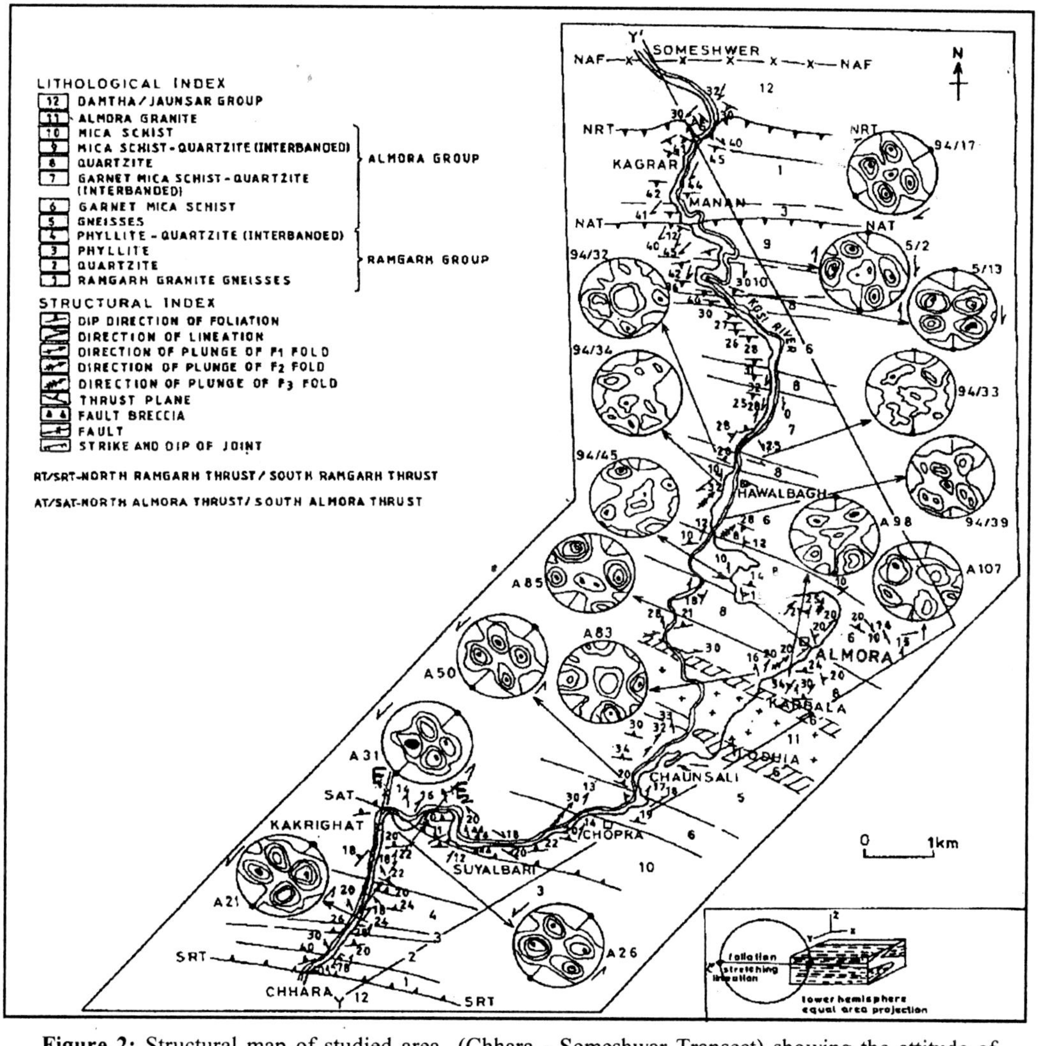

Figure 2: Structural map of studied area (Chhara - Someshwar Transect) showing the attitude of structural elements and quartz C - axes fabric diagrams.

Formation show well developed F_1- and F_2- tight to isoclinal folds plunging around 25° in NNE direction (Fig 3). The axial planes of these folds dip NNE at low angles (<30°). The F_1 folds are gently plunging reclined to recumbent type folds, (Fig 3). The earliest foliation (S_1) is developed parallel to the axial planes of F_1 folds and appears to be genetically

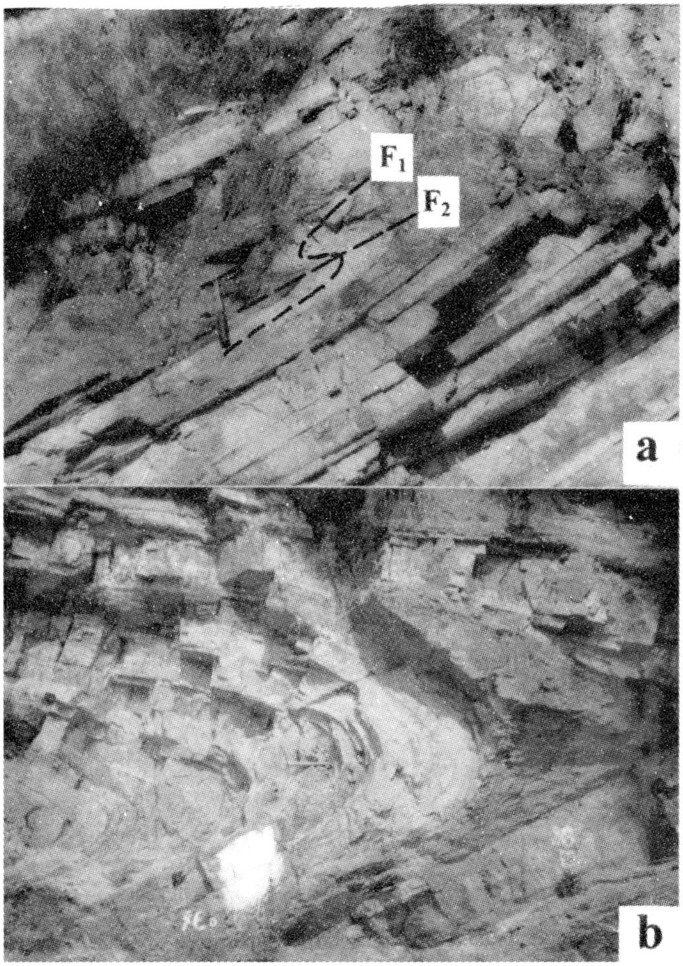

Figure 3: Field photographs

3 a: The F1- fold axis being folded by F2- fold axis. F1- folds are isoclinal folds while the coaxial F2- folds are tight folds in Saryu Formation rocks. Loc : Near Khatyari, Almora.

3 b: The F2- tight folds in unmylonitized quartzites of the Saryu Formation. Loc : Near Khatyari, Almora.

related to the latter. The F_2 folds are coaxial with the tight F_1 folds. The bedding (S_0) is folded by F_1 folds while the F_2 folds affect the oldest foliation (S_1) and axial planes of F_1 folds. The F_3 open folds with chevron type hinges and sub - vertical axial planes show plunges between 10 – 40° in NNW directions (Fig 4). The S_2 – foliations for the metamorphics unaffected by mylonitization and exposed between Chaunsali and Hawalbagh (in the central unmylonitized part of Almora Nappe) on the chosen traverse have been plotted on lower hemisphere equal area projection (Fig 4). The F_3- and F_4- fold axes orientations deduced from the diagram show the orientations of 40°,N 38°W for F_3- fold and 20°,N 70°E for F_4- fold. Both these interpreted values closely correspond to the field measurements for the F_3- and F_4- folds suggesting the mesoscopic folds are small scale manifestations of the regional folding. The F_1- and F_2- folds observed in the central parts of the nappe have escaped shearing both on mesoscopic (outcrop) scale and on the microscopic scale. This is evidenced by the absence of shear sense indicators, viz. shear bands, asymmetric porphyroclasts on the outcrop scale while the mica fish, pressure shadows, grain shape

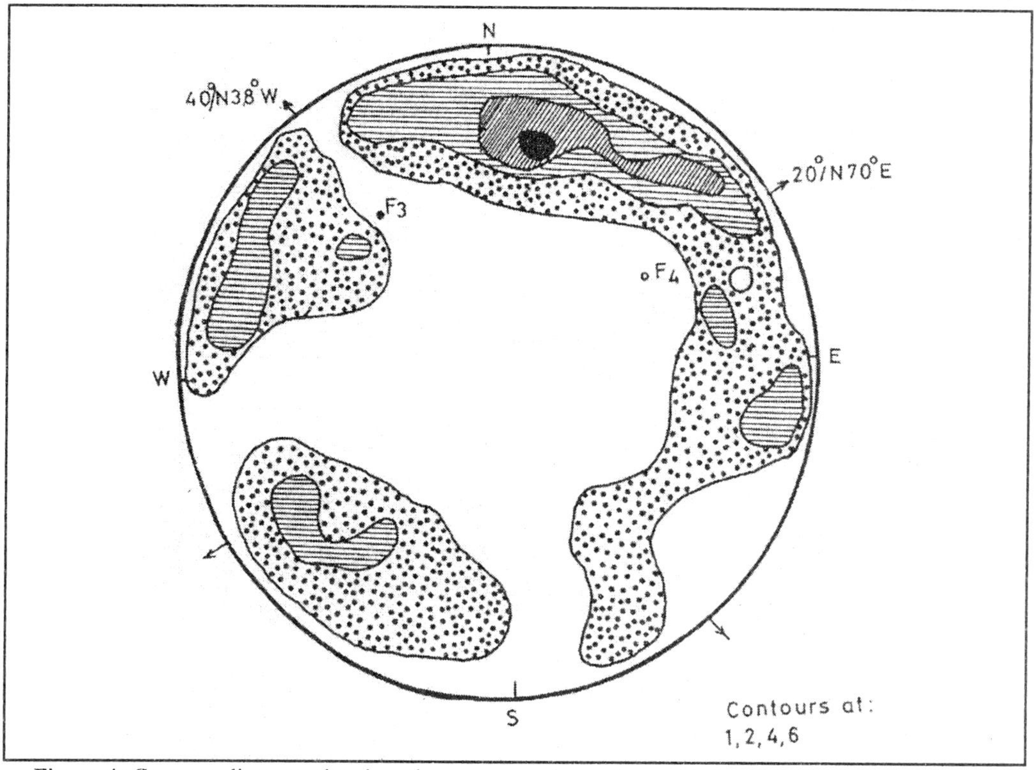

Figure 4: Contour diagram showing the macroscopic analyses of S_2- foliation of Almora Nappe (N = 90) plotted on lower hemisphere equal area projection. The F_3- and F_4- fold axes orientations deduced from the diagram show plunge of 40°/ N38°W for F_3- folds and plunge of 20°/ N70°E for F_4- folds.

preferred orientation and quartz Lattice Preferred Orientations (quickly checked by the technique after Simpson, 1986) are characteristically absent under the microscope.

Syn - shear zone structures encompass a range of shear sense indicators, viz. shear bands, asymmetric porphyroclast systems (s_a-, s_b- and d type), mica fish, pressure shadows and the characteristic Lattice Preferred Orientation (LPO) of quartz.

The preferred orientation of micas define the shear band cleavages (White, 1979; Gapais and White, 1982). The C - and C′- type extensional crenulation cleavages of Platt and Vissers (1980) are observed in the mylonites of both the northern and southern flanks of Basal Shear Zone of the Almora Nappe on megascopic and microscopic scale. The C′- type shear band cleavages are developed during the late stages of shear zone activity and make an angle with the earlier foliation as well as with the shear zone boundary (White, 1979; Platt and Vissers, 1980) (Fig 5a). The shear bands observed in the mylonites of the SASZ (Chhara – Chaunsali sector) and the NASZ (Someshwar – Hawalbagh sector) generally indicate a top - to - south movement. However in the NASZ some shear bands show top - to - north movement. The internally monoclinic symmetry shown by the asymmetric

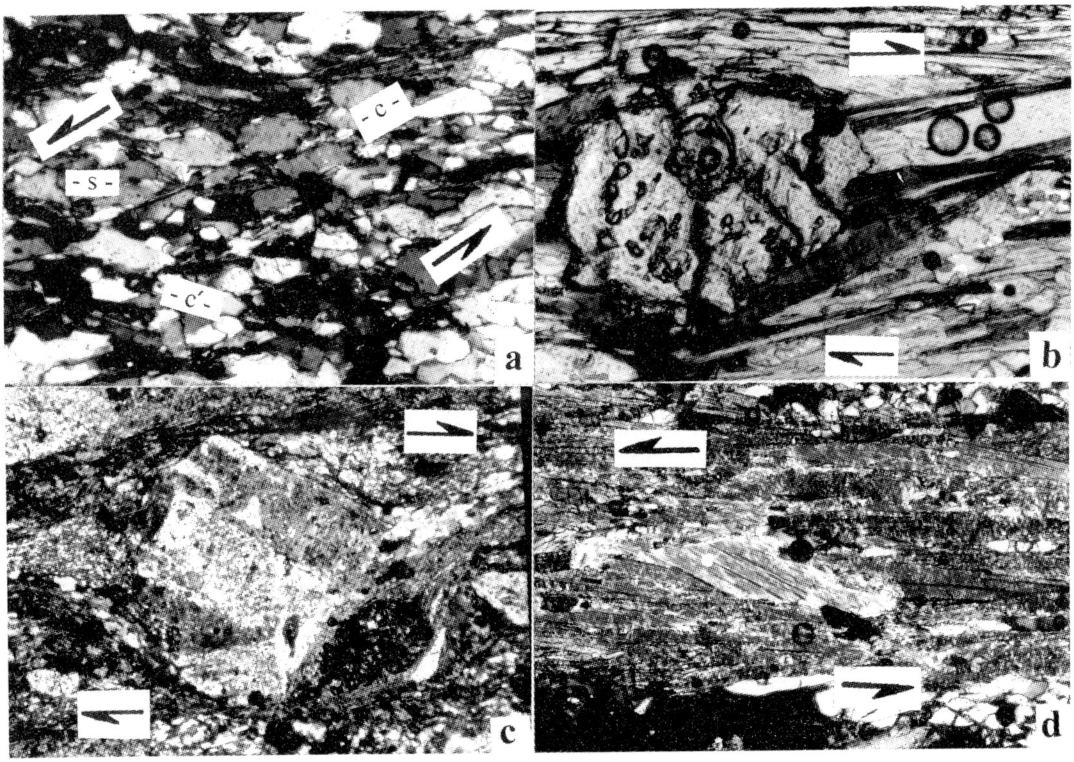

Figure 5: Photomicrographs

5 a: The S - C - C′ band defined by mica flakes in Ramgarh Group of rocks. The C′- band is oblique to both the S- and C- band. The shear sense movement direction is top - to - south. Loc: South of Kakrighat (SASZ).

5 b: The close up of a a- type asymmetric pressure shadow tail of garnet porphyroclast in which biotite and muscovite have recrystallized. Also notice that the crude alignment of inclusions (Si) in garnets is oblique to the external schistosity (Se) due to rotation of garnet. Shear sense movement direction is top - to - south. Loc : Chopra.

5 c: A b- type porphyroclast system in Ramgarh Group of rocks belonging to the northern flank of Almora Nappe. Notice the shear sense movement direction is top - to - north. Loc : North of Manan.

5 d: A mica fish in Almora Group of rocks. Notice the shear sense movement direction is top - to - south. Loc : South of Chopra.

porphyroclast systems in the X - Z sections appear as tails around porphyroclasts and is generally defined by wedges of dynamically recrystallised material characterizing the mylonites. Equidimensional ellipsoidal varieties of s_a - porphyroclasts (Passchier and Simpson, 1986) (Fig 5b) are quite common. The ultramylonites after mica schists of the Nathuakhan Formation of Ramgarh Group are characterized by the development of s_b - type porphyroclasts (Fig 5c) in narrow zones of high shear strain (C - planes). Both s_a - and s_b - type porphyroclasts in the thin sections of rock samples from the protomylonites and mylonites of Ramgarh and Almora Groups, exposed in both the northern and southern flanks of nappe show top - to - south movement but in some cases a reverse sense of movement i.e. from top - to - north is discerned in the mylonites of the northern flank. The reversed sense of movement shown by the s_a - type porphyroclast systems of the NASZ is due to non-coaxial deformation during second stage of shearing movement. However, during the second shearing movement, the movement direction was also from top - to - south similar to the first movement shown by the s - type asymmetric porphyroclast systems in the SASZ. The mica fish (Eisbacher,1970; Lister and Snoke,1984) considered as a special case of s - type porphyroclast systems (Simpson,1986) developed in mylonites of Saryu Formation of Almora Group as a result of combined brittle and crystal-plastic processes affecting the pre-existing mica lepidoblasts (Fig 5d).The shear sense movement directions indicated by mica fish are top - to - south for both the northern and southern flanks of the Basal Shear Zone of the Almora Nappe in the area studied.

The Lattice Preferred Orientation studies were carried out on 16 oriented samples for the Chhara - Someshwar transect across the Almora Nappe. Two thin sections (viz. XZ and XY sections) were studied for each rock sample. The Lattice Preferred Orientation (LPO) of C - axes of at least 300 grains of quartz were determined on a five axes universal stage for each thin section. All the measured data for the two thin sections were fused in the XZ plane, by rotating the measured data from the XY plane by 90^0 to the XZ plane.

The details of quartz C - axes orientation diagrams for area studied are presented in fig. 2. The interpretation of these quartz C - axes orientation diagrams is summarized below :

1. The last ductile movement as deduced from the geometry of the crossed C - axes type – I girdles (Simpson,1986; Passchier and Trouw, 1996) was directed from top - to - north for the mylonites of the NASZ and top - to - south for the mylonites of the SASZ.

2. By comparing the LPO patterns of quartz C - axes with the flinn diagram of Lister and Hobbs (1980), it can be concluded that the sample no. A.83, A.85, A.98, A.107, 94/45, 94/32, 94/33 and 94/34 were subjected to plain strain to flattening strain-conditions during their last deformation. All these samples belong to the schists and gneisses of the Saryu Formation of the Almora Group, exposed in the central parts of the nappe.Thus it is clear that these rocks have totally escaped the intense shear deformation to which the basal parts of the Almora Nappe have been subjected.

3. The girdles show strong clustering of C - axes close to periphery and around the Y - axes, suggesting that the LPO patterns developed by a dominant basal <a> slip.

At low temperatures, the girdles developed by basal <a> slip while at medium temperature conditions the girdles developed by prism <a> slip (Wilson,1975; Bouchez,1977; Lister and Dornsiepen,1982; Law,1990). Thus it can be concluded that the last ductile deformation responsible for shearing occurred under low to medium temperature conditions.

Both the NASZ and SASZ have been affected by different types of post shear zone brittle faults. A N – S striking fault (F_1 - F_1) with the eastern block up thrown (Rawat,1988 Joshi,1999) is observed in the Khairna - Suyalbari sector, along which the river Kosi flows in an unusual straight course (Fig 2). The northwestern block is up thrown along another brittle fault (F_2 - F_2) with NW - SE strike direction (fig 2) which appears to be a neotectonic rejuvenation subparallel to the SAT.

CONCLUSIONS

The structural elements in the Almora Nappe can be distinguished into the (a) pre-shear zone structures, (b) syn - shear zone structures and (c) post - shear zone structures with respect to the Basal Shear Zone (BSZ) of the Almora Nappe (Joshi,1999). A complex three stage tectonic history and kinematic evolution of the Basal Shear Zone (BSZ) of the Almora Nappe has been deduced from these structures. The first stage movements of the Almora Nappe were top - to - south ductile movements of very strong intensity. This was the movement which brought the Almora Nappe from the Higher Himalaya to its present Lesser Himalayan locale to accommodate the crustal shortening by thrusting as a consequence of the Himalayan continental collision. The second stage of movement was ductile to brittle - ductile in nature and of strong intensity. The direction for the second movement were top - to -south for the SASZ and top - to - north (reversed) for the NASZ. The third and last recorded movements were moderately strong and largely brittle in nature. The directions for the third movements were similar to those of the second movement.

The shear deformation during the Tertiary orogeny was restricted only to the basal parts of the Almora Nappe (see fig. 2). The structurally higher region, now comprising the central parts of the nappe remained largely unaffected by mylonitization. Only few brittle faults generally running parallel to the strike of the nappe (and a few running across) have affected the central parts of the nappe. The last shearing movements occurred under low to medium temperature conditions.

This fact is clearly documented by the quartz C - axes diagrams for the central parts of the nappe which show plain strain to flattening strain conditions (see previous sections) unlike the cross girdles which characterize a shear zone.

The movement direction for the last two movements (second and third) were reversed due to the continuing northward movement of the Indian plate. The overlying Almora Nappe was subjected to stresses of lesser magnitude as compared to the sedimentaries underlying the Ramgarh Thrust. As a consequence there was some sort of a relative decoupling along the still ductile mylonites zone comprising the Basal Shear Zone of the Almora Nappe, whereby the Almora Nappe was relatively free to move both northwards and southwards as compared to the over - compressed underlying sedimentaries.

ACKNOWLEDGEMENTS

We gratefully acknowledge the financial assistance by the DST for the project No. ESS/23/132/91. We are also grateful to the DST for using the computer facilities provided under Project No. ESS/16/091/97. ANT is thankful to the UGC for financial assistance as a Senior Research Fellowship.

REFERANCES

Bhargava, O.N., 1972 : A reinterpretation of the Krol belt. Him.Geol., 2, 47-81.

Bhattacharya, A.R., 1987 : "A ductile thrust" in the Himalaya. Tectonophysics, 135, 37-45.

Bouchez, J.L., 1977 : Plastic deformation of quartzites at low temperatures in an area of natural strain gradient. Tectonophys., 39, 25-50.

Eisbacher, G.H., 1970 : Deformation mechanics of mylonitic rocks and fractured granites in Cobequid Mountains, Nova Scotia. Can. Bull. Geol. Soc. Am., 81, 2009-2020.

Gairola, V.K. and Joshi, M., 1980 : Structure of a part of Dudatoli - Almora Crystalline Thrust sheet around Thalisain, District Pauri Garhwal, Uttar - Pradesh. Him. Geol., 8, 379 - 398.

Gansser, A., 1964 : Geology of the Himalayas. Interscience Publishers, London, 273 p.

Gapais, D. and White, S. H., 1982 : Ductile shear bands in a naturally deformed quartzite. Text Microstruct., 5, 1- 17.

Heim, A. and Gansser, A., 1939 : Central Himalaya, Geological observations of the Swiss expedition in 1936. Mem. Soc. Helv. Sci. Nat., 73, 1-245.

Jain, A. K., 1971 : Stratigraphy and tectonics of Lesser Himalayan region of Uttarkashi, Garhwal, U. P. Him. Geo., 2, 188-205.

Joshi, Mallickarjun., Singh, B. N. and Goel, O. P., 1994 : Metamorphic conditions of the aureole rocks from Dhunaghat area, Kumaun Lesser Himalaya. Cur. Sci., 185-188.

Joshi, Mallickarjun., 1999 : Evolution of the Basal Shear Zone of the Almora Nappe, Kumaun Himalaya. Gondwana Research Group Mem., 6, 69 – 80.

Kumar, G. and Agrawal, N. C., 1975 : Geology of Srinagar - Nandprayag area (Alaknanda Valley), Chamoli, Garhwal and Tehri Garhwal District, Kumaun Himalaya, U. P. Him. Geol., 5, 29-59.

Law, R. D., 1990 : Crystallographic fabrics : a selective review of their applications to research in Structural Geology. In : Knipe R J, Rutter E H (eds.) Deformation mechanism, rheology and tectonics, Geol. Soc. Spec. Publ., 54, 335 - 352.

Lister, G. S. and Dornsiepen, U. F., 1982 : Fabric transitions in the Saxony granulite terrain. J. Struct. Geol., 41, 81 - 92.

Lister, G. S. and Hobbs, B. E., 1980 : The simulation of fabric development during plastic deformation and its application to quartizite : the influence of deformation history. J. Struct. Geol., 2, 355-371.

Lister, G. S. and Snoke, A. W., 1984 : S - C mylonites. J. Struct. Geol., 6, 617 – 638.

Mehdi, H. S., Kumar, G. and Prakash, G., 1972 : Tectonic evolution of Eastern Kumaun Himalaya : a new approach. Him. Geol., 2, 4 81 - 501.

Merh, S. S., 1968 : A preliminary note on the structural history of Kumaun Himalaya. Bull. Geol. Soc. Ind., 5, 1-7.

Merh, S. S. and Vashi, N. M., 1965 : Structure and metamorphism of Ranikhet area of Almora District, Uttar Pradesh. Indian Mineralogist., 6, 55-66.

Misra, R. C. and Sharma, R. P., 1972 : Structure of the Almora Crystallines, Lesser Himalaya : An interpretation. Him. Geol., 2, 330-341.

Molnar, P. and Tapponier, P., 1975 : Cenozoic tectonics of Asia : effects of a continental collision. Science., 189, 419-426.

Passchier, C. W. and Simpson, C., 1986 : Porphyroclast systems as kinematic indicators. J. Struct. Geol., 8, 831-841.

Passchier, C. W. and Trouw, R. A., 1996 : Microtectonics. Springer - Verlag Berlin Heidelberg., 289p.

Pilgrim, G. E. and West, W. D., 1928 : The structure and correlation of the Simla rocks. J. Struct. Geol. Surv. Ind., 53, 1-140.

Platt, J. P. and Vissers, R. L. M., 1980 : Extensional Structures in anisotropic rocks, J. Struct. Geol., 2, 397-410.

Powar, K. B., Gairola, V. K. and Dixit, P. C., 1969 : Relation between volcanism, Plutonism, regional metamorphism and tectonism in Himalayan orogeny as related by rocks of a part of Kumaun Himalayas. Pub. Centre Adv. Stud. Geol. Punjab Univ., Chandigarh., 6, 1-7.

Rawat, J. S., 1988 : Geomorphology and tectonic history of the South Almora Thrust zone near Kakrighat, Nainital District (U.P.), India. Singh, S. and Tiwari, R. C. (ed.) for Allahabad Geogr. Soc., 325-336.

Saklani, P. S., 1993 : Geology of the Lower Himalaya (Garhwal). International Books and Periodicals, Delhi, 246p.

Saxena, S. P. and Rao, P. N., 1975 : Does Almora Nappe exists ? Him. Geol., 5, 169-184.

Simpson, C., 1986 : Determination of movements sense in mylonites. J. Geol. Education., 34, 246-261.

Singh, B. N., Goel, O. P., Joshi , Mallickarjun. and Shereton, J. W., 1993 : Geochemistry and petrogenesis of the Champawat Granitoids occurring around Dhunaghat, Distt. Pithoragarh, U.P. J. Geol. Soc. Ind., 42, 289-302.

Valdiya, K. S., 1962 : A study of the Champawat Granodiorite and associated metamorphics of the Lohaghat Subdivision, District Almora, U.P. with special reference to petrography and petrogenesis. Indian Mineralogist, 3, 6-37.

Valdiya, K. S., 1976 a : Himalayan Transverse faults and folds and their parallelism with subsurface structures of North India Plains. Tectonophys., 32, 353-386.

Valdiya, K. S. 1976 b : Structural set up of the Kumaun Lesser Himalaya. 'Ecologie et Geologie de 1' Himalaya., 268, 449-462.

Valdiya, K. S.., 1978 a : Extension and analogues of the Chail Nappe in the Kumaun Himalaya, India. Indian Jl. Earth Sci., 5, 1-19.

Valdiya, K. S., 1978 b : Outline of the structure of Kumaun Lesser Himalaya. In : Saklani, P. S. (Ed.), Tectonic Geology of the Himalaya., 1-14.

Valdiya, K. S., 1980 : Geology of Kumaun Lesser Himalaya. WIHG, Dehradun,291p.

White, S. H., 1979 a : Grain and sub-grain size variations across a mylonites zone. Contrib. Mineral. Petrol., 70, 193-202.

White, S. H. 1979 b : Large strain deformation : report on a tectonic studies group discussion meeting held at Imperial College, London; Introduction., J. Struct. Geol. 4 : 333-339.

Wilson, C. J. L., 1975 : Preferred orientation in quartz ribbon mylonites. Bull. Geol. Soc. Am., 86, 968-974.

POP-UP KLIPPEN IN THE MUSSOORIE SYNCLINE, LESSER HIMALAYA: EVIDENCE FORM FIELD AND MODEL DEFORMATION STUDIES

A.K. Dubey* and R. Jayangondaperumal*.*

Wadia institute of Himalayan Geology
33 General Mahadev Singh Road, Dehradun-248001
. Present Address: Indian Bureau of Mines. Chennai

ABSTRACT

The Satengal and Banali klippen occur in the core of the Mussoorie Syncline, Lesser Himalaya. Their geographical position has led many geologists to consider them as demonstrating the maximum translation amongst the Himalayan thrusts. However, field observations made by a large number of geologists during the last 116 years (Middlemiss, 1887 to the present authors) suggest a small amount of displacement along the klippen thrust. Based on field observations, and supported by model deformation studies, an alternate model is proposed for the structural evolution of the area. The model explains the occurrence of older leaser Himalayan rocks over younger rocks by a combination of fault bifurcation, foreland propagating thrust, back thrusts, and subsequent superposed folding. The Main Boundary Thrust (MBT) bifurcates into two thrusts; the first is exposed as a combination of frontal and oblique thrust ramps, where as the second was initially a blind frontal thrust ramp which was later partially exposed during the superposed deformation to form the klippen structures. The klippen are described as pop-up or parautochthonous klippen because they lie over their roots

INTRODUCTION

The Mussoorie Syncline, a part of the Krol belt, occurs as a doubly plunging syncline and extends in the NW-SE direction (Fig.1). It lies between the Main Boundary Thrust (MBT) MBT in the south and North Almora Thrust (NAT) in the north.

Middlemiss (1887) suggested an autochthonous model for the structural evolution of the area on the basis of field observations. Half a century later, the concept of large-scale translation of thrust sheets became popular in the Scottish Highlands and in the Alps. Influenced by this, Pilgrim and West (1928) proposed a large-scale displacement along thrusts in the Simla Lesser Himalaya. The area of Garhwal Lesser Himalaya was assigned by the Geological survey of India to J. B. Auden who proposed a revised litho-tectonic set-up (table1) and a geological cross-section of the area (Fig. 2) (Auden, 1937). The Garhwal Lesser Himalaya was classified into two dominant structural units, namely, Krol Nappe and Garhwal Nappe. The Garhwal Nappe was depicted as occurring in the from of detached klippen over the Krol Nappe with displacement of nearly 80km southward form its root in the Central Crystalline of the Higher Himalaya. The klippen occur in the core of Mussoorie Syncline (Fig. 1). Most of the later workers (e.g., Gansser, 1964; Rupke, 1974; Valdiya, 1980) have followed the Auden's work despite the fact that there is a marked contradiction between his stratigraphic observations (Table 1) and the structural interpretation (Fig. 2). The Table describes the uppermost horizon of the Garhwal Nappe with Chandpur Fm. whereas the cross-section brings the klippen rocks from the Higher Himalaya where Chandpur rocks have no existence. Jain (1972), while confirming that the klippen rocks resemble the older lesser Himalayan formations, stated that, "Both tectonic units of Garhwal Thrust Unit seem to be of more of local origin, developed probably somewhere from near the base of the Krol Thrust Sheet in the north. It is difficult to visualize their large-scale movement from the base of the Central Himalaya. Nor does it seem tenable to connect Garhwal Thrust with the Main Central Thrust". Later, Saklani (1993) mapped the rocks as Chandpur Phyllite. Bali and Bhattacharya (1998) have also envisaged an autochthonous model for the adjoining region of Garhwal Synform.

It is evident from the above discussion that the earlier workers have clearly identified the klippen rocks as the Lesser Himalayan formations but the rocks were assigned different names in order to explain the regional structure in the terms of large scale displacement along thrust faults, from an assumed root zone. This has created a major confusion regarding the structure of the region. Though Auden (1937) commented that "the problem remained as Middlemiss left it for exactly fifty years", fact remains that it is no different than what he himself left sixty-six years ago!

In order to solve this long standing problem, the present investigation is an attempt to integrate field model deformation studies.

GEOLOGICAL FIELD STUDY

Geological mapping was carried out on 1:50,000 scale. Additional information on lineaments and lithological boundaries were obtained form the computer compatible tape of IRS LISS II/A2 (path 28 Row 26 Data of coverage 7/8/1989). These results were incorporated in the structural map (Fig.1). More than 6000 structural data were recorded. A revised geological cross-section of the area is shown in figure.3, and a revised tectonostratigraphy is shown in Table 2. The present field and petrographic studies confirm that rocks of the klippen units are the lesser Himalayan rock formations (i.e. Mandhali and Chandpur formations). A brief summary of the structural features is made here.

South of the Mussoorie Syncline, the MBT is shown as displaced by a strike-slip fault which trends in the NS direction (Valdiya, 1980). The stratigraphic horizons follow the trend of the fault. The fault was regarded as youngest in the tectonic history because it displaces the MBT. However, it is to be noted that: (i) the fault is confined to offset thrust segments, i.e. the displacement along the fault is equal to the length of the fault - a condition most improbable for strike slip faults; (ii) it does not displace

Table 1

Tectonic Unit		Formation	
	Amri Nappe	Chandpur (Metamorphosed)	
		-------------Thrust-------------	
Garhwal Nappe		Nagthat Chandpur	(Little metamorphosed)
	Bijni Nappe	Boulder beds, slates and limestone of uncertain Stratigraphical horizon occur in one outlier below Metamorphosed Chandpurs.	
		-------------Garhwal Thrust---------	
Krol Nappe		Nummulitic	
		Tal Formation	
		Krol Formation	
		Blaini Formation	
		Nagthat Formation	Metamorphosed
		Chandpur Formation	and unmetamorphosed.
		-----------Krol Thrust----------	
Autochthonous		Dagshai, Nummulitic/ Siwalik / Simla slates	

Table 2

Formation	GrossLithology	~Thickness in meters
Chandpur Formation	Low grade, greyish- green phyllite	?
Mandhali Formation.	Blue, greyish fawn arenaceous limestone, yellowish to white, gritty and slaty quartzite Phyllite.	?
.........Kathu-ki-chail Thrust..........		
Subathu Formation	Ferruginous purple to dark grey shale with lateritic/ limonitic nodules	200
...............unconformity.................		
Tal Formation	Orthoquartzite and calc-arenite with pebbly quartzite, brownish yellow siliceous limestone, dirty white to yellowish quartzite, greenish- grey shale, dark black phosphoritic and cherty layers	1552
Krol Formation	Dark bluish- grey dolomitic limestone, fine grained carbonaceous limestone, and shale alteration	2027
Blaini Formation	Slate and muddy quartzite, conglomerate, greywacke and limestone	517
Nagthat Formation	Orthoquartzites (local pebbliferous) with subordinate shale	546
Chandpur Formation	Olive green and grey phyllite with subordinate slate	1782
Mandhali Formation	Blue, greyish -fawn arenaceous limestone, yellowish to white gritty and slaty quartzite Phyllite.	>345
.........MBT(Krol Thrust)		
	Siwalik/Dagshai/Simla slates	

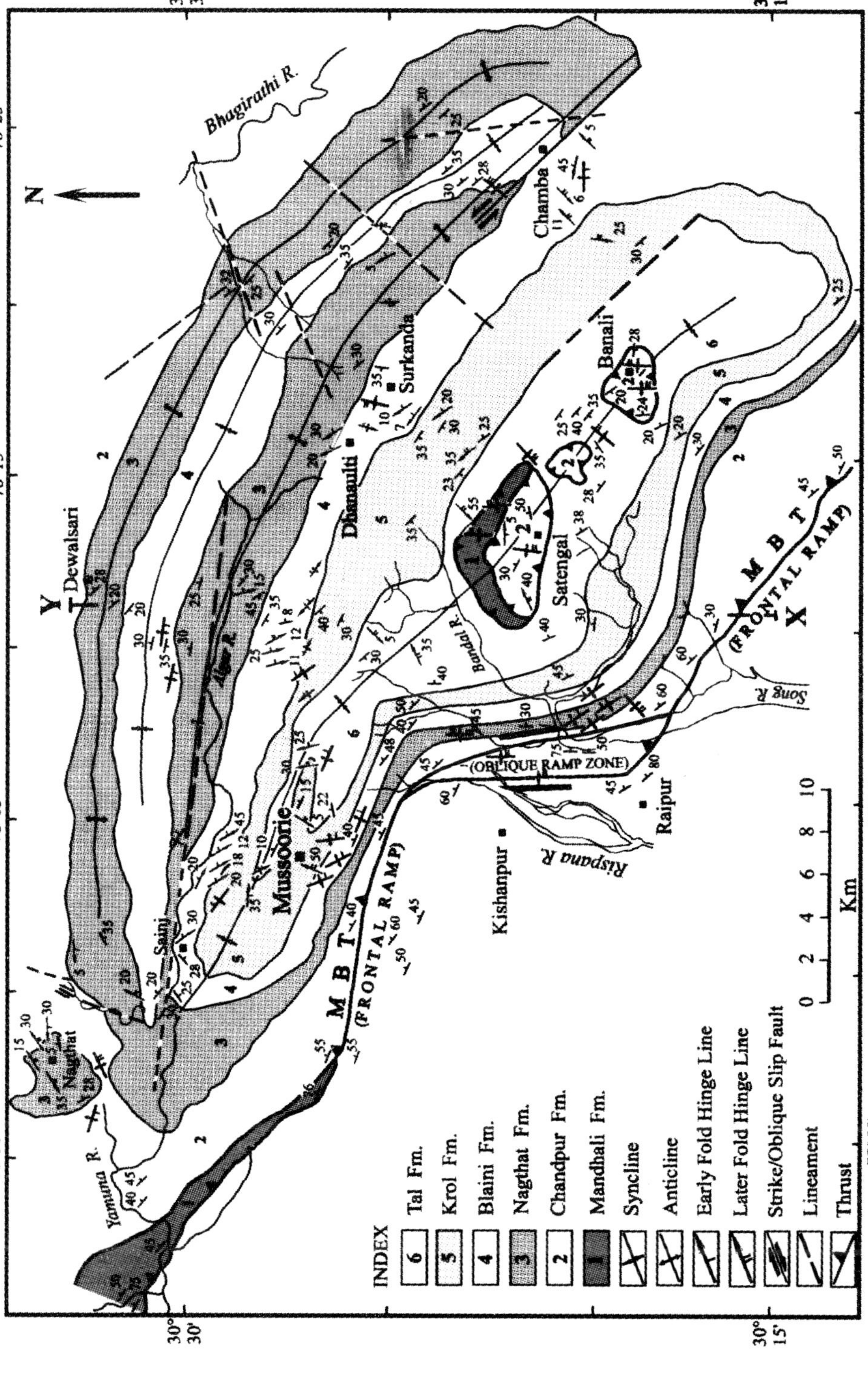

Figure. 1: A geological and structural map of the Mussoorie Syncline, Lesser Himalaya (modified after Valdiya, 1980).

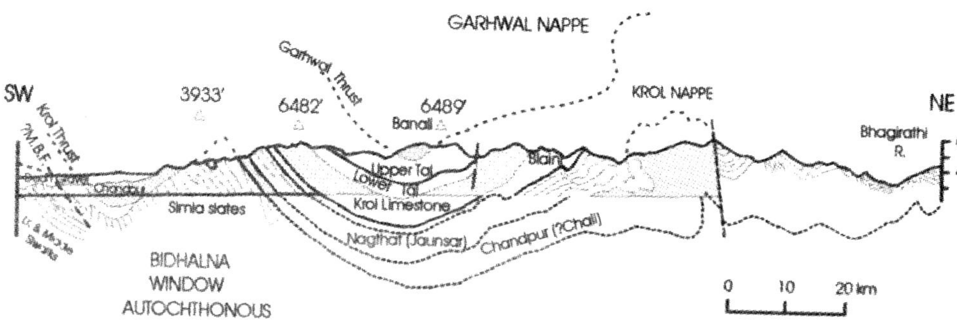

Figure. 2: A geological cross-section of the Mussoorie Syncline (after Auden, 1973).

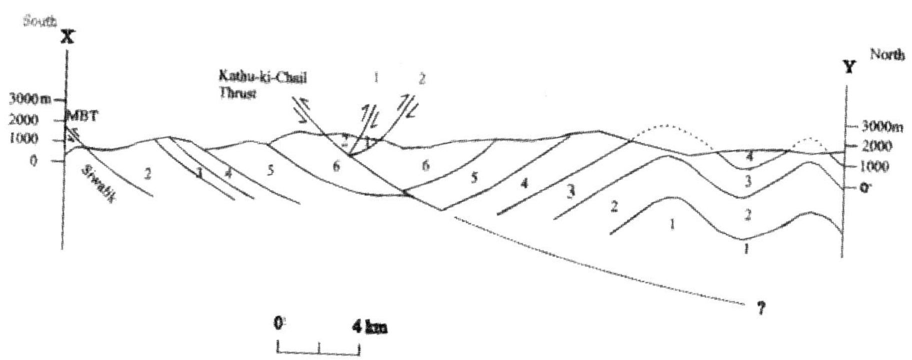

Figure. 3: A revised geological cross-section of the area along X-Y marked on Fig. 1.

stratigraphic horizons but, the stratigraphic horizons follow the trend of the fault, and (iii) fold hinge lines are not traceable across the fault. The above features suggest that the fault is most likely an oblique thrust ramp rather than a strike slip fault (Dubey, 1997). The fault may have formed as an oblique ramp normal fault during the earlier tensional regime in the region (Dubey and Bhat, 1986). The fault is well exposed east of Raipur (Fig.1). The oblique ramp gradually tapers towards the trailing ramp and merges with the MBT frontal ramp, NE of Kishanpur. The tapering and merger are evident near Rajpur where intense crushing is observed in field.

The rock units are involved in two phases of folding episodes (Fig. 1). The first one was developed simultaneously with the development of the prominent Himalayan thrusts and the second was developed after locking of the thrusts (Dubey and Bhat, 1986). The early folds show a wide variation in geometry from open to isoclinal, occasionally recumbent. The trend of the fold hinge lines varies from NW-SE to E-W with gentle plunge in both the directions. The superposed folds are mostly upright or asymmetric chevron folds with varying trends from NE-SW to N-S and gentle plunge towards N or S. The minor folds are most prominent in the Chandpur Formation inside the klippen. The fold hinge line curvatures can be explained by simultaneous development of frontal thrust ramps, oblique thrust ramp and non cylindrical folds (Dubey, 1997). As a result of fold interference between the early and superposed folds, the dips of the klippen rocks are not always centripetal. The inward dips were observed in the surrounding rock outside the klippens and superposed synformal hinge lines separate the Satengal, Darak (Ringalgarh), and Banali klippen (Fig.1).

Minor normal faults are prominent in the southern (near Main Boundary Thrust) and central part of the area. The continuity of some foliations and displacement of others across the faults indicate that these faults were formed at different periods. The earliest synsedimentary faults have been observed south of Dewalsari (near Magra Rest House) in the Blaini Formation. Their characteristic feature is that they do no displace the early foliation, which indicates that they were formed during post-rift subsidence and sedimentation (Dubey and Bhat, 1986). The youngest faults displace all the foliation planes present in the area. Hence normal faults were formed at different periods of tectonic history of the region.

In the northern part of the area a major E-W trending fault can be seen along the Aglar River. Here the Nagathat Formation abruptly terminates against the Blaini Formation. The fault can be seen South of Dewalsari (Magra-Dewalsari Section) as well.

A few strike slip faults and other lineaments that were traced trend NW-SE to NE-SW to N-S, without displacing the rock types. However, at few places, e.g. southeast of Nagathat village, the N-S trending lineament shows displacement of the Nagthat formation (Fig.1)

The principal susceptibility directions (i.e. KMAX AND KMIN) form different sites were derived form the anisotropic magnetic susceptibility data (Jayangondaperumal and Dubey, 2001) and contoured on equal area net (Fig.4a) with majority orientations in the NE-SW quadrant with gentle plunge. The orientations around the E-W axis may be attributed to the early subhorizontal compression that resulted in the formation of early folds whereas the orientations around NW-SW can be attributed to the later superposed deformation and related superposed folding. The concentration in the central part indicates reactivation along the oblique ramp and normal faulting movement during the superposed deformation (fig. 10 in Dubey, 1977) The KMIN axis (Fig. 4b) shows concentration in the central part of the diagram indicating steep plunge. Two orientations, i.e. N-S and E-W are representative of the two folding episodes present in the area. Steep orientations of the KMIN are indicative of the predominant oblique slip displacement during the two folding episodes.

TWO -DIMENSIONAL MODEL

The existing model for the structural evolution of the area fails to explain the occurrence of the Lesser Himalayan formations as klippen. Hence, an alternate evolutionary model is proposed. The model is based on field data and tested by model deformation studies. However, before presenting a three dimensional evolutionary model, an attempt is made here to understand the occurrence of the older Lesser Himalayan rocks as klippen using a two dimensional model.

The different layers designated as A to F (Fig. 5) roughly correspond to the thickness of the Lesser Himalayan formations (Table 2). 'A' represents Mandhali, 'B' represents the next younger Formation Chandpur, 'C' Nagthat, 'D' Blaini, 'E' and 'F' represent the Tal Formation.

The initial fracture with listric geometry (Fig. 5a) was probably formed during the tensional regime in the region (cf. Dubey and Bhat, 1986; Coward, 1996). Normal fault displacement, which took place along the fracture in the subsequent stage (Fig. 5b), was accompanied by anti-clockwise rotation of the hangingwall block and tilting of the beds opposite to the dip direction of the fault. This was followed by initiation of a conjugate set of faults (Fig. 5c). The faults initiated either at a late stage of normal fault displacement or at an early stage of the compressional phase, which followed the tensional regime.

The compressional phase during the Tertiary Himalayan orogeny resulted in reactivation of the earlier listric normal fault as listric thrust fault. After crossing the null point (Williams et al., 1989) the hangingwall beds were tilted in the dip direction of the thrust fault. The conjugate set of faults behaved like back thrusts (referred as 1 and 2 in Fig. 5d). The main listric thrust brought the Chandpur formation at higher elevations along with the back thrusts. At a subsequent stage of the compressional phase (Fig, 5d), a larger displacement along back thrust 2 brought the Mandhali formation at upper horizons and in thrusted contact with the Chandpur Formation (cf. Fig. 3). Thus, a combination of the main listric thrust and thrusts brought the Chandpur and Mandhali formations as Klippe (cf. fig. 1 in Coward, 1983). The later shortening resulted in an open synformal structure.

The Kathu-ki-chail Thrust represents the main thrust (Fig. 3) and the two back thrusts occur east of Anand chowk (near Satengal). The geometry of the cross-section suggests that if the dip of fracture is steep the relative translation may be a of a higher order of magnitude. However, it is to be noted that a steep fault dip will result in an increased friction and an early fault locking. Taking these constraints into account it is proposed that

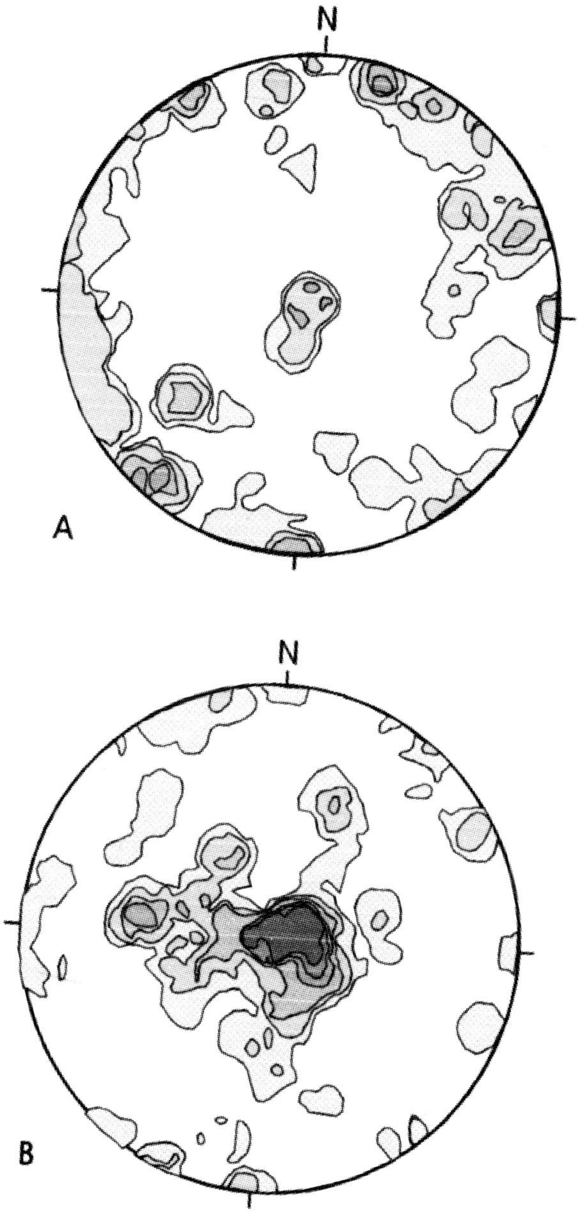

Figure 4: A a. The orientations of KMAX of specimen from different sites (equal area lower hemisphere projection). Contoured at intervals of 0,2,4,6,and 8 (No.152). b.The orientations of KMIN of specimen from different sites (equal area lower hemisphere projection). Contoured at intervals 0,2,4,6,8,10, and12 (No. 152).

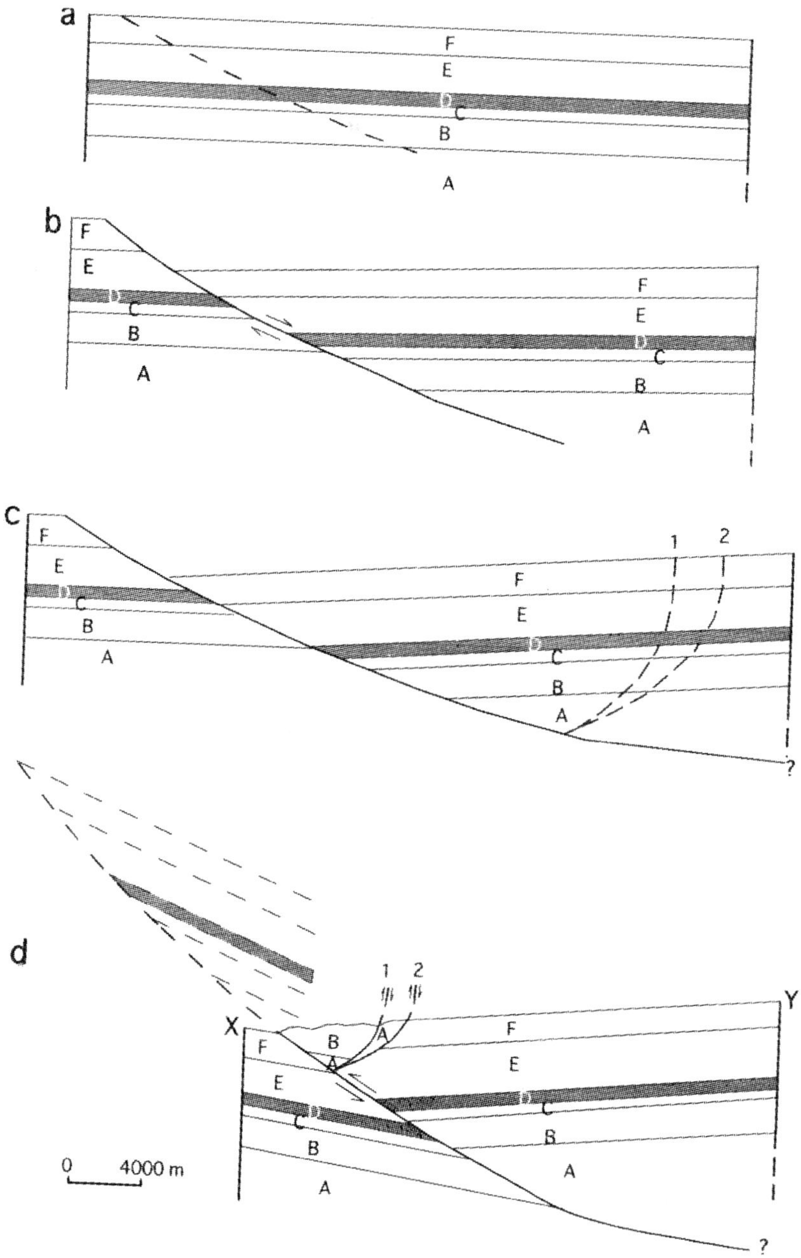

Figure 5: A schematic diagram showing sequential development of Klippe in the Mussoorie syncline. a. Initiation of a fracture during tensional regime in the region. b. Normal fault displacement along the fracture .c. Initiation of conjugate set of fractures. d. During the later compressional regime, a large displacement along the main thrust and minor displacements along the back thrusts leading to the development of Klippe. X-Y represents the present erosion level (scale approximate).

the maximum translation, that may have taken place along the Kathu-ki-chail thrust is~18-20 Km. The geometrical estimation is considerably less than the earlier estimated figure of 80-90 Km (Auden, 1937; Heim and Gansser, 1939; Valdiya, 1980).

THE EXPERIMENTS

In order to understand the present map pattern (Fig. 1) and structural evolution of the area in three dimensions, a series of experiments were performed using multilayer modelling clay models. The external dimensions of the models were 15x11x7 cm. Five layers of modelling clay, each approximately 1 mm thick, formed the multilayer packet. The multilayer packet was sandwiched between two slabs of modelling clay.

The models were cut to a predetermined geometry for simulation of thrust faults that were initiated as normal faults during the earlier tensional regime. The multilayers showed two parallel frontal ramps joined by an oblique ramp (e.g. Dubey, 1997) (Fig. 6a). The fault geometry was different in the basement, as the trailing frontal ramp extended to the other end of the model and the leading frontal ramp terminated in the oblique ramp, which acted as a connecting splay with the trailing frontal ramp (fig. 6b). The upper slab was homogeneous except in one experiment (Experiment two) where the multilayer fault extended into the slab. The initial angle between the basement oblique ramp and the axis of maximum compression was 50^0 and the initial dip of the fault was 30^0

The models were deformed in a biaxial press under the general strain boundary condition i.e., the models were allowed to extend along the horizontal as wall as vertical axis of the press when the axis of maximum compression acted along the layering in a horizontal direction. After a few stages of deformation, the axis of maximum compression acted horizontally but normal to the earlier axis of compression, in order to develop superposed folds.

Two experiments performed under slightly different conditions of deformation are described here.

Experiment one

In this experiment, the multilayer fault did not extend into the upper modelling clay slab. It was anticipated that absence of the fault in the upper slab would restrict that thrust displacement.

The early deformation resulted in formation of cylindrical asymmetric folds as fault propagation folds in the multilayers. Deformation and folding was concentrated near the thrust. The blind frontal thrust ramp was reflected in the orientation of fold hinge lines in the cover (Fig. 7a). Non cylindrical buckle folds did not initiate because the stresses were released on the fault surface. The geometry of cross-sections on two side of the deformed model was different and revealed that the amount of displacement varied along the two frontal ramps. The leading frontal ramp showed a larger displacement as compared to the trailing frontal ramp. The model was compressed up to 12% overall shortening during the first phase of deformation and then it was compressed in the horizontal direction normal to the earlier maximum compression direction under the general strain boundary conditions.

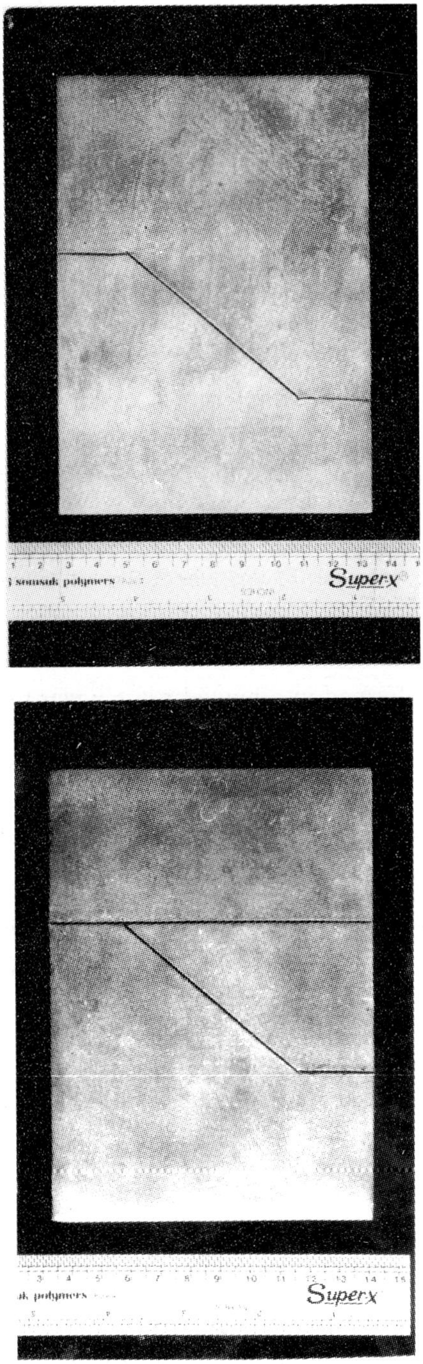

Figure 6: a. Initial fault on a layer surface showing frontal and oblique ramp geometries. b. Initial fault in the basement. Note a continuation of the trailing frontal and oblique ramps in the basement.

The superposed deformation resulted in reversal of displacement along the oblique ramp and formation of superposed folds (Fig. 7a). In contrast to the early folds, these folds were mostly buckle folds and initiated both, on the hangingwall and the footwall, irrespective of the location of the thrust. Polyharmonic fold geometry and interference effects (Dubey, 1997) between the superposed folds and the thrust were observed. After 17% shortening, a horizontal slice was cut from the deformed hangingwall layers to expose the internal structures above the blind thrust (Fig. 7b). Non cylindrical geometry of the antiformal superposed folds revealed closed outcrop patterns marked by partial Klippe structures (i.e. older rock completely or partially encircled by younger rock and part of the contact is a thrust; Dubey and Paul, 1993). The traced structural features are shown in fig. 7c.

Experiment two

In this experiment, fault geometry in the basement and multilayers remained similar to Experiment one, but in order to facilitate a larger thrust displacement both the forntal ramps (i.e. leading and trailing) and the oblique ramp extended into the upper slab. The blind thrust remained confined to the basement. A large thrust displacement inhibited the formation of buckle folds and prominent asymmetric cylindrical folds were observed only above the propagating thrusts (Fig. 8a). After 25% overall shortening of the early phase, the axis of maximum compression was changed to an orthogonal direction to produce superposed folds. A polyharmonic superposed antiformal fold was prominent in the central part of the layers after 23% shortening (Fig. 8a). Penetration of the blind thrust into the multilayers is also visible in the figure. A horizontal slice was then cut of the hanging wall layers to expose the internal structures (Fig. 7b, c). Partial inliers were observed above the initial blind thrust and the oblique ramp. A comparison with Experiment one indicates that formation of partial inliers in the vicinity of an oblique ramp depends upon the amount of vertical component of displacement along the oblique ramp. A large displacement along blind thrust and small displacement along oblique ramp may not result in the formation of partial inliers near an oblique ramp.

THREE-DIMENSIONAL MODEL

Based on the field study an experimental work, a three-dimensional model is proposed for the structural evolution of the area (Fig. 9). The frontal and transfer fault geometries prior to the compressional phase are depicted in Figure 9a. The leading and the trailing frontal ramps are marked a 'L' and 'T', respectively. The trailing frontal ramp bifurcates into two, the first is an oblique ramp (OR) and the second is a blind thrust (BT). The blind thrust is confined to the basement and supports a back thrust with arcuate strike.

During the early stage of deformation, when the maximum compression was normal to the trend of the frontal ramps (Fig. 9b), fault displacement resulted in the formation of fault propagation folds at the propagating thrust tips. The arcuate back thrust resulted in the formation of arcuate noncylindrical folds (cf. Dubey and Bhat, 1986). With increase in shortening, a number of other noncylindrical folds also appeared on the layer surface. As compared to the leading frontal ramp, the displacement was less along the blind thrust and this was reflected in the intensity of folding in the corresponding fault propagation fold

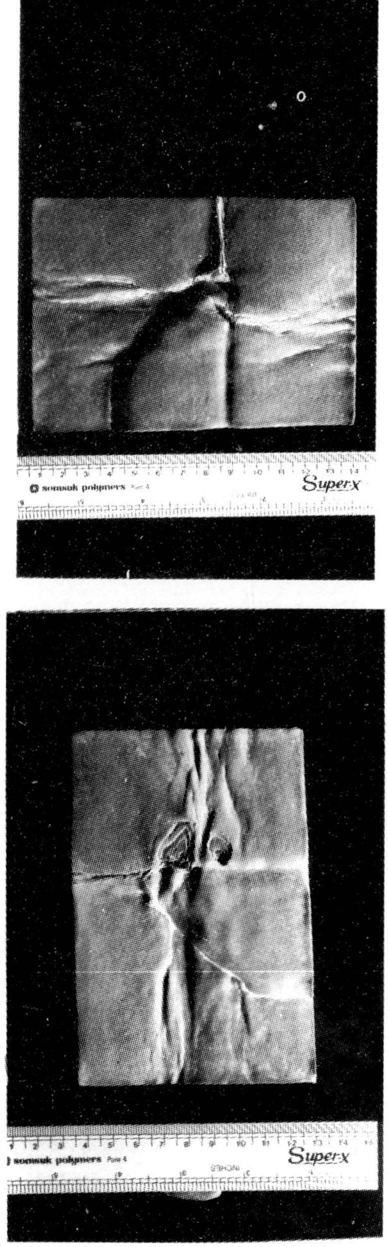

Figure 7: a. Development of early folds (fold hinge lines trending left to right) and superposed folds (trending top to bottom) on a deformed layer surface (after 12% early shortening and 17% superposed shortening). b. Removal of a horizontal slice from the upper part of the model, revealing a partial Klippe (left) and an inlier (right) .

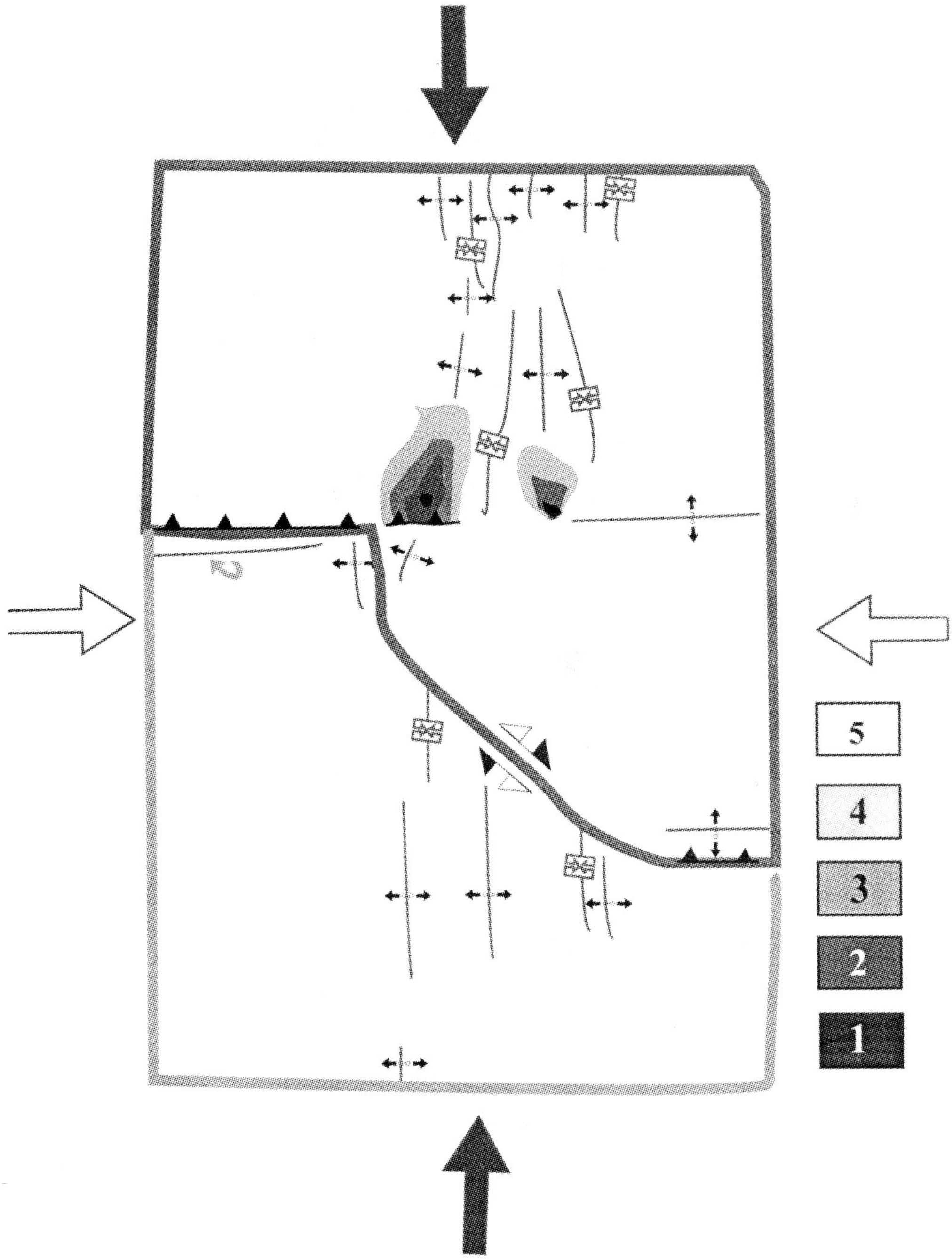

Figure 7c. The structural features at stage b. 1to5 represent successively overlying layers in the multilayer packet. Solid arrows: axis of maximum compression during early deformation; open arrows: axis of maximum compression during superposed deformation.

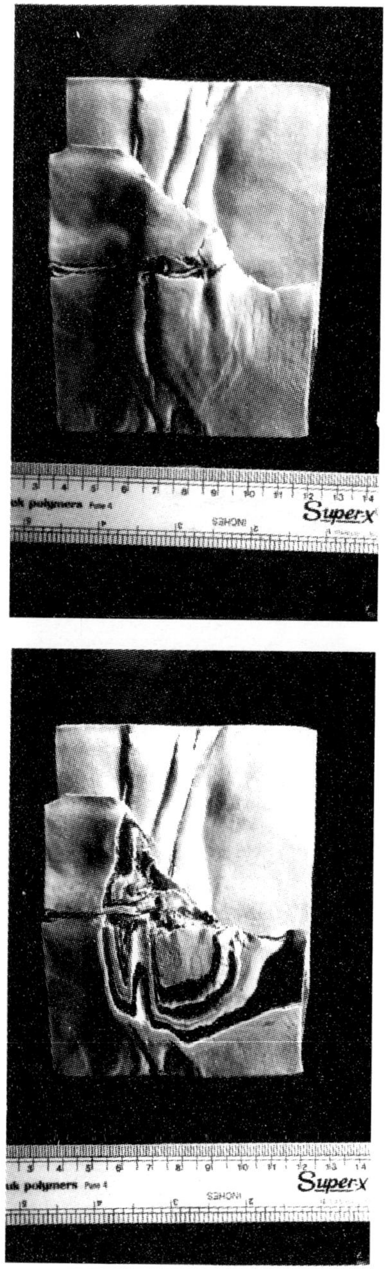

Figure 8: a. Development of early folds (fold hinge lines trending left to right) and superposed folds (trending top to bottom) on a deformed layer surface (after 25% early shortening and 23% superposed shortening) . b. Removal of a horizontal slice from the upper part of the model revealing partial klippen.

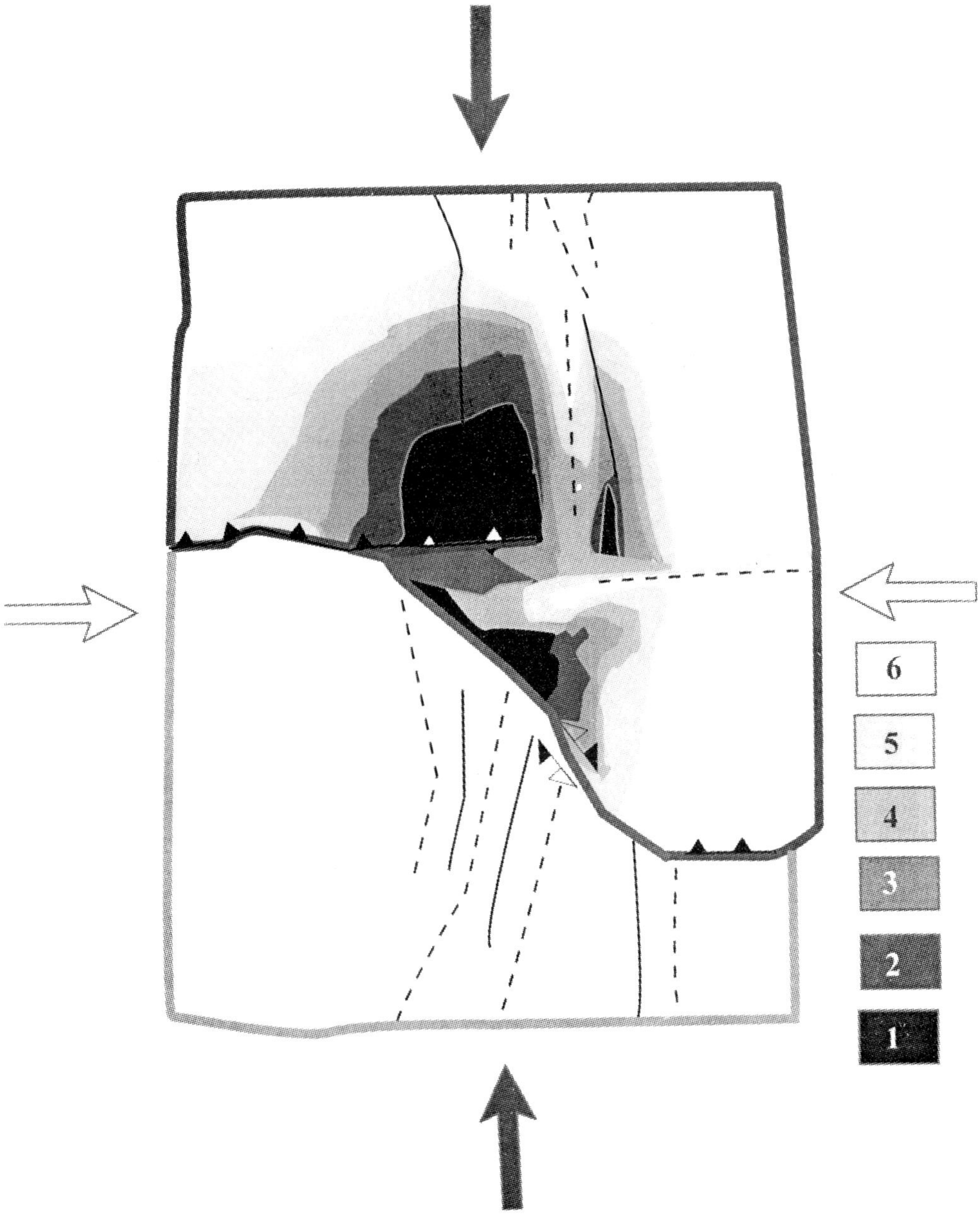

Figure 8c. The structural features at stage b. 1is basement. 2 to 6represent successively overlying layers in the multilayer packet. Solid arrows: axis of maximum compression during early deformation; Open arrows: axis of maximum compression hinge lines.

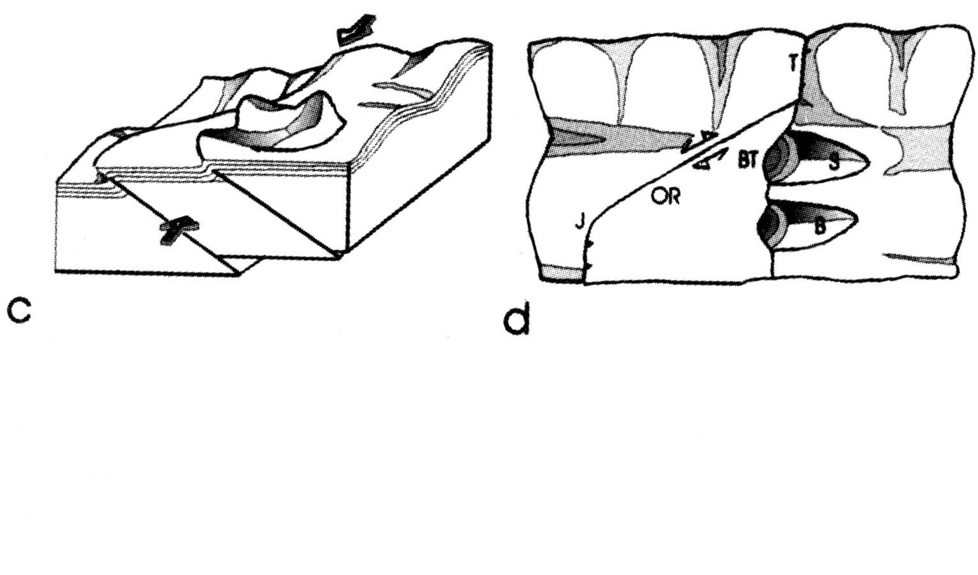

Figure 9: A schematic diagram showing sequential development of klippen.

a. Geometry of initial basement faults at the onset of compressional phase.
b. Thrusting and development of thrust propagation folds above the leading frontal ramps during early deformation. Large arrows indicate the maximum compression direction.
c. Interference of early and superposed folds during the superposed deformation. Note a change in the compression direction. Prominent folds occur above the blind thrust.
d. Erosion of hanging wall with exposure of older rocks surrouded by younger rocks.

(Fig. 9c). During the superposed deformation, the maximum compression axis changed its orientation to the perpendicular direction (Fig. 9c). A number of superposed folds initiated on the surface and interfered with the early folds (Dubey, 1997). Since the early and superposed fold hinge lines were perpendicular to one another, dome and basin were the main interference patterns. The initial thrusts and the arcuate fold were also folded during the superposed deformation. Figure 9c shows prominent dome and basin pattern in the hanging wall above the blind thrust. A subsequent erosion of the hangingwall led to the exposure of older rocks surrounded by younger rocks (Fig. 9d) thereby forming the klippen structures. The two klippen are separated by a synformal hinge line.

The proposed model can be compared with the structural map (Fig. 1). The label 'J' in Fig. 9d represents the junction between the leading frontal ramp of the MBT and the oblique ramp observed northeast of Raipur. The label 'OR' represents the oblique ramp along the Song River and 'T' represents the trailing frontal ramp northeast of Kishanpur. Labels S and B represent the present position of Satengal and Banali Klippen.

The above model is based on inversion tectonics. Since the amount of initial extension in unknown, there is displacement out of the tectonic transport plane, and subsurface data north of the Main Boundary thrust are not available and hence no effort can be made to balance the section.

CONCLUSIONS

In the present tectonic set-up, the Main Boundary Thrust bifurcates into an oblique ramp and a blind thrust. The blind thrust supports two arcuate back thrusts and is partially exposed as a result of dome and basin pattern thereby forming the klippen in the Mussoorie Syncline.

Rock distribution and structural features similar to the Mussoorie syncline area occur in other part of the lesser Himalaya as well (e.g. Sexena and Rao, 1975; Kumar and Dhaundial, 1997; Bali and Bhattacharya, 1988). It is suggested that some of these klippen might have formed by the similar process. Hence the field data and the map patterns should be critically reviewed in the light of the present work in order, to understand the geometry of thrusts and amount of maximum displacement.

Acknowledgements- Several discussions were made with S.J. Sangode, M.I. Bhat, T.N. Bagati, N.S. Gururajan, and S.S. Bhakuni. The project was funded by the Dept. of Science & Technology, Government of India. Help received from the India Bureau of Mines is also acknowledged.

REFERENCES

Auden, J.B., 1937. The Structure of the Himalaya in Garhwal. *Records Geological Survey of India* 71, 407-433.

Bali, R. and Bhattacharya, A.R., 1988. Geological and Structural studies of the rocks of the Dwarhat- Chaukhutia area, Kumaun Lesser Himalayas with special reference to the north Almora fault. *Geoscience Journal* 9, 215-230.

Coward, M.P., 1983. Thrust tectonics, thin skinned or thick skinned, and the continuation of thrusts to deep in the crust. *Journal of Structural Geology* 5, 113-123.

Coward, M.P., 1996. Balancing sections through inverted basins. In Modern Developments In *Structural interpretations. Validation and Modelling*, P.G. Buchanan and D. A. Nieuwland (eds.) *Geological Society Special Publication*, 99, 51-77.

Dubey, A.K., 1997. Simultaneous development of noncylindrical. Fold frontal ramps, and transfer faults in a compressional regime: experimental investigations of Himalayan examples. *Tectonics* 16, 336-346.

Dubey, A.K., and Bhat, M.I., 1986. The role of reactivation of pre-rift basement listric faults in the structural evolution of the Himalaya: an experimental study. In *Current Trends in Geology, Himalayan Thrusts and Associated Rocks*, P.S. Saklani (ed.), 9, 265-290.

Dubey, A.K. and Paul, S.K. 1993. Map patterns produced by thrusting and subsequent superposed folding: model experiments and example from the NE Kumaun Himalayas. *Eclogae Geologicae Halvetiae* 86-839-852.

Gansser, A., 1964. *Geology of the Himalayas*, Interscience, New York.

Heim, A. and Gansser, A., 1939. Central Himalayas: geological observation of the swiss expedition in 1936. *Mémoires de la Socie'te' Helev'tique des Sciences Naturelles* 73, 1-245.

Jain, A.K. 1972. Structure of Bidhalna - Pharat Windows and Garhwal Thrust Unit, Garhwal, U.P., *Himalayan Geology* 2, 188-205.

Jayangondaperumal, R. and Dubey, A.K. 2001. Superposed folding of blind thrust and formation of klippen: results of anisotropic magnetic susceptibility studies from the Lesser Himalaya. *Jour. Asian Earth Sci.*, 19, 713-725.

Kumar, G. and Dhaundial, J.N. 1979. Stratigraphy and structure of "Garhwal Synform", Garhwal and Tehri Garhwal districts, U.P.: a reappraisal. *Himalayan Geology* 9, 18-41.

Middlemiss, C.S. 1887. Crystalline and metamorphic rocks of the Lower Himalaya, Garhwal and Kumaun. *Memoires Geological Survey of India*, 20, 134-143.

Pilgrim, G.E. AND West, W.D., 1928. The structure and correlation of the Simla rocks. *Memoir Geological Survey of India* 53, 1-140.

Rupke, J., Stratigraphy and Structural evolution of the Kumaun Lesser Himalayas, *Sedimentary Geology* 11, 81-265.

Saklani, P. S., 1993. *Geology of the Lower Himalaya*. IBS, Delhi

Saklani, P. S., Khanduri, H.C. and Singh, S., 1977 A *reappraisal* of geology of a part of Mussoorie Syncline around *Satengal*, Garhwal Himalaya. *Chayanika Geologica* 3, 95-103.

Saxena, S.P. and Rao, P.N., 1975. Does Almora Nappe exist ? *Himalayan Geology* 5, 169-184.

Valdiya, K.S., 1980. *Geology of the Kumaun Lesser Himalayas*, Wadia Institute of Himalayan Geology, Dehra Dun 291 pp.

Williams, G.D. Powell, C. M. and Cooper, M.A., 1989 Geometry and kinematics of inversion tectonics. In: *Inversion Tectonics*, (ed.) M.A. Cooper and G.D. Willaims, 44, pp. 3-15, Geological Society Special Publication.

DEVELOPMENT OF THRUSTS IN GARHWAL HIMALAYA

Arun K. Shandilya

Centre of Advanced Studies in Geology, Dr.H.S.G.University,
SAGAR, M.P.

ABSTRACT

An attempt has been made to suggest the classification of the Himalyan thrusts, on the basis of their relationship to the asymmetric fold limbs which they cut, Further it has been classified on the basis of hanging/ footwall zones. The various stages of development of thrusts in the Himalaya and tectonic model for thrusting have been described.

INTRODUCTION

The major structures of the Himalayan rocks in Garhwal Himalaya are characterised by folding, faulting and thrust system. The thrust tectonics separates the rocks of two different geological ages and at places separates the rocks of higher grade metamorphics from the low grade. The important various type of thrusts from south to north in Garhwal Himalaya are as -

(1) Main Boundary Thrust (MBT) (2) Krol Thrust (KT) (3) Garhwal Thrust (G.T.) (4) North and South Almora Thrusts (NAT & SAT) (5) Chail Thrust (CT) = Ramgarh Thrust (6) Jutogh Thrust = Main Central Thrust (MCT) = Munsiari Thrust (7) Vaikrita Thrust etc.

Although the study of these thrust of Himalayan belt has been carried out by a group of pioneer workers such as Heim and Gansser (1939), Saklani (1971-1993), Valdiya (1978, 1980), Jain (1971, 1987), Pati and Rao (1981, 1983), Kumar and Agrawal (1975), Agrawal and Kumar (1973), and reviewed by Fuchs and Sinha (1978), Raina (1978), and Shandilya and Saklani (1986). Saklani and Bahuguna (1983) and Thakur and Choudhary (1983), and author himself Shandilya (1987). Since that time, these concepts of Himalayan thrusting have been applied to numerous other parts of Himalayan belts and with this has come a plethora of descriptions of structures, to such on extent that no longer exists a commonly used terminology. These thrusts have been locally named according to the choice of workers in the various parts of Himalayan belt (Table 1).

An attempt has been made in this paper to correlate all the various thrust planes of the Himalaya, including various sectors and the process of their development has been

Table 1 : Correlation of Thrusts in Garhwal HImalaya.

E of Tons Area Dhoundial and Ali (1967)	Purola Area Pati et al (1981)	NE of Tons Ranga Rao (1972)	Bhagirathi valley. Saklani (1972)	Bhilangana Mandakini valley Pati & Rao (1983)	Alaknanda valley Kumar & Agrawal (1975)
Purola sheet	Crystalline Belt = Purola F mn.	Purola Central crystalline	Muliagaon Fmn Ramoli Group	Gangi Kedar Nath Fmn.	Central Cryst. Group Joshimath Fmn
Central Himalayan Thrust	Jutogh Thrust Dharmigad Fmn.	MCT	Mukhen Thrust	Ponwali Thrust	MCT
Nawgaon sheet or Barahat window zone		Chamoli window	Pratapnagar thrust sheet	Bhilangana F mn.	Chamoli Fmn.
Thrust Deoban	Chail Thrust /MC Nangaon Fmn Mandali Fmn Shali Fmn	Badian Thrust Nangori Syncline - Deoban & Rama Sarai sheets	Pratap nagar T. Kotga Balali group	Tharali Thrust Deoban Group	Garhwal - Pipal koti Fmn. or Calc.Group Ghat Fmn.
	Fault (CHAMI)	Tons Thrust	Dharkot Thrust	NAT	NAT
Simla Slate Morar - Chakrata	Morar - Chakrata Fmn.		Chandpur Group SAT	Chandpur Group SAT	Chandpur Group SAT
Krol fmn. Krol.Thrust Tal fmn. Krol Thrust	Krol fmn. Krol.Thrust Tal fmn. Krol Thrust	Krol fmn. Krol.Thrust Tal fmn. Krol Thrust	Krol fmn. Krol.Thrust Tal fmn. Krol Thrust	Krol fmn. Krol.Thrust Tal fmn. Krol Thrust	Krol Fmn. Krol.Thrust Tal Fmn. Krol Thrust
MBT	MBT	MBT	MBT	MBT	MBT
Siwalik	Siwalik	Siwalik	Siwalik	Siwalik	Siwalik

discussed. A recent review by McClay (1981), although not exhaustive, provided some useful definitions. In accordance with McClay's proposal, a thrust is considered to be a constrictional fault and the rock mass carried upon it is called a thrust sheet. In some what broader term 'slide' meaning a Synmetamorphic detachment surface or zone, carries no implication of displacement sense. This present paper is more concerned with the development of thrusts and their correlation in the entire Himalayan belt. Author has tried in this paper to apply the geometric principle, in deciphering Three Dimensional Thrust surfaces of Garhwal Himalaya. The principal lithounits in Garhwal are shown in Table-2.

Contd...

Table 2: Principal Lithounits in Garhwal Himalaya

Siwalik Group

——— Main Boundary Thrust ———

Krol Unit

——— Krol Thrust ———

Tal fmn.

——— Tal Thrust ———

——— Krol Thrust ———

Simla Slate / Blaini boulder/Chandpur fmn.

——— Garhwal Thrust ———

Nagthat fmn.

——— Amri Thrust ———

Schistose fmn.

——— Bijni Thrust ———

Lansdowne crystallines

——— Bijni Thrust ———

Schistose fmn.

——— Amri Thrust ———

Damtas

——— South Almora Thrust (SAT) ———

Dudatoli/Chandpur Group

——— North - Almora Thrust (NAT) ———

Garhwal Calc Group/Shali-Deobans Group

——— Pratap Nagar Thrust (PNT) ———

Pratap Nagar Quartizite / Berinag fmn

——— Main Central Thrust / Vaikrita Thrust (MCT) ———

Central Crystalline Group

——— Munsiari Thrust ———

Tethyan Sediment

Indus Suture Zone

Thrust surfaces

The process of thrusting, is characterised by movement of one body of rocks over another along a tectonic plane or surface. This was reported by Pilgrim and West (1928) and Heim and Gansser (1939), from Himachal and Garhwal-Kumaon Himalaya, e.g. Main Central Thrust and Main Boundary Thrust/fault.

The rocks which overlie a particular thrust surface are said to occur in the hangingwall and the rocks below it are in the footwall. The thrusts must cut up section in the transport direction (Dahlstrom, 1970, p.342) but they rarely do so, as a smooth plane, and in most cases they follow a stair case trajectory made up of "Ramps" and Flats (Fig.1). A flat is that part of a thrust surface which was horizontal at the time of fault initiation. When a thrust sheet is transported over a previously undeformed sedimentary sequence a flat will be parallel to the bedding. Flats are connected by relatively steep ramps along which thrusts climb up section in the trasport direction. Ramps are classified not only by their orientation (Dahlstrom, 1970, p.345) but also as they cut off the hangingwall or footwall of a thrust surface. Those ramps present in the hanging wall are termed as Hanging wall ramps and those of the foot wall are Foot wall ramps (Fig.2,3,4).

It should be realised that although thrusts climb up section in the transport direction

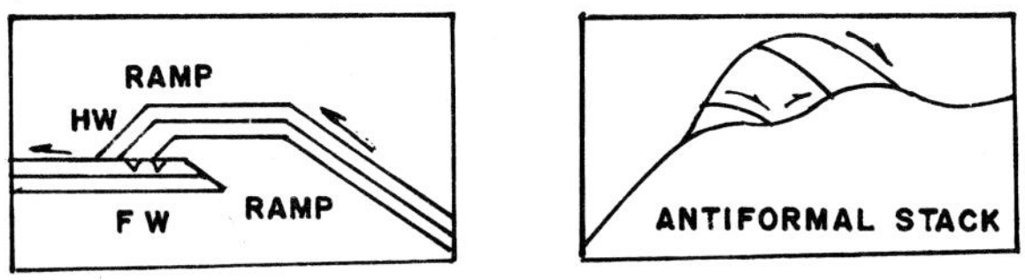

Fig. 1. Ramps and antiformal stack.

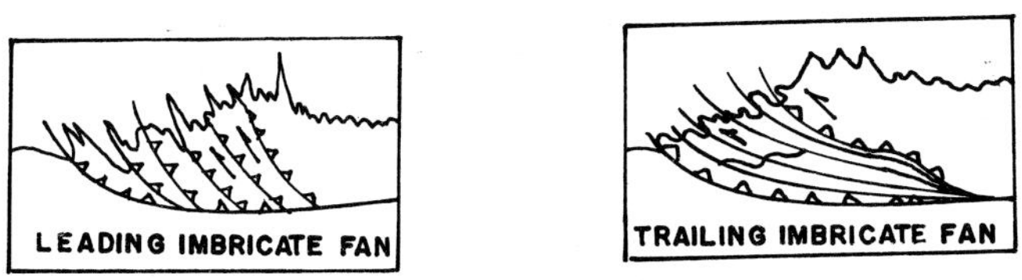

Fig. 2. Leading imbricatefan and trailing imbricate fan.

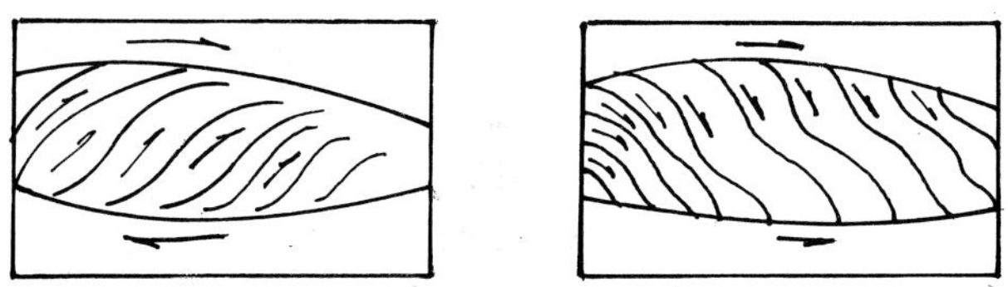

Fig. 3. Imbricates

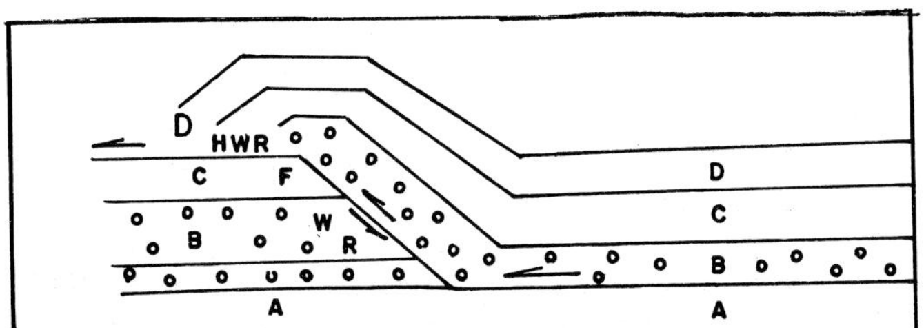

Figure 4: Relation between frontal hanging wall (HWR) and Footwall (FWR), Beds are in ascending order A to D.

yet they can migrate both to higher and lower levels across the transport direction (Dahlstrom, 1970, p.343, Eliott, 1976, Elliott and Johnson, 1980, p.70). When thrusts do not project up to the topographic surface they are referred to as 'Blind thrusts' (Thompson, 1981, p.452)

Thrust sequence in Garhwal Himalaya

There are two possible propagations and displacement sequences to explain the development of any array of thrust surfaces which diverge in the direction of tectonic transport. One arises if a younger thrust develops in the foot wall of an older thrust, the other arises if a younger thrust develops in the hanging wall of an older thrust. In the first situation the older movement plane will be carried by further displacement on the lower thrust. This is termed "Piggy back thrust" propagation (Dahlstrom, 1970, p.349) and in an array of thrusts the highest will represent the earliest displacement, while the lower ones will represent the last displacement. (Dahlstrom, 1970, p.354, Cooper, 1981, p.228).

The Piggy back thrust sequences result from new thrusts being developed in the footwall of what was called previously an active thrust. This occurs by progressive failure of footwall ramps (Cooper, 1981, p.228) and abandonment of a part of the old thrust surface. If the failure continues an *imbricate stack* or *Schuppen structure* (Peach et al., 1970) will be developed.

A series of horses together with their bounding thrusts are collectively known as a "Duplex" (Dahlstrom, 1970, p.353).

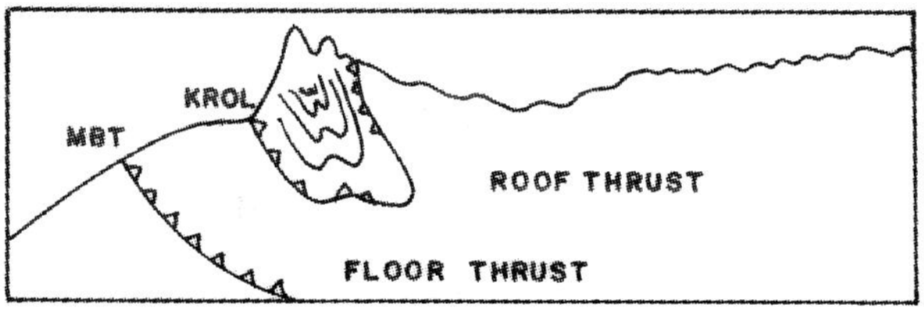

Figure 5: Rishikesh area (roof and floor thrusts).

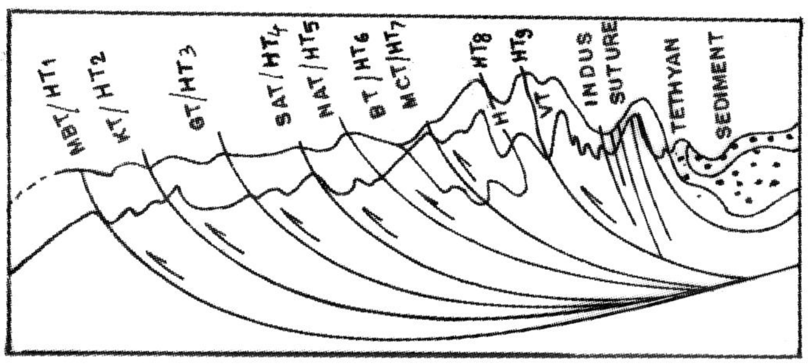

Figure 6: Cross Section of Imbricate fan at two levels of erosion in Garhwal Himalaya.

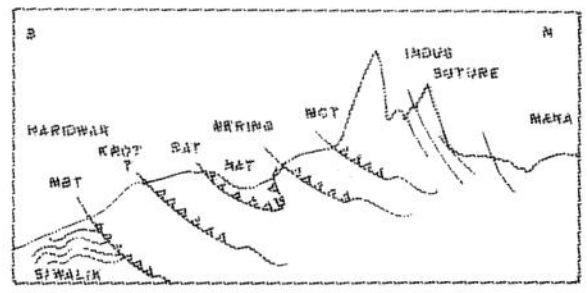

Figure 7: Haridwar-Badrinath-Mana Section showing various thrusts in Garhwal

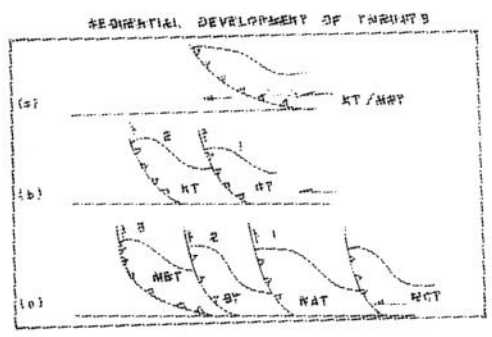

Figure 8: Piggy back thrust sequence

The highest thrust surface is termed as the 'Roof thrust', and the lowermost retaining thrust forms the 'Floor thrust' (Dahlstrom 1970, p.357). The various Himalayan Thrusts (Fig. 5,6,7,8) in Garhwal are classified as -

1. Crystalline/ Ramgarh Duplex - Vaikrita Thrust (Roof Thrust)
   Munsiari Thrust (Floor Thrust)
2. Dudatoli /Almora Duplex - Pratap Nagar Thrust - Roof Thrust
   NAT & SAT - Floor Thrust.
3. Lansdowne Duplex - Bijni Thrust - Roof Thrust
   Amri Thrust - Floor Thrust
4. Krol Duplex - Tal Thrust (Roof Thrust)
   Krol Thrust (Floor Thrust)
5. Bhilangana/ Mandakini Imbricates - Jutogh Thrust
   Chail Thrust
6. Alaknanda /Pinder Imbricate
   Nagnath Pokhari Imbricate -
7. Yamuna Imbricates - Tons thrust
8. Bhagirathi Imbricates - MCT=Mukhem Thrust
   Partap Nagar Thrust
9. Khirshu Duplex - Khirshu Thrust - Roof Thrust
   NAT - Floor Thrust
10. Ganga valley - Krol Thrust - Root Thrust
    Main Boundary Thrust-Floor Thrust
11. Garhwal Tectonic Window -

Major Thrusts in Garhwal

Though a difference of opinion prevails regarding the nomenclature, majority of workers including Saklani (1971), Valdiya (1980), Schwan (1980), Saklani and Bahuguna (1983) and Shandilya (1986) described following thrust sheets in Garhwal Himalaya :

		Munsiari Thrust
Crystalline Thrust Sheet	1. Central Crystalline =	
		Vaikrita Thrust
		NAT
	2. Almora = Dudatoli	
		SAT
	3. Garhwal = Ramgarh (=Chails)	
Metasedimentary thrust Sheet	4. Berinag Thrust	
	5. Krol Thrust	
	6. Jaunsar Thrust	
	7. Main Boundary Thrust (MBT)	

CENTRAL CRYSTALLINE THRUST SHEET

The term Main Central Thrust (MCT) was introduced by Heim and Gansser (1939) for the tectonic plane, which separates the epi, meso, and kata zonal metamorphics of the Greater Himalaya from the less metamorphosed sedimentary sequence of the lesser Himalaya. According to them the so called Central crystallines represented about 10-20 km. thick mass of metamorphosed sediments and intrusive bodies. Present author (Shandilya, 1984) considered that the present line of this tectonic dislocation represents the root zone of an extensive sheet of crystalline rocks. The crystalline rocks of Almora, Baijnath, Dudatoli, Lansdowne in Kumaon and in Garhwal and Narkanda in Himachal were considered to be the parts of the thrust slices from the north.

The MCT of Garhwal and Himachal Himalaya has been delineated and studied by many workers e.g., Valdiya (1978), Saklani (1971), Saklani and Bahuguna (1983), Raina (1978), Pati and Rao (1983), Agarwal and Kumar (1980) and Jain (1971) and Shandilya (1984) etc. In the MCT zone two tectonic planes are recognised along which the Central Crystalline rocks are thrust over the allochthonous Berinag (=Pratapnagar) quartzite of Dudatoli (=Chandpur) group and which rest over the Deoban = Garhwal group. Out of these two tectonic planes, i.e., the lower one wherever present separates the Central Crystalline Group from the Deoban=Garhwal, while higher tectonic plane divides the former into two tectonic units; the lower and the upper, which are continuation of Chail and Jutogh formations of H.P. Himalaya, respectively. Hence the higher tectonic plane at the base of the Jutogh Fmn. can be correlated with 'Jutogh Thrust' and the lower at the base of the Chail tectonic unit can be correlated with the Chail thrust. In the Tons, Yamuna, Bhagirathi, Bhilangana, Mandakini, Alaknanda and in Pinder valleys the two tectonic planes have been traced. The general trend of these thrusts is almost WNW-ESE to NNW-SSE showing northwardly dips varying from 15° to 23°.

In the south of area, the 'higher tectonic unit overlaps the lower one and as a result Jutogh Fmn. directly rests over the Deoban (autochthons). Between the Bhagirathi and Bhilangana valleys the high angle Balganga fault offsets the lower tectonic contact (=Chail thrust). From Chirpatiyakhal upto the Mandakini it trends in ENE-WSW and dips 20°-30°. The Jutogh thrust (=MCT) trending NW-SE direction from the Bhagirathi to Mandakini river dips towards the NE. This tectonic plane, separating the Crystalline Group from the Deoban Group of rocks trends roughly E-W between the Mandakini and Alaknanda rivers. From Alaknanda river it takes a turn and its strike is almost NNW-SSE along the Pinder river and dips at low to moderate angles due NE. From Pinder river onwards again it takes NE-SW direction upto the Kali river in Kumaun-Himalaya. In the Kumaun Sector, Valdiya (1980) suggested a new name, Munsiari and Vaikrita Thrusts to MCT and differentiated these two on the basis of the grade of metamorphism.

Almora Thrust

Heim and Gansser (1939) have designated the two flanks of the asymmetrically synformal thrust sheet as North Almora Thrust and South Almora Thrust (NAT and SAT) respectively, Their roots lie at the base of the Great Himalayan ranges implying that it is the prolongation of the Munsiari Thrust. This thrust sheet defines the tectonic base of the great Almora nappe built up of medium grade metamorphics and granitoids. In the Kumaun region the South Almora Thrust has been worked out by the various workers.

In the northern flank of this thrust sheet e.g. the North Almora Thrust has also been described as the Saryu Thrust by Valdiya, (1962a), Srinagar Thrust by Kumar et al., (1974), Ghosh (1973, 1974), Kumar and Agrawal (1975), and Shandilya (1984). This thrust sheet is described here as a floor thrust. Another thrust is the Pratapnagar Thrust (=Berinag/ Khirsu thrust) which brings the quartzites over the rocks of the Garhwal Calc (=Deoban) Group. This thrust represents the Roof Thrust.

Lansdowne/Garhwal Thrust

The rocks of the Lansdowne area are characterised at their base by the presence of the Amri Thrust and the Bijni Thrust. The Garhwal Duplex is characterised by the presence of the Amri thrust as Floor Thrust and Bijni Thrust as the roof in Lansdowne, Satengal-Banali areas from the frontal chain and the Askot-Baijnath-Nandprayag klippen make the chain exposed between the Almora Nappe and its root at the base of the Great Himalayan range. Amri thrust was recognised by Auden (1937) in the Lansdowne hill, which separates the medium grade mica schist capped by the gneissose Lansdowne granite from epi grade metasediments of Bijni thrust sheet. Bijni nappe is thrust over the schists which was described as Garhwal Thrust by Auden (1937). In the Kumaun area these thrusts have been renamed as Baijnath and Askot Thrusts forming Baijnath duplex in the Kumaun Himalaya.

Ramgarh Thrust

The rocks of Ramgarh group, thrust over the Nagthat formation of Krol belt has been designated as the 'Ramgarh Thrust' in Kumaon Himalaya (Pandey, 1950) in the north of Chukka on the Kali river section, through Diuri on the Tanakpur-Champawat bridle path. Valdiya (1962) described this thrust as Ladhiya Thrust in the Ladhiya valley. This thrust has been equated with the Chail Thrust of Pilgrim and West (1928) in the western most extension of the Ramgarh-Chail Thrust. In the Central Nepal this thrust delimiting the base of the Chail Nappe-2 of Fuchs and Sinha (1978), or the "Lower Midland Formation (Lefort, 1975, Pecher, 1976, Hashimoto et al., 1973) is analogous to the Bhatwari (=Chail) Thrust. In the Darjeeling area it has been described as Daling Thrust, which represents the eastern most extension of the Chail/Ramgarh thrust, while in Bhutan it is called as the Samchi Thrust.

Krol Thrust Sheet

Auden (1934) designated this tectonic plane at the base of the Krol succession as the 'Krol Thrust', which brought various litho-stratigraphic units of the Krol Units (namely, Mandhali, Chandpur, Nagthat, Blaini, Krol, and Tals) over the three Tertiary formations, namely Subathu, Dagshai, and Kasauli on the Siwaliks. In the Kumaon Himalaya, excepting the short stretch in the west of Yamuna, where a narrow horizon of the Subathu intervenes, everywhere the Siwaliks are overridden by the six units of the Krol belt. In a way, there is a thrust which serves as the boundary of Siwalik sub province. At some places the Chandpur Group rocks are brought by the M.B.T. over the Tertiary rocks in Garhwal. The Krol thrust has overlapped the concealed M.B.T. in the Yamuna valley and in Kotdwar section in Garhwal.

Main Boundary Thrust

A tectonic contact between the Pre-Siwalik or Pre-Tertiary with Siwalik and the Siwalik sedimentation has been described as Main Boundary Thrust by Middlemiss (1888). This thrust fault that has brought the lower Eocene Subathu Formations over the younger lower Siwalik sediments exposed in between Dakpathar-Dehradun-Rishikesh-Kotdwar in Garhwal called as MBT/MBF. This thrust is the lower most thrust in a succession of a large number of lesser Himalayan thrusts, it is overridden by the higher Krol thrust (Roof Thrust). A majority of workers including myself have however described the MBT=Krol thrust where Krol/Subathu/Chandpur /Blaini rocks etc.are thrust over the Siwalik rocks, in Garhwal-Rishikesh-Kotdwar areas (Fig.5).

EVIDENCES OF THRUSTING

(1) *Older rocks resting over younger rocks.* The Central Himalayan Crystalline rocks (1800 M.Y., Bhanot et al., 1976) rest over the stromatolite bearing Tejam/Deoban formation of late Pre-cambrain to early cambrian age. This can be possible only due to thrusting. (2) *Increase in the grade of metamorphism.* (3) *Injection of Basic to acid igneous rocks.* (4) *Abrupt modification of lithology* and (5) *Cataclasis,* at Sainj in Bhagirathi, and at Thayeli, in Bhilangana, at Kund in the Mandakini, at Helang in Alaknanda and South of Kunwari khal in Loharkhet areas, and Munsiari area near Bauna and Pyan in Kumaun. Along the thrust contacts the quartzites are highly sheared and pulverised. The rocks are sheared and marked by elongated augen and bent mica streaks. The development of mylonite and phyllonite is also observed. The feldspars are flattened. The thrust zone is

marked by the development of 'a' lineation plunging towards NE to NNE. (6) *Variation of Mesofabric elements* - (a) Truncation of strike trend (b) Abutment of large scale fold (c) Abutment of faults (7) *Geomorphic evidences :* breaks of slopes, nickpoints well developed all along the thrust contacts. Landslides and neotectonic activities were observed commonly along the thrusts. and (8) *Tectonic elimination of litho units.*

DEVELOPMENT OF THE HIMALAYAN THRUSTS

The tectonic framework of the Himalaya can not be treated as one event, when there were numerous evidences to reveal the orogenic episodes prior to the Cretaceous-Tertiary orogeny. The Himalayan orogeny can not be worked out in isolation and with a bias, particularly when geological evidences strongly point to a conclusion that the Indian shield was very close to its present position since the Upper Proterozoic. Even the Gondwana flora once considered to be the indigenous to Gondwana land, is a very strong evidence in favour of drift tectonics, and is seen in Tibet much closer to the Protagonists of the drift and plate tectonics, as well as to the Indus Suture Zone (ISZ), where subduction is supposed to have taken place to form the northern boundary of the Indian plate. The collision of the Indian plate with Tibetian plate (a part of Eurasia) is supposed to have formed the Himalaya According to Dewey and Bird (1970); Gilluly (1971), the folded thrust sheets / nappes and wide spread shallow seismic activity are in support of the collision of the plates.

However the gravity data have been successfully applied by Qureshi et al., (1974) in explaining the causes of the vertical block uplift of the Middle-Central Himalaya. The suggestions pertain to the thickening of a basaltic layer having roots beneath the Central Himalaya as a possible cause. A belt of positive isostatic anomalies are also characterised in the Central Himalayan region.

In the start of Himalaya mountain the part of northern Indian Peninsula was subiected to the orogenic activity and the northern rocks of Main Central Crystallines and part of Peninsular India were folded and thrusted on the parautochthonous metasedimentaries along the MCT. The sedimentary rocks deposited along the sea of Tethys, got folded in the later phase of Himalayan orogeny. The recumbent folding in older low to high grade metamorphic rocks were also thrust in the form of Almora and Dudatoli Crystalline Duplexes in the middle Himalaya of Garhwal and Lansedowne/ Satengal nappes in the lower Himalyan region. The rocks of Blaini, Krol, Tal, Chandpur, are thrust over the Siwalik rocks along the MBT/MBF in the outer Himalaya.

AGE CORRELATION OF THE THRUSTS

Structural analysis of the thrusts in Himalaya with the overlying phanerozoic sequence of this Himalayan zone and radiometric data on a regional scale indicate that the Central Crystallines represented the Pre-Cambrian Basement (1800-2000 m.y). These thrusts have been reactivated during 500 m.y. and 12-18 m.y. old orogenic phases. Sarkar (1980) dated the Pre-Himalayan granitic gneiss and granites as old as 2000 m.y. and 500 m.y. respectively, which suggest that the Central Crystallines of the Himalaya represent the northern most extension of the basement of the Indian shield, incorporated into the Himalaya orogeny. As the Himalayan rocks document the four phases of deformation (D_1, D_2, D_3 and D_4) the over printing of four generations of fold, F_1 (E-W), F_2 (NW-SE), F_3 (NE-SW) and F_4 (N-S) and related major and minor structures on the Central Crystallines and Para-autochthonous rocks, indicate the crustal shortening.

The regional metamorphism produced leuco-granite. The Main Central thrust, a post D_2 and Pre D_3 structure which was initiated as a ductile crustal shear in the basement and later developed into a southward moving major thrust was juxtaposed over the Central Crystalline belonging to zone of imbrication. The environment of the development of the folds in the Central Crystalline, Ramgarh, Almora crystalline, Chandpurs, Lansdowne crystalline rocks etc. were at the deeper levels accompanied by metamorphic changes, whereas the sedimentaries of Lesser Himalayan zone were folded into large folds by buckling of layers in the semibrittle environment at shallower depth. The MCT, Chail, NAT, Jutogh and SAT thrusts of Middle Miocene age, resulted to Miocene deformation of the Himalayan orogeny.

CONCLUSION

The rocks in the Garhwal Himalaya were subjected to four phases of deformation during the Himalayan orogeny The Central Cyrstalline rocks all along the Himalayan Central axis are parts of the extension of the northern part of Peninsular India. It is equivalent to the Precambrian granitic basement 1800-2000 m.y. The tectonic pressure from the Eurasian plate on the Sea of Tethys developed the recumbent folds in the crystalline rocks which are thrust over the metasedimentaries in Garhwal. By the orogenic process and Himalayan uplift the various thrusts in the area during the Himalaya were formed.

The compression caused by the northward movement of Indian plate formed the Krol-Thrust. The Almora thrust got folded in the later phase of deformation. Amri and Bijni Thrusts in Lansdowne area are also the southern most extensions of Almora thrust.

ACKNOWLEDGEMENT

The critical discussions and comments of Prof. P.P.Roday, of the University, Sagar (M.P.) and Dr.A.K.Dubey, Scientist W.I.H.G. Dehradun were quite encouraging, and the author is thankful to both of them. Author is thankful to UGC, for providing financial assistance to carry out this work.

REFERENCES

Agarwal, N.C., and Kumar, G., (1973) Geology of Upper Bhagirathi and Yamuna valleyes, Uttarkashi district, Kuman Himalaya Himalayan Geology, 3, 1-23.

Ashgirie, G.D. (1972) The problem of the origin of continental earth crust., Int. Geol. Cong. Section. 3, Tectonics Montreal, 3-9.

Auden, J.B., (1937) Structure of Himalayas in Garhwal. Rec. Geol. Surv. Ind., 71.

Bhanot, V.B., Bhandari, A.K., Singh, V.P. and Bansal, A.K. (1976b). Rb-Sr whole rock isochron age for Bandal granite, Kulu Himalaya H.P., Him. Geol. Sem. New Delhi (1976).

Boyer, S.E. and Eliott David (1982) "Thrust system". Am. Asso. Petrol. Geol. V.66 No.9. p.1186-1229.

Butler, R.W.H. (1981) Terminology of structures of thrust belt, Geol. Soc. Lond.

Butler, R.W.H. (1983) Hangingwall strain : A function of duplex shape and footwall topography - Tectonophysics.

Cooper, M.A. (1981) The internal geometry of Nappe Criteria for models of emplacement. In thrust and nappe Tectonics, Eds. McClay, K.R. and Price, N.J. special publs, Geol. Soc. Lond, 9, 225-234.

Dahlstrom, C.D.A. (1970) Structural Geology in the Eastern margin of the Canadian Rocky Mountains. Bull Can. Petrol. Geol. 18, p.332-306.

Dewey, J.F. and Bird, J. (1970) Mountain belts and new global tectonics, Jour. Geophys. Res., 75, 2625-2647.

Elliott, D. and Johnson, M.R.W. (1980) Structural evolution in northern part of Moine thrust belt, N.W., Scotland. Trans. R. Soc., Edinb : Earth Science, 71, 69-96.

Elliott, D. (1976) The energy balance and deformation mechanism of thrust sheet, Phil, Trans. R. Soc. A283, 289-312.

Fuchs, G. and Sinha, A.K. (1978) The tectonics of Garhwal Kumaun lesser Himalaya. Jb. geol. B.a., 121(2), 219-224 Wien.

Ganguly, S. Jagtap, P.K. (1966) A note on structure of Garhwal Himalaya, Foot Hills, Science and Culture, 32.

Ghosh, A. (1973) Tectonic Evolution of Lesser Himalaya of Kumaun U.P. Proc. Seminar Geodyna. Him. Region. NGRI. Hyderabad, 136-147.

Ghosh, A. (1974) Structural and metmorphic-history of Almora group. Him. Geol. 4, 171-194.

Gilluly, J. (1960) A folded thrust in Nevada - Inference as to time relationship between folding and faulting. Am. Journ. Sci. 258 A.P. 68-79.

Hashimoto, S. Ohta Y. Akiba (1973) Geology of Nepal Himalaya (eds) Him. Comm. Hokkaido Uni. Sapporo. Japan.

Heim, A. and Gansser, A., (1939) Central Himalayan geological observation of swiss expedition, 1936, Mem. Soc. Helv. Sci. Nat., 73(1), 246, Zurich.

Jain, A.K. (1971) Stratigraphy and tectonics of lesser Himalayan region of UP. Garhwal Himalaya, Him. Geol., 1, 25-28.

Kumar, G. and Agarwal, N.C. (1975) Geology of Srinagar-Nandprayag area, Alaknanada valley, Chamoli & Tehri Garhwal, Distts. of Kumaon Himalaya. Him. Geol. 5, 29-59.

Lefort, P. (1975) Himalayas the collided range. Am. Jour. Sci. 275A, 1-44.

Mc Clay, K.R. (1981) What is thrust ? what is Nappe ? In. Thrust & Nappe tectonics (Ed. Mcclay K.R. and Price, N.J. Special Publ. Geol. Soc. Lond., 9. 7-9.

Middlemiss, C.S. (1988) Section III Rec. Geol. Surv. India 21, 11-128.

Pati, U.C. and Rao, P.N. (1983) Geotectonics of M.C.T. of UP. Himalaya. In: Saklani, P.S. (Ed.) Himalayan shears, 68-83, Himalayan Books, N. Delhi.

Pilgrim, G.E. and West, W.D. (1928) The structure and correlation of Simla rocks. Mem. Geol. Surv. Ind. 53, 1-140.

Powell, C. Mc, A. and Conaghan, P.J. (1973) Polyphase deformation in Pheneorzoic rocks of Central Himalayan Gneiss, N.W. India, Jour. Geol. 81(2) 127-140.

Qureshi, M.N. (1971) Geophysical investigation in the Himalaya. Him. Geol., 1, 165-177.

Qureshi, M.N., Venkatachalam, S. and Subramanyam, C. (1974) Vertical Tectonics in middle Himalaya, an appraisal from recent gravity data. Geol. Soc. Amer. Bull. 85, 921- 926.

Raina (1978) A review of stratigraphy and structure of U.P. In: Saklani P.S.(Ed.). Tect. Geol. of Himalaya, TTP, New Delhi, 79-111.

Ranga Rao, A. (1972) Traverses in Himalaya of UP. Misc. Pub. Geol. Surv. Ind. 15, 31-44.

Saklani, P.S., (1971) Structure and tectonics of the Pratapnagar Area, Garhwal Himalaya, Him. Geol. N. Delhi, 1, 75-91.

Saklani, P.S., and Bahuguna, V.K. (1983) Main Central Thrust Zone and associated Imbricate structures in Chhatera Area, Garhwal Himalaya, In : Saklani P.S., (Ed.), Himalayan Shears, 1-9, Himalayan Books, New Delhi.

Sarkar, S.N., Reddy, U.S. and Nair, P.K.R. (1965) Tectonic pattern of part of Almora Nappe zone around Almora, Dr.D.N. Wadia Comm. volume, Min, Metal. Instt. Ind. Calcutta, 668-691.

Saxena , S.P. and Rao, P.N. (1975) Does Almora Nappe exist ? Himalaya Geol., 5, 169-184.

Schwan, W. (1980) Shortening structures in Eastern and NW Him. rocks. Today and Tommorow Publ. New Delhi 62.

Shandilya, A.K., (1984) Studies on structures and morphotectonics between Srinagar thrust and Alaknanada fault, Garhwal Himalaya India. (Unpublished D.Phil thesis), Garhwal University, Srinagar. Garhwal, (U.P.)

Shandilya, A.K., (1987) Lithostratigraphy and structure of Krol thrust sheet Dugadda-Satpuli area. Garhwal Himalaya. Conf. on Krol. Geol. Garhwal University, Srinagar. 71-88.

Shankar, Ravi, and Ganeshan, T.M. (1973) A note on Garhwal nappe. Him. Geol., 3, 72-83.

Thompson, R.I., (1981) The nature and significance of large blind thrust within the northern rocky mountains of Canada. In . thrust and Nappe tectonics (ed. McClay, K.R. and Price, N.J.) Spec. Publ. Geol. Soc. London.

Valdiya, K.S. (1980) Geology of Kumaun Lesser Himalaya, W.I.H.G. Dehradun.

West, W.D. (1939) Structure of the Shali Window near Simla. Rec. Geol. Ind. 74, 133-163.

METAMORPHIC AND DEFORMATION HISTORY OF THE CENTRAL CRYSTALLINES OF THE UPPER ALAKNANDA VALLEY, GARHWAL HIMALAYA, AND ITS BEARING ON INVERTED METAMORPHISM

Surya Pratap Singh*, Prem Swarup Saklani, Alan Bruce Thompson*****

*Department of Geology, Bundelkhand University, Jhansi - 284128 (U.P.)
**Department of Geology, Delhi University, Delhi - 110007,
***Inst. Min. Pet., Dept. Erdwissenschaften, ETH Zurich, CH-8092, Switzerland

ABSTRACT

The metamorphic rocks of the Vaikrita Group of the Badrinath area, witnessed three episodes of metamorphism and five phases of deformation. The mineral assemblages suggest a medium-pressure Barrovian-type metamorphism (M1) which began at D1 deformation , and reached its culmination at D2- deformation. This was followed by granitisation, partial melting, and formating of augen-gneisses in the late stage of metamorphism. The D3- deformation is related to large scale folding after the formation of rocks of M1 metamorphic . The second episode of metamorphism (M2) was mainly retrograde and localised only along major shear zones at the late stage of D3- deformation. However along the Main Central Thrust (MCT) and in a few shear zones a prograde nature of metamorphism was noticed on the basis of growth of porphyroblastic garnet. The poorly developed contact aureole (M3) around Tertiary granite in the Central Crystalline zone is post-tectonic to D3 deformation and regional metamorphism (M1). The folding of the MCT thrust plane and development of transverse faults were caused by D4 and D5 phases of deformations. The last two phases of deformation however did not initiate crystallisation of new minerals but were related to late stage of retrograde metamorphism where coarse-grained apatite, tourmaline, andalusite-chlorite-biotite -bearing schist developed along shear zones. The inverted metamorphic zonation near the MCT in the Alaknanda Valley section is due to the cumulative result of polyphase metamorphism (M1 and M2) and deformation (D1 to D2) without the course of inverted thermal gradients.

INTRODUCTION

The metasedimentary rocks of northern part of the Central Crystalline zone (CCZ) along the upper part of the Alaknanda valley belong to Vaikrita Group.These rocks generally represent the highest grade of metamorphism in the area and consist of granitoids, migmatites, gneisses, calc-silicates and metamorphics. The southern part of the Vaikrita Group shows thrusted contact i.e., Vaikrita thrust (Valdiya 1980) or Main Central Thrust-I of (Saklani et al., 1991) associated with low-to medium-grade metamorphics of the Munsiari Group (Fig.1, Table 1). The Vaikrita Group is intruded by two episodes of granitoids (Late Precambrian and Tertiary). Both the granitoids are nearly unmetamorphosed and occur in the northern

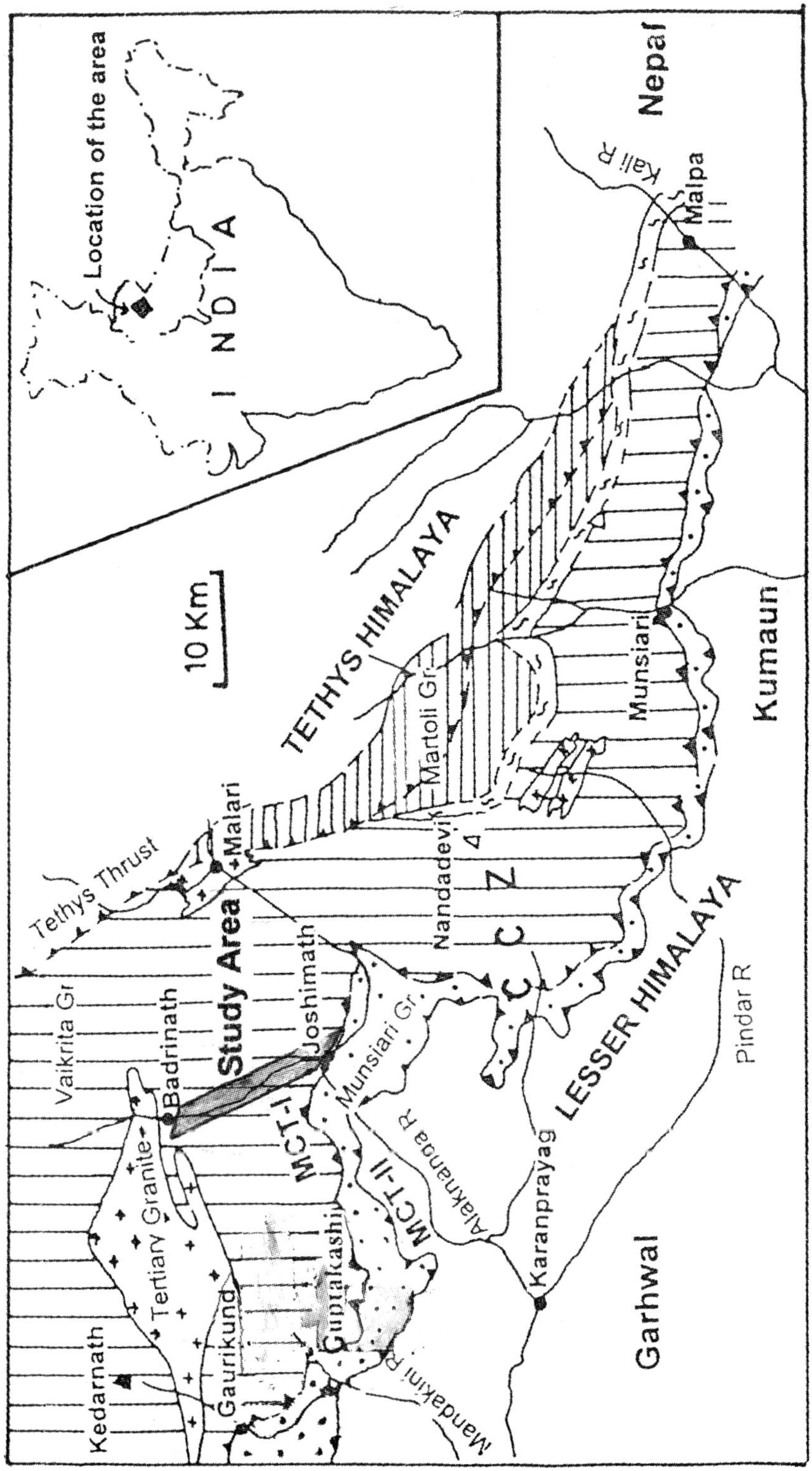

Figure1: General tectonic map of the Himalaya, after Valdiya (1980).

Table 1: Tectonic setting, Structures and Metamorphism of in the upper Alaknanda valley, Garhwal Himalaya, India

Group/Fm.		Rock Type	Metamorphic Grade	Structure
TETHYS THRUST				
VAIKRITA GROUP	Panduke-shwar Formation	Quartzites, micaceous-quartzite, metapelite band within quartzite	Low-grade (Bio and Gt Zone), Greenschist facies.	Sometimes sedimentary features (viz. Cross-bedding) present. Different kinds of fold in pelitic bands are present.
	Joshimath = Badrinath Formation	Gneisses: Gt-Bio-gneisses, Ky-Sill gneisses, migmatite, calc-silicates and quartzite-migmatites Schists: Ky-Mica schist, Gt-Mica schist, Mica schist, Quartzites	High-grade (Ky, $Sill_I$ and $Sill_{II}$ Zone) of M_1 metamorphism.	Thin quartzite band, pelitic band, calc-silicates from high-grade metamorphics have prominent S_2 schistosity, poor S-C fabrics, cross folds, hook folds.
	Jharkula Formation	Schists: Gt-Mica schist, Mica schist, Chlorite schist, Quartzites, mylonites and cataclasites	Low-grade, (but high-grade M_1 as almond shape within shear zone) M_2 metamorphism (up to Gt Zone).	Predominated by S-C and S-C' shear fabrics, and different kind of sheath folds are present.
MAIN CENTRAL THRUST - I (VAIKRITA)				
Munsiari Group		Chlorite schist, Mica schist, Quartzites, Chl-Mus Schist. Augen gneiss, Streaky gneiss, Porphyritic gneiss, migmatite, schist and quartzites	Low-grade metamorphics superimposed on grani toids, gneisses and schist. Recrystallisation along shear zone.	Mainly mylonites, ductile, brittle ductiles and brittle conditions. sheath folds, S-C and S-C' prominent.
MAIN CENTRAL THRUST - II (MUNSIARI)				
Garhwal Group		Phyllite, Quartzite, Slate, Carbonates	Very low-grade metamorphism.	Primary and secondary structures,S-C fabric

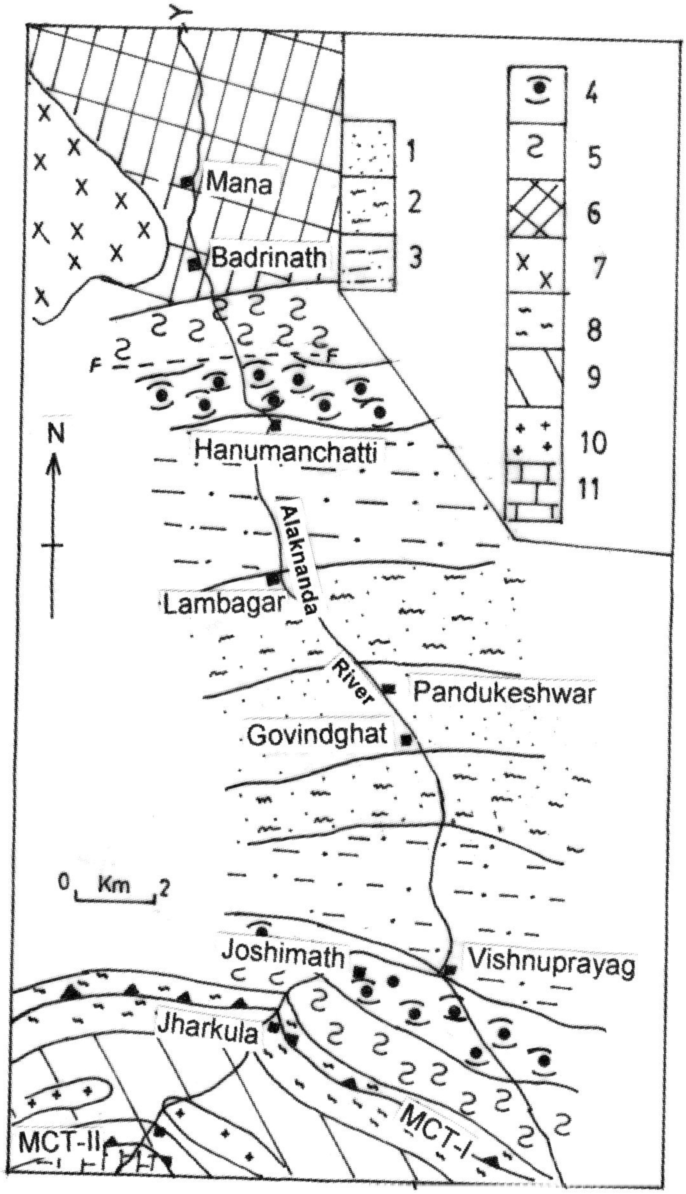

Figure 2: Geological map of the Upper Alaknanda valley, Garhwal Himalaya, India. 1: massive quartzite with thin pelitic bands; 2: micaceous quartzite with thin pelitic bands; 3: quartzite and schist bands; 4: gneisses and migmatite; 5: kyanite-sillimanite gneisses and garnetiferous gneisses; 6: calc-silicates with cordierite -sillimanite gneisses; 7: Badrinath granite; 8: MCT shear zone: kyanite-garnet-staurolite-mica schist and garnet-mica schist; 9: mica schist, chlorite-biotite schist, chlorite schist; 10: porphyritic granitoid and gneisses; 11: carbonates. (1-3: Pandukeshwar Fm; 4-6: Joshimath/Badrinath Fm; 8: Jharkula Fm; 9-10: Munsiari Group and 11: Garhwal Group)

part of the Central Crystalline zone (Singh et al., 1997). The Vaikrita Group is tectonically overlain by a thick sequence of Tethys sediments ranging in age from Late Precambrian to Mesozoic in the Garhwal Himalaya and adjoining areas (Valdiya, 1981; Sinha, 1989; Thakur, 1992).

It was firstly suggested by Heim and Gansser (1939), and developed further by Gansser (1964), that the Central Crystallines were evolved at greater depth and were brought to the surface along the Main Central Thrust (MCT), during intracontinental horizontal shortening in Tertiary time. The data from Rb-Sr, K-Ar, and Pb-Pb geochronology of granitoids and gneisses of some areas of the Garhwal Himalaya which are equivalent to the Central Crystallines, reveal three strong tectono-thermal events (1850 ± 50; 1150 ± 50; 500 ± 25 Ma) of Precambrian age and an event of Tertiary (25 ± 5 Ma) age (Singh et al., 1997 and Sorkhabi & Stump, 1993). Although the data are too scanty to reveal any generalised conclusion nevertheless it gives good evidences of the Precambrian and Tertiary orgenesis. The observation of clasts and conglomerates of metamorphic rock and granitoids from base of Tethys sediments (Fuchs 1968, 1992; Srikantia et al., 1978; Azami, 1994) also indicate that Precambrian metamorphism can not be ruled out in the Himalaya.

The polytectonic and polymetamorphic character of the Himalayan Crystallines is disputed but the records are not always well preserved due to intense shearing or thrusting during phases of deformation. The Joshimath-Badrinath area is perhaps a unique terrain where the relationship between deformation and metamorphism is traceable inspite of the influence of the superposed folding and crystallisation phases. It is feasible to establish a relationship between time of deformation and growth of porphyroblasts in the geochronological sequences from the study between the fabrics of the rocks and internal structure of the porphyroblasts.

REGIONAL GEOLOGY AND STRUCTURE

The metasediments belonging to the Vaikrita Group (Fig. 2) comprise the main lithology of quartzites, pelitic and semipelitic schists, gneisses, migmatites, amphibolites, calc-silicates and granitoids along the Alaknanda river section and can be divided into two formations (i) Pandukeshwar Fm and (ii) Joshimath (=Badrinath) Fm. The regional tectonic setting of the area is given in Table 1. The Vaikrita Group Crystallines in general show 30 to 40^0 dip towards NNE or NE; whereas rocks of mesograde (Jharkula Fm), associated with MCT-I shear zones show 20 to 30^0 dip towards N or NNE. The shear zones also comprise of small-scale folds, strain slip cleavage, sheath folds, S-C fabrics and silicified quartz veins.

The quartzites of Pandukeshwar Fm are usually coarse-grained and at places show graded and cross-beded structures (Virdi, 1986). The low-grade metamorphics of pelitic rocks sequences have been also rarely noticed from these quartzites. The quartzites of the Pandukeshwar Fm show gradational contact with pelitic and quartzofeldspathic schist and gneisses (Joshimath/Badrinath Fm) in the south and north of the Pandukeshwar village (Fig. 2). The high-grade schists, gneisses and migmatites are found mainly around Joshimath and Badrinath and have thrusted contacts (Vaikrita=MCT-I) with Jharkula Fm and Munsiari Group towards the south. The quartzite and pelitic bands have equal proportion (each about 10m thick) in kyanite-sillimanite zone. A thick shear zone (Jharkula Fm) at the base of the

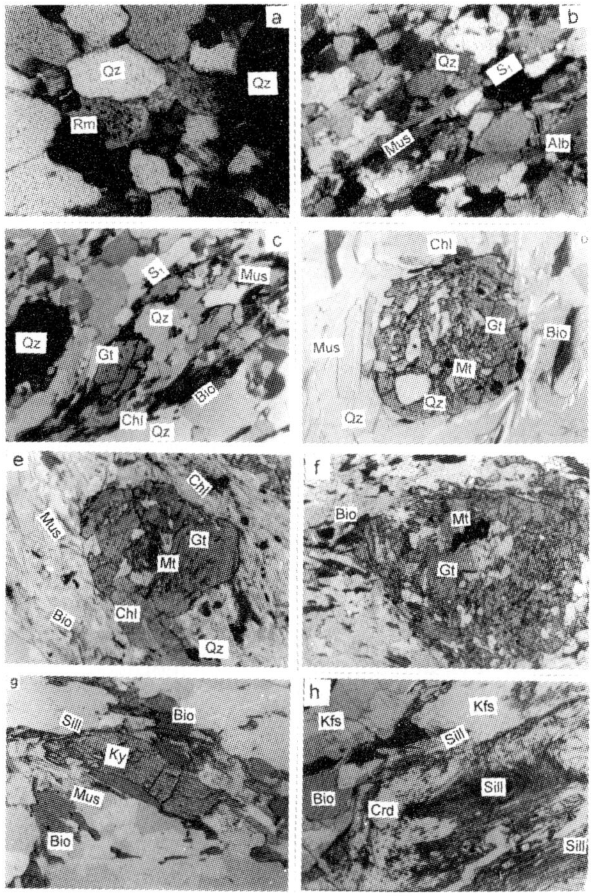

Figure 3: **(a)** Rounded to sub-rounded quartz (Qz) grain recrystallised in argillaceous matrix (Rm) and overgrowth quartz (Qtz) (300m north of Pandukeshwar). in crossed nicol 40x. **(b)** Overgrowth quartz (Qz) and crystallisation of muscovite and albite (Alb) from micaceous quartzite along S1 schistosity (north of Pandukeshwar). in crossed nicol 40x. **(c)** The crystallisation of garnet (Gt) on the S1 schistosity due to prograde reaction (1) (see text), (2 km south of Govindghat), under the gypsum plate 40x. **(d)** Sieved garnet (Gt) shows the reaction (1) from the pelitic band of Pandukeshwar Fm, nearly 2.5 km south of Govindghat containing inclusion of coarse rounded to angular quartz (Qz), chlorite (Chl) and magnetite (Mt). The muscovite (Mus) and biotite (Bus) along S2 schistosity. in PPL 40x. **(e)** Medium to coarse-grained quartz (Qz) and magnetite (Mt) inclusion in garnet (Gt) present in the matrix of chlorite (Chl), muscovite (Mus), biotite (Bio) and iron oxides on S2 schistosity. 3km south of Govindghat. in PPL 40x. **(f)** Xenoblastic, medium to coarse-grained quartz (Qz), magnetite (Mt) and chlorite (Chl) in the core of the porphyroclastic garnet (Gt) from pelitic band of quartzite. 4km south of Govindghat. in PPL 40x. **(g)** Coarse grained fractured kyanite (Ky) with rim of fibrolite (Sill) related to reaction (2) from Joshimath Fm. 500m north of Vishnuprayag. In PPL 40x. **(h)** High-grade metamorphic assemblage shows fibrolite (Sill) intermingled with cordierite (Crd), biotite (Bio) and K-feldspar (Kfs) from the gneisses of Badrinath Fm, near Badrinath. In PPL 40x.

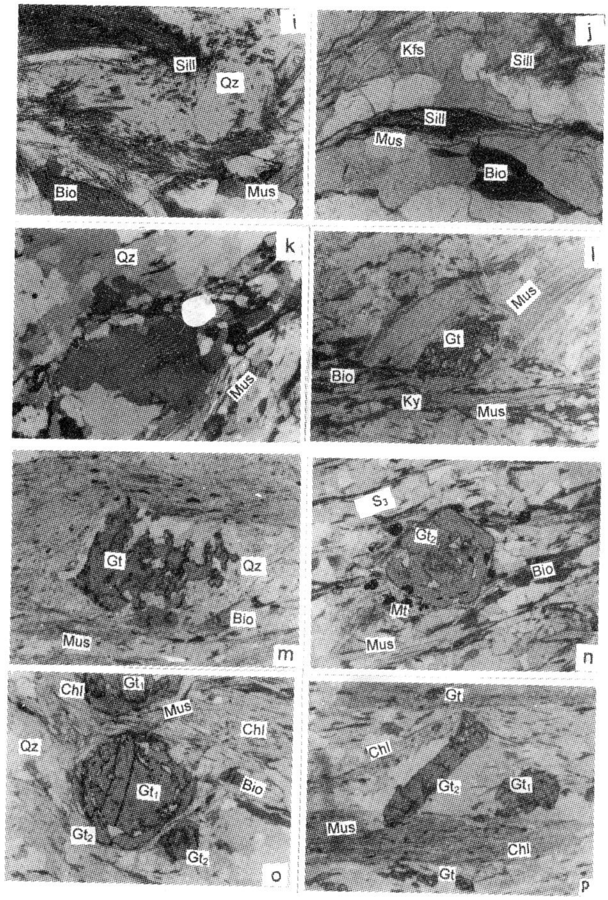

Figure 3: **(i)** Fibrolite sillimanite (Sill) in the matrix of biotite (Bio), muscovite (Mus) and quartz (Qtz), 3km south of Badrinath. The development of sillimanite is related to reaction (3)(see text). In PPL 40x. **(j)** Rim of muscovite (Mus) around sillimanite (Sill) in the matrix of quartz and K-feldspar (Kfs) is related to reversal of reaction (3), 1 km south of Badrinath in PPL 40x. **(k)** Flame perthite (Pert) and dynamic recrystallisation of porphyroclastic feldspar of high-grade gneisses of Joshimath Fm. The shear plane is defined by elongated muscovite (Mus) and quartz (Qtz). 2.5 km south of Joshimath. In crossed nicol 40x. **(l)** S-C fabric defined by xenoblastic sieved garnet (Gt), chlorite, muscovite, biotite on C surface and kyanite (Ky), biotite (Bio), muscovite on S-surface from Jharkula Fm. near MCT-I, in crossed nicol 40x. **(m)** Skeletal garnet (Gt) embedded in the matrix of chlorite, muscovite (Mus), biotite (Bio), quartz (Qtz) and iron oxides of shear zone, in PPL 40x. **(n)** Spiral garnet (Gt2) with ideoblastic rim from the matrix of biotite (Bio), muscovite (Mus), magnetite (Mt), and chlorite of shear surface (-C=S3) showing syn to post-M2 metamorphism of D3 deformation (Jharkula Fm), in PPL 40x. **(o)** Coupled garnet in the matrix of chlorite (Chl), muscovite (Mus), biotite (Bio), quartz (Qtz) from the Jharkula Fm. Earlier garnet (Gt1) either chloritised or rimmed by garnet (Gt2) See text, in PPL 40x. **(p)** Elongated garnet (Gt2) showing oblique to shear surface (C=S3). The medium to fine-grained garnet (Gt) embedded in the matrix of muscovite (Mus), chlorite (Chl) of shear plane. The garnet (Gt1) is xenoblastic and chloritised at the rim, in PPL 40x.

Vaikrita Group is interesting - which comprises of quartzites, garnet-chlorite schist, mica schist and chlorite-mica schist. These are coarse-grained and sometimes the high-grade metamorphics have been found in small, lensoidal almond shaped bodies within this shear zone. The shears near the Vaikrita thrust (MCT-I) are very wide (~500m), frequent and closely spaced where coarse-grained garnets are commonly embedded and show a late stage growth. The shears gradually become faint and thin and finally disappear nearly 3 km north of MCT-I.

The Vaikrita Group Crystallines in general show 30 to 40° dip toward NNE or NE whereas rocks of mesograde, associated with shear zone show 20 to 30° dip towards N or NNE. The shear zone also comprises of small scale cross folds, strain slip cleavage, sheath folds, S-C fabrics and silicified quartz veins.

The first generation of folds (F1) formed during D1 deformation are present as small scale structures and are characterised by tight isoclinal to hook type folds in gneisses. The plunge is 15° NE and SW direction. The second type of F2 folds formed during D2 deformation are coaxial to F1 folds. The rocks formed during D1 and D2- deformation are characterized by the S2-schistosity (parallel to S1-foliation) of regional metamorphism (M1) (Fig. 3a, b, c). The F3-folds are large-scale anticlines to synclines of recumbent-type (characterized by folding of S2-penetrative schistosity; Fig.4), and were possibly developed during D3 deformation. The axis of the major synclinal trends in the E-W direction plunge at low angle (15-20°) to SW. This deformation was brittle-ductile shearing at different levels and was characterised by S-C structures. Subsequent to D3, an effect of renewed movement related to D4 deformation was identified on microscopic and mesoscopic scale, and characterized by new shear fabrics (S-C structures). Due to this deformation at large scale dome and basin structures developed and the MCT zone was folded (plunge 5^0 to 15^0 Northward). with generation of folds (F4). The final phase of deformation (D5) was characterised by brittle deformation, extensional tectonics, kink structures and development of cataclasites around MCT-I and major shear zones. This caused local fault, kink folds, etc. but did not initiate recrystallisation and development of new minerals in the rocks.

METAMORPHISM

The medium- to high-grade pelitic schists intercalated with quartzites and characterised by biotite, almandine garnet, kyanite-sillimanite-muscovite, biotite-sillimanite- K-feldspar-bearing assemblages exhibit Barrovian type of regional metamorphism. Their spatial distribution consists of Gt-Chl, Ky-Mus, Sill Mus and Sill- Kfs zones (Fig. 2 and 4). The isograds patterns are simple than the complex regional structure indicating thereby that the regional metamorphism (M1) was subsequent to D2 deformation. The macroscopic folding marked by repetition of metamorphic zones caused the D3 deformation after M1 metamorphism. The low-grade rocks (chlorite schist, mica schist, garnet-chlorite schist, garnet-mica schist) occurring along the shear zones are evidences of the M2-metamorphism. The presence of different rotational garnets with idioblastic rim from mylonite zone or shear zone at various levels of Vaikrita Group indicate a syn- to post-tectonic event of M2 metamorphism. Moreover, the presence of Badrinath Tertiary Leucogranite and presence of a very poorly developed contact aureoles around Badrinath further suggest the effect of low-pressure metamorphism (M3) over the regional metamorphites (M1). In the present discussion these episodes have been described separately with deformation.

Progressive Regional Metamorphism (M1) in relation to Deformation

The regional metamorphism (M1) is almost contemporaneous with the early F1 folding phase during which chlorite, muscovite, and biotite recrystallised in the pelitic schists. The interleaved muscovite and biotite grew parallel to the axial plane schistosity (S1=S2) in pelitic schist suggesting their synkinematic growth during D1 deformation (Fig. 3b). The textural study also points that the growth of garnet started during late D1 deformation after the F1 folding (Fig. 3c). The porphyroclasts of garnet grown during early stage, were rotated subsequently by F2 folding as ascribed by curved or spiral internal schistosity (Si) occurring discordant to external schistosity (Se) in the pelitic rocks of Pandukeshwar Fm (Fig. 3d, e). The trails of inclusions consisting of opaque and lenticular quartz in the garnet (Fig. 3e), suggest the following reaction:

Fe-Chl + Mus + Qtz == Gt + Bio + Mt + H2O

The kyanite porphyroblasts ware grown over S2 schistosity. Flakes of muscovite, biotite, quartz and magnetite were common minerals associated with kyanite. The evidences of synkinematic (D2) growths were developed in specimens of kyanite +/-staurolite + biotite zone of amphibolite facies (near iron bridge; 1.5 km north of Vishnuprayag). Textural and structural study reveals that this metamorphism (M1) continued and culminated in the post-D2 deformational episode with the growth of garnet, kyanite, sillimanite and staurolite - bearing assemblages in the different pelitic rocks (Fig. 3f, g, h). The growth of idioblastic garnet followed F2 folding episode which is characterised by a xenoblastic texture. Garnets are homogeneous in amphibolite facies of rocks. The folding of elongated cleavage in kyanite and its rotation was contemporaneous to D3 deformation, This the kyanite of regional metamorphism grew prior to D3 deformation.

The presence of small needles of sillimanite present at the trail and fracture cleavage of kyanite (Fig. 3g) indicates an advanced grade of metamorphism. This record can be explained by "first sillimanite isograd"reaction

Kyanite == Sillimanite ...... (2)

Fibrolite and needles of sillimanite have been noticed in rocks exposed towards north of Hanumanchatti (Fig. 2). The fibrolite is found along the cleavages and at margins of muscovite in various pelitic rocks. The presence of sillimanite, xenoblastic muscovite along with K-feldspar and quartz (Fig. 3i, j), points to the decomposition of muscovite into sillimanite and thus the "second sillimanite isograd" reaction was:

Mus + Plag + Qtz == Sill + Kfs + H2O ...... (3)

There are a few evidences for the presence of kyanite porphyroblasts (folded and fractured) due to the above reaction (3) in thin section. Singh et al., (1996, 97) found similar texture in rocks of the adjoining areas and suggested that this univarient to divariant reaction (3) was controlled by fluid activities. The behaviours of kyanite and low fluid activity caused the presence of a narrow zone of kyanite-sillimanite-muscovite -bearing assemblages within the sillimanite-muscovite zone, which is also represented from a narrow divariant field in the AKF-petrogenetic grid (Fig. 5).

The complete absence of muscovite + quartz association from Badrinath area (Fig.4) indicates a progressive metamorphism towards high P-T conditions. Cordierite associated with Bio + Sill was found in a few pelitic rocks (Fig. 3h). The appearance of cordierite can been explained by following univarient reaction in KFMASH system by the presence of all of the phases.

Bio + Sill + Qtz == Crd+ Kfs + H2O ...... (4)

The textural study indicates that cordierite was formed after the sillimanite-orthoclase stage of the amphibolite facies. It is also an evidence for decompression subsequent to regional Barrovian metamorphism (M1).

The petrographic study suggests that biotite (early D1) grew earlier than garnet (late D1) and was followed by kyanite, sillimanite, cordierite - bearing assemblages (D2) during the M1 progressive metamorphism. The zonal distributions of index mineral from quartzites to pelitic rocks along Alaknanda valley were possibly affected by progressive metamorphism in space as well as in time. The feldspathisation of medium-grade pelitic schists at Vishnuprayag and their transformation into migmatites and gneisses without a tectonic break in area (north to south of Pandukeshwar village) suggests of regional metamorphism. The migmatisation, granitisation and cordierite - sillimanite gneisses commonly intermingled with each other. The mechanism of appearance of cordierite (by reaction 4) is also an evidence for decompression after tectonic loading at the end of culmination episode of M1 metamorphism (Fig. 6).

All the prograde reactions involved in the metapelites of Vaikrita Group (Fig. 4 and 5) are schematically explained in the K2O-FeO-MgO-Al2O3-SiO2-H2O (KFMASH) system by the petrogenetic grids of Thompson (1982) and of Grant (1985). The involved phases viz. garnet, biotite, kyanite, sillimanite, muscovite, K-feldspar and cordierite with extra phases (quartz, H2O and plagioclase) display various univariant reaction in the AKF diagram from [Cordierite], [K-feldspar], [Muscovite] stable invariant points. The phase compatibility and textural relationships of gneisses and schists of Joshimath/Badrinath Fm suggest that M1-metamorphic (prograde metamorphism) path proceeded through the univariant curves (Kfs), (Gt), (Mus) of the [Cordierite]-absent invariant point. The appearance of cordierite in a few pelitic bands may be related to a decompression path passing though the curve (Mus) and (Kfs) of the [Garnet]-absent invariant point (Fig. 5).

Second phase (M2) metamorphism in relation to Deformation

The evidences of retrograde metamorphic assemblages (Fig. 3k, 1) formed by prograde (M1) sequences are widely distributed in the investigated area and are mainly confined to the MCT-I (Vaikrita thrust) and some other major shear zones of Vaikrita Group. Mylonitisation occurred along the shear zone varying from a few centimeters to 500 meters in width. The ductile shears were prominent towards 1 to 2 km north of Vaikrita Thrust while the basal part of this thrust sheet (~600 to 800 m thick) formed under ductile-brittle conditions. The shears lying at the base of the high-grade rocks comprised of low-grade (chlorite-, garnet-mica, mica-schists) metamorphic rocks associated with small enclaves of high-grade schist rocks.

INVERTED METAMORPHISM DUE TO TECTONIC SUPERPOSITION

A few proposals i.e. recumbent folding and structural juxtaposition, divergence of isograde, frictional heat due to shearing along MCT, transfer of heat along MCT, S-C fabrics in the CCZ and superposed metamorphism along MCT zone were responsible for inversion of grade of metamorphism across the Main Central Thrust (e.g., Windley, 1983; Hodges & Silverberg, 1988; Metcalfe, 1993; Jain, 1993; Mukhopadhyay et al., 1996). However, the "inversion" is due to the tectonic superposition of low-grade metamorphism (M2) on regional high-grade (M1) metamorphism. Singh et al., (1996, 1997) pointed out that development of MCT shear zone started under the low -grade (green schist facies) conditions (14-16 km/ 400-500^0 C). Garnet from shear zones shows different kind of textures with progressive deformation. Our textural study reveals six types of garnet from shear zone i.e., i) skeletal/ fractured garnet, ii) spiral garnet, iii) zoned garnet iv) elongated garnet, v) rotated garnet and vi) massive garnet.

Skeletal/Fractured Garnet

It is medium-grained, highly fractured, sometimes, sheared, showing xenoblastic texture (Fig. 3m). it contains inclusion of biotite, muscovite, kyanite and quartz. The retrogression into chlorite, took place along fractures and rim. It was pre-tectonic to D3 deformation.

Spiral Garnet

Medium- to coarse-grained, porphyroblastic garnet is of two types- (i) core part showing spiral inclusion with an idioblastic rim (Fig. 3n) and (ii) spiral type. The earlier variety shows sinistral and dextral coiling. The inclusions are mainly of chlorite, muscovite flakes and angular fragments of quartz and magnetite. The rim part contains the inclusion of apatite, tourmaline and sericite. Its growth was post-tectonic (D3) (Fig. 3n).

Zoned Garnet

It is medium to coarse -grained. The core of this garnet is massive (occasionally contains inclusions of biotite, quartz, magnetite) but the rim is characterised by idioblastic inclusion of apatite, chlorite, tourmaline, magnetite and quartz (Fig. 3o).

Elongated Garnet

It is rarely found. and is coarse-grained and massive but contains inclusions of magnetite/ilmenite and quartz in the core part (Fig. 3p). It also shows pressure shadows. Sometimes it is present inclined to the shear plane (C- surface).

Rotated Garnet

It is idioblastic. The internal schistosity of this garnet is straight and is characterised by quartz grains. The external schistosity shows oblique relation with the internal schistosity. The pseudo pressure zone of it is asymmetrical.

Massive Garnet

It is medium- to fine-grained, and is disseminated in chlorite-muscovite-biotite schist. The chlorite or mica is not present in textural contact. This garnet contains rarely the inclusions of apatite, It was post-tectonic (D4) in origin.

Other retrograde reactions in the shear zones

In addition to garnet, are the minerals of shear zones chlorite, muscovite, biotite, andalusite, apatite, sericite, plagioclase, microcline, perthite etc. are the minerals of the shear zones. Sometimes, tiny crystals of kyanite (M2 episode) near garnet or andalusite are also noticed but no phase transformations of aluminosilicate was observed. Interestingly tourmaline, apatite, ilmenite are found frequently and are aligned along the shear planes. Occasionally plagioclase porphyroblasts, containing remnants of lenticular quartz with specks of mica and laths of tourmaline, indicate their crystallization during the shearing. As a result of intense shearing and alteration, feldspars were sericitised, and garnet and kyanite showed pseudomorphic texture.

Inverted metamorphism

The textural, structural and mineral paragenesis observed from rocks of MCT-I shear zone (Jharkula Fm) reveals two metamorphic zones of greenschist facies (i) Gt-Chl-Bio zone and (ii) Bio-Chl zone. The lower grade rocks are exposed mainly towards the southern part of the Vaikrita Shear (Munsiari Group). Both zones were delineated at 500m northwards from the base of Jharkula Fm. and their minerals were formed by the following reaction:

Qtz + Chl + Mus == Alm (Gt) + Bio + H2O .... (5)

It is therefore evident that M2 metamorphism and formation of schist along shear zones of Jharkula Fm occurred during the late D3 deformation and probably continued until early synkinematic D4 deformation (Fig. 6).

The garnet formed during the M2 metamorphism (from Jharkula Fm) was prograde. Some of the garnets contain inclusions of chlorite, tourmaline, etc. This was also reported by Hubbard (1989). Moreover, Singh et al (1997) have discussed the presence of high temperature at shallow depth. The presence of idioblastic thick rims on the spiral garnet points towards the progressive growth and was a part of thermal reaction of M2 metamorphism. It is fact that the Jharkula Fm (MCT-I shear zone) was retrograded to a low-grade. So retrograde origin for inverted metamorphism can be a possibility.

THERMAL / CONTACT METAMORPHISM

The Tertiary granitoids (leucogranite) were noticed from many places of Garhwal viz. Gangotri, Malari, Kedarnath and Badrinath. All these granites are plutonic and intrusive into the Central Crystallines. The Tertiary granitoids possibly were emplaced around 20 Ma (Sinha, 1989; Sorkhabhi & Stump 1993; Stern et al., 1989). No data on thermal metamorphism are available. In the present paper, the studies carried out were only on a few samples. The study of thermal metamorphism (M3) did not reach above hornblende-hornfels facies.

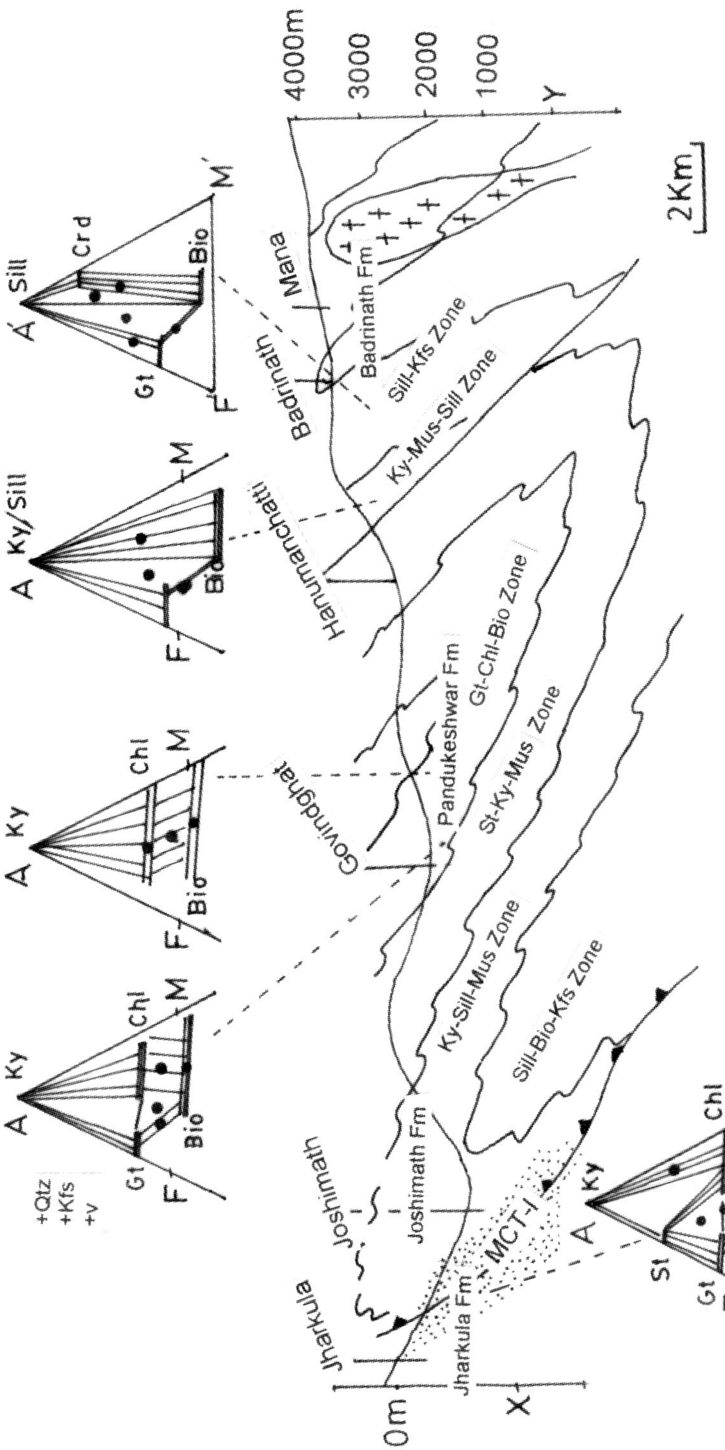

Figure 4: Different metamorphic zones in the Upper Alaknanda valley along X-Y cross section (Fig.2) are described in the projection diagrams. The solid circles in the triangular diagram represent observed assemblages of M1 and M2 episode.

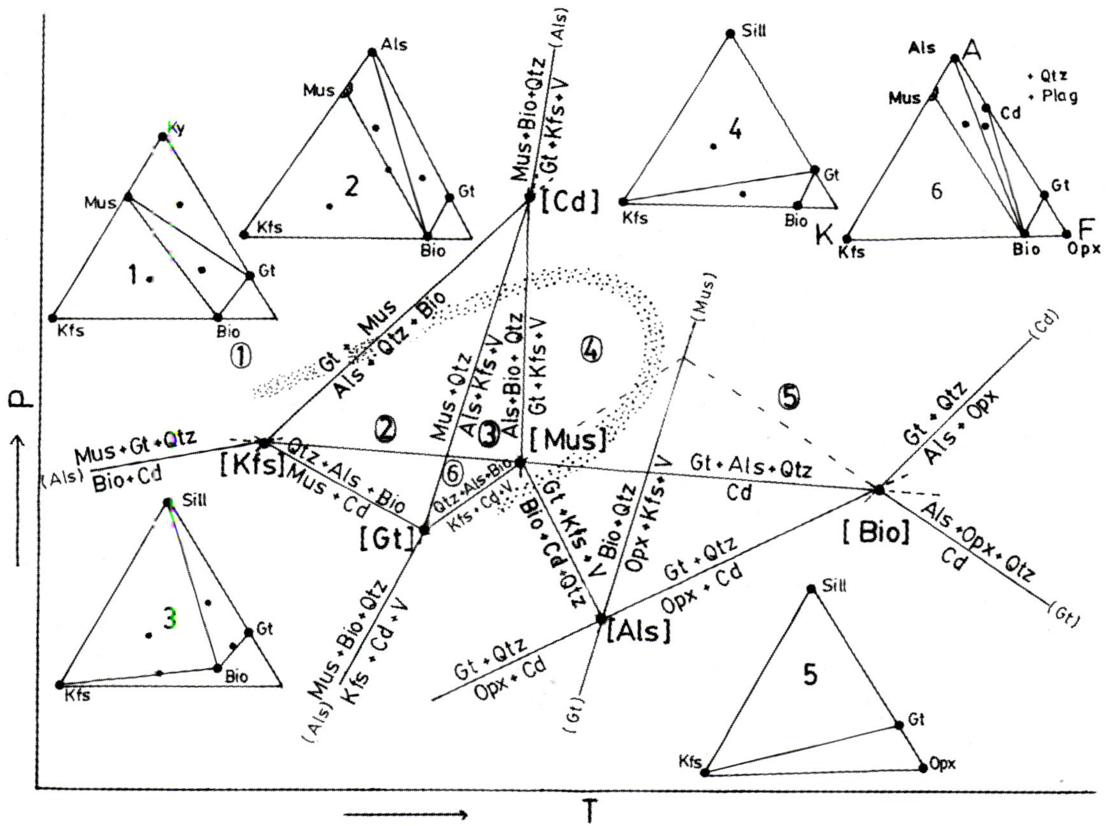

Figure 5: Petrogenetic grid in KFMASH system based on mineral assemblages and reaction textures (modified after Grant, 1985). The arrow indicates prograde path of the Central Crystalline Zone. In one thin section of rock from the adjoining area (30-km west from the present one) an orthopyroxene intergrowth with K-feldspar has been observed. Our petrogenetic grid has included that aspect also. Other petrologists are also realising the presence of incipient granulite facies metamorphism in the Himalaya.

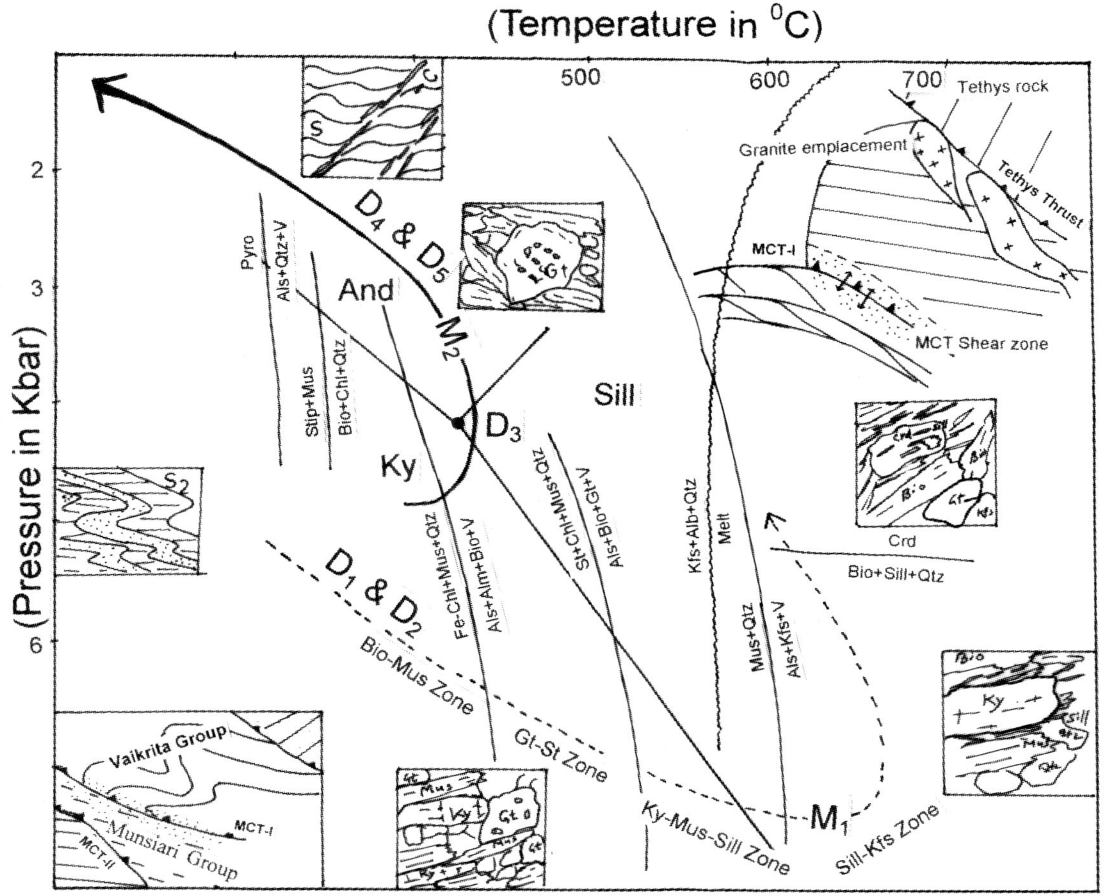

Figure 6: P-T path of M1 and M2 episodes of metamorphism along the Alaknanda valley. D1, D2 represents Precambrian events while D3, D4 and D5 related to M2 episode and Tertiary. The squares sketch out the important texture and structure noticed in different metamorphism.

Limited constraints on Timing and P-T conditions of the Thrust

The Main Central Thrust zone represents by a major intracontinental shear zone where high-grade rocks of the Vaikrita Group were thrust over the low-grade metamorphics of Lesser Himalayan (Valdiya, 1980; Singh et al, 1997; Sinha, 1989; Thakur, 1992). Even though it is considered that MCT evolved around 30 Ma (Searle et al., 1988), yet the limited geochronological data, shearing and recrystallisation do not show very conclusive evidences for the evolution of Central Crystallines along the MCT-I.

Detailed geochronology of Crystallines reveals three tectonothermal events 1850 ± 50 Ma, 1150 ± 50 Ma, 500 ± 50 Ma; (Trivedi et al., 1984; Singh et al., 1986; Le Fort et al., 1986; Scaillet et al, 1990) and one of Tertiary (25 ± 5 Ma) event proposed by Singh et al (1998). The whole rock and mineral data the for Vaikrita Crystallines in Garhwal Himalaya reveal the following points (1) shear zone rocks yielded 30 to 20 Ma (2) age of intrusive was~500 Ma and ~20 Ma granites and 3) age gneisses and migmatites was 1850 ± 50Ma. The detailed structural studies of the Garhwal Himalaya reveals five phases of deformation in Central Crystalline Zone (Schwan 1980,81; Saklani 1993), four phases of deformation in Garhwal Group (Gairola, 1992), and two phases of deformation in Siwalik Group (Thakur, 1992). The Precambrian sedimentaries of Garhwal Group rocks (south of CCZ) are Late Precambrian to Cenozoic.

The conventional geothermobarometry used to calculate the P-T conditions of the MCT shear zone of the Vaikrita Group revealed 550 °C - 700 °C / 6-9 Kbar in Garhwal and adjoining areas (Metcalfe, 1993; Gairola & Ackermand, 1988; Hodges et al., 1988). However unusual P-T conditions of metamorphism within the MCT-I shear zone was found by many workers (Singh et al., 1997). The abrupt change in P-T was partially due to disequilibrium of garnet and biotite mineral phases, (II) geothermobarometry was applied without any correction of mixing models of garnet, and (III) scattered values of distribution coefficients KD between garnet and biotite (Fe/Mg), Singh et al., (1997) pointed that P-T conditions were around 500+50 °C / 4-5 Kbar in the shear zone,e.g., (higher temperatures at shallower depths) and suggested that this reflected the supply of heat from deep-seated magma. These P-T data also suffered the mineral assemblages viz. And-Chl-Mus-Bio-Qtz and rarely with Ky-Gt-Mus-Bio-Qtz noticed near MCT at M2 episode Jowhar and Verma (1995) studied Si-Al order in the alkali-feldspar of gneisses, pegmatites and granite from the Badarinath area and suggested that the high degree order in feldspar held slow cooling conditions but they did not ascribe to slow cooling for Tertiary metamorphism.

CONCLUSIONS

The deformation and textural studies suggest that the inverted metamorphic zonation near the MCT in the Alaknanda valley section was due to cumulative result of polyphase metamorphism (M1 and M2) and deformations (D1 to D5). The M1 episode event was Pre-MCT while M2 was contemporaneous to it during Tertiary. The first metamorphism (M1) essentially was synkinematic with respect to F1 folding, initiated crystallisation of chlorite, biotite and muscovite. The last two minerals were intergrown and defined axial plane schistosity (S2). Following this, grade of metamorphism in creased with formation of porphyroblastic garnet, biotite, kyanite and muscovite which continued subsequent to D2 deformation. At the end, the rocks were decompressed and were characterised by of cordierite + sillimanite migmatisation, and partial melting. Thus, during M1 metamorphism a series of different isograds developed across the terrain.

The phase relationship and identified reaction isograds in the Alaknanda valley suggest a repetition of metamorphic zones (Fig. 4 and 6). The metamorphism was of lowest grade (i.e. greenschist facies) around Pandukeshwar and was highest at Badrinath and Joshimath (Fig. 2). The structural study revealed that the D3 deformation was responsible for the repetition of metamorphic zones and was post-kinematic to M1 episode. At the advanced stage of folding (F3), the rocks were thrusted and translated towards the south. The rotated porphyroclasts of garnet, feldspar and kyanite were formed due to the strong shearing under ductile to brittle-ductile conditions (Fig. 3 l).

The M2 metamorphism restricted to shear zones and crystallisation was contemporaneous to thrusting demonstrated by rotated garnets in footwalls of the MCT-I (Fig.3 m, n, o, p). The metamorphic grade recorded from the MCT shear zone of M2 episode was in the garnet zone defined by Gt + Chl + Ky+Bio+Qtz assemblage at 500m from the MCT-I. The asymmetrical shearing marked by rotation of garnet during M2 episode pointed another phase of deformation (D4) after the M2 metamorphism. This deformation was defined S-C structures with occurrence of chlorite-muscovite-quartz. This metamorphism was of retrograde stage of M2 metamorphism (Fig. 5 and 6). The thermal (contact) metamorphism (M3) developed due to emplacement of Badrinath leucogranite and was pre-kinematic to D4 and post-tectonic to D3 deformation.

Thus, the inverted metamorphic zonation near the MCT in the Alaknanda Valley section is due to the cumulative result of multiple metamorphism (M1 and M2) and deformation (D1 to D5).

ACKNOWLEDGEMENTS

ABT is grateful the ETH Schulleitung for sabbatical support.

REFERENCES

Azmi, R.J.,(1994). A Pre-Vandian orogeny and existence of the Himalayan Central Crystalline zone in the NW Himalaya. -9th Himalaya-Karakorum-Tibet workshop Kathmandu, (abstract), 10: p14.

Fuchs, G., (1968). The geological history of the Himalaya- 22nd Int. Geol. Cong. New Delhi, pp 161-174.

Fuchs, G., (1992). Pre-Alpine and Alpine orogenic phases in the Himalaya. In: Sinha, A.K, (ed.): Himalayan Orogen and Global Tectonics, pp. 19-34.

Gairola, V.K., (1992). Structure and tectonics of the Garhwal synform. In: Sinha, A.K, (ed.): Himalayan Orogen and Global Tectonics, 91-106.

Gairola, V.K., Ackermand, D., (1988). Geothermobarometry of the Central Crystallines from the Garhwal Himalaya. J. Geol. Soc. Ind., 31 pp. 91-106.

Gansser, A.,(1964). Geology of the Himalaya. Wiley Interscience, London, 289p.

Grant, J.A. (1985). Phase Equilibria in partial melting of pelitic rocks, In: J.R. Ashworth (ed.), Migmatites, pp. 86-144.

Heim, A. and Gansser, A., (1939), Central Himalaya, geological observations of the Swiss expedition 1936. Mem. Soc. Helv. Sci. Nat. pp. 73- 245.

Hodges, K.V. and Silverberg, D.S., (1988). Thermal evolution of the Garhwal Himalaya, India, Tectonic 7 pp. 583-600.

Hodges, K.V., Hubbard, M.S., Silverberg, D.S., (1988). Metamorphic constraint on the thermal evolution of the Central Himalaya orogen. Phil. Trans. Roy. Soc. Lond on A326, pp. 257-280.

Hubbard, M.S., (1989). Thermobarometric constraints on the thermal history of the Main Central Thrust zone and Tibetan Slab, eastern Nepal Himalaya. J. Met. Geol. 7, pp. 19-30.

Jain, A.K., Manikavasagam, R.M., (1993). Inverted metamorphism in the intercontinental ductile shear zone during Himalayan collision tectonics, Geology 21, pp. 407-410.

Jowher, T.N. and Verma, P.K. (1995). Alkali feldspar from the Badrinath Crystalline complex and their bearing on the Himalayan metamorphism. The Indian Mineralogist 29, pp. 1-12.

Le Fort, P., Deobon, F., Pecher, A., Sonet, J., Vidal, P., (1986). The 500Ma magmatic event in Alpine southern Asia episode at Gondwana scale. Sci. de la Terre, Mem. 47, pp. 191-209.

Mukhopladhyay, D.K., Bhadra B.K., Ghosh, T.K., Srivastava, D.C. (1996). Evidence of the thrust emplacement of the "Lesser Himalaya" Chur granite, Himachal Pradesh. Proc.Indian Acad. Sci (Earth Planet. Sci) 105, pp. 157-171.

Metcalfe, R.P., (1993). P-T-t constraints on metamorphism across the MCT zone and High Himalaya slab in the Garhwal Himalaya. In: Treloar, P.J. and Searle, M.P. (eds.): Himalayan Tectonics. Geol. Soc. London, Sp. Publ., 74, pp. 485-509.]

Saklani, P.S., (1993). Geology of the Lower Himalaya (Garhwal). International Books and Periodicals, Delhi, 246p.

Saklani, P.S., Nainwal, D.C. and Singh, V.K., (1991). Geometry of the composite Main Central Thrust (MCT) in the Yamuna valley, Garhwal Himalaya, India. N. Jb. Geol. Paläont. Mh., pp.364-380.

Scaillet, B., France-Lonard, Ch. and Le Fort, P., (1990). Bhagirathi-Gangotri plutons (Garhwal, India): petrological and geochemical evidence for fractionation processes in a high Himalayan leucogranite. J. Volcanol. Geotherm. Res., 44, pp.163-188.

Schwan, W., (1980). Shortening structures in Eastern and North Western Himalayan rocks. Saklani, P.S. (ed.), 62p.

Schwan, W., (1981). Key structures of minor tectonics, major tectonic forms, and shortening kinematics in the Himalaya. In: Sinha, A.K. (ed.): Contemporary Geoscientific Researches in Himalaya, I, pp.13-20.

Searle, M.P., Cooper, D.J.W., Rex, A.J., (1988). Collision tectonics of the Ladakh-Zansker Himalaya. Phil. Trans. R. Soc. Lond on A 326, pp.117-150.

Singh, R.P., Singh, V.P., Bhanot, V.B., Mehta, P.K., (1986). Rb-Sr ages of the gneiss rocks of Rihee-Gangi, Bhatwari, Hanumanchatti and Naitwar areas of the Central Crystalline zone of Kumaon Himalaya (U.P.) Ind. J. Earth Sci. 13, pp.197-208.

Singh, S.P., Singh. V.K. and Saklani, P.S., (1997). Metamorphic evolution of the Central Crystallines of Higher Himalaya along Dhauliganga valley, Garhwal Himalaya, India. N. Jb. Geol. Paläont. Abh., 206, pp.249-275.

Singh, V.K., Singh, S.P. and Saklani, P.S., (1996). Metamorphic and Geodynamic evolution of the Central Crystallines of the Garhwal Himalaya, India. N. Jb. Geol. Paläont. Abh., 199, pp. 89-109.

Sinha, A.K., (1989). Geology of the Higher Central Himalaya. John Wiley & Sons., 219p.

Sorkhabi, R.B. and Stump, E., (1993). Rise of the Himalaya: A Geochronologic approach. GSA Today 3, pp. 87-92.

Srikantia, S.V., Ganesan, T.N., Rao, P.N. Sinha, P.K. and Tirkey, B. (1978). Geology of Zanskar area Ladkah Himalaya. Him. Geol. 8, pp.1009-1033.

Stern, C.R., Kligfield, R., Schelling, D., Virdi, N.S., Futa, K., Peterman, Z.E. and Amini, H., (1989). The Bhagirathi leucogranite of the High Himalaya (Garhwal, India): age petrogenesis and tectonic implications. In: Malinconico, L.L. and Lillie, R.J. (eds.): Tectonics of the Western Himalaya. Geol. Soc. Am. Sp. Pap., 232, pp. 33-45.

Thakur, V.C., (1992). Geology of Western Himalaya. Pergamon press, London, 355p.

Thompson, A.B., (1982). Dehydration melting of pelitic rocks and generation of H2O- undersaturated granitic liquids, Am. J. Sci. 282, pp.1567-1595.

Trivedi, J.R., Gopalan, K. and Valdiya, K.S., (1984). Rb-Sr age of granitic rocks within the lesser Himalayas Nappe, Kumaun, India. J. Geol. Soc. Ind. 25, pp.641-654.

Valdiya, K.S., (1980). Geology of Kumaun Lesser Himalaya. Wadia Inst. Him. Geol., Dehradun, 219p.

Valdiya, K.S., (1981). Tectonics of the central sector of the Himalaya, In: Gupta, H.K., Delany, F.M. (eds), Zagras-Hindukush-Himalaya geodynamic evolution. Am. Geoplhy. Union Geodyn. Ser. 3, pp. 87-110.

Virdi, N.S., (1986). Lithostratigraphy and structures of the Central Crystallines in the Alaknanda and Dhauliganga valleys of Garhwal, U.P., In: Saklani, P.S. (ed.), Himalayan Thrust and associated rocks, Today and Tomorrow's Publ., New Delhi, pp. 155-166.

Windley, B.F., (1983). Metamorphism and Tectonics of the Himalaya, J.Geol. Soc. London, 140, pp.849-865.

STRUCTURAL AND METAMORPHIC STUDIES OF ROCKS ASSOCIATED WITH THE CHAIL UNIT AND MAIN CENTRAL THRUST ZONE OF THE GARHWAL HIMALAYA

V. K. Singh

Department of Geology, Bundelkhand University, Jhansi - 284128 (UP),

ABSTRACT

The Chail Unit is composed of mainly quartzite, phyllite, slate, carbonaceous slate, and schists along with intrusive porphyritc granites while the Main Central Thrust (MCT) zone is characterised by sheared granite-gneisses along with greenschist facies of rocks. The MCT zone is directly thrust over the Chail Unit along MCT-II and its northern limit is known as MCT-I. Structural studies show five phase of deformation (D1 to D5), textural relationships and mineral parageneses indicate two episodes of metamorphism in the present area. The early phases of deformation (D1 and D2) were restricted to limited zones while D3 deformation produced a strong schistosity overprinting the D1 and D2 structures. The M1 prograde regional metamorphism is related to emplacement of porphyritic granitoids, while the prograde M2 (low-grade) is superimposed on the M1 (higher-grade)

The F1 folds are very rare and F2 folding is represented by tight and isoclinal folds plunging towards ENE. or WSW. The F3 folds formed during D3 deformation, are recumbent, frequently overturned and plunging towards NW. The F4 chevron folds were developed during D4 and D5 deformation phases and they plunge NE or SW, and sometimes in N or S direction.

The minerals formed during prograde metamorphism (M1) include biotite, muscovite, garnet, kyanite (rarely), sillimanite (rarely), plagioclase and K-feldspar (D1 and D2 deformation). The transformation of kyanite into sillimanite (~620° C/6 Kbar) during the prograde event was followed by the formation of andalusite from sillimanite (520° C/4 Kbar) and suggests a clockwise P-T trajectory. The succeeding event of M2 metamorphism was characterised by mainly low-grade metamorphic assemblages during D3 and D4 deformations. The D5 deformation episode produced transverse faulting in the area. The metamorphic conditions of the rocks reveal that the metamorphism (M1) took place at 600° C to 700° C at 6-7 Kbar pressure during D1 and D2 deformations, while M2 metamorphism took place at 350 - 400° C/ 4-5 Kbar.

INTRODUCTION

Broadly, the structural architecture of Garhwal-Kumaon Himalaya is outlined by (i) Tethys Sedimentary Zone, (ii) Central Crystalline Zone, (iii) Lesser Himalayan Zone (parautochthonous and allochthonous units of metasediments of Precambrian) and (iv) Outer Himalaya (Sedimentaries of Tertiary) from north to south.

The Main Central Thrust (MCT) Zone lies between the Central Crystalline Zone (CCZ) and Lesser Himalayan Zone (LHZ) which is about 10-15 km thick in the Garhwal region and comprises mainly the metamorphic series of different grade and intrusive of granitoids which range in age from 2200 Ma to 8 Ma (see references in Singh et al., 1997). The northern part of the MCT zone is thrusted (Vaikrita thrust of Valdiya, 1980 and MCT-I of Saklani et al., 1991). It is in contact with the Central Crystallines (Vaikrita Group). The southern part of the MCT zone has a thrusted contact (Main Central Thrust of Heim and Gansser, 1939; Munsiari Thrust of Valdiya, 1980 and MCT-II of Saklani et al., 1991) with the Precambrian sedimentary of LHZ (Chail Unit, Garhwal Group and Krol Group). The Chail Unit comprises of quartzites and schists of low-grade and has also a sheared contact with the MCT zone as well as with the Garhwal Group (Fig. 1).

The structural and petrological studies of associated rocks of Chail unit and MCT zone point to polyphase deformation and metamorphism. Various metamorphic zones and isograds related to Tertiary metamorphism have been identified and it is argued that there is an increase in grade of metamorphism to north of MCT (Hodges and Silverberg, 1988; Metcalfe, 1993 and Singh et al., 1996).

Narayan Das et al., (1981), Trivedi et al., (1984), Bhattacharya et al., (1984), and Singh et al., (1986) obtained significant geochronological data from different parts of Garhwal Himalaya, and suggested that the granite-gneisses of MCT zone and Central Crystallines associated with the metamorphics are in general of Precambrian age. However, the appearance of Tertiary ages (Sorkhabi and Stump, 1993) of the CCZ suggests that the older granite-gneisses as well as metamorphics were possibly remobilised and recrystallised during the Tertiary. So, the polyphase deformation and metamorphism of the MCT zone range in age from Precambrian to Tertiary and, thus, the Chail Unit may be of structural and petrological

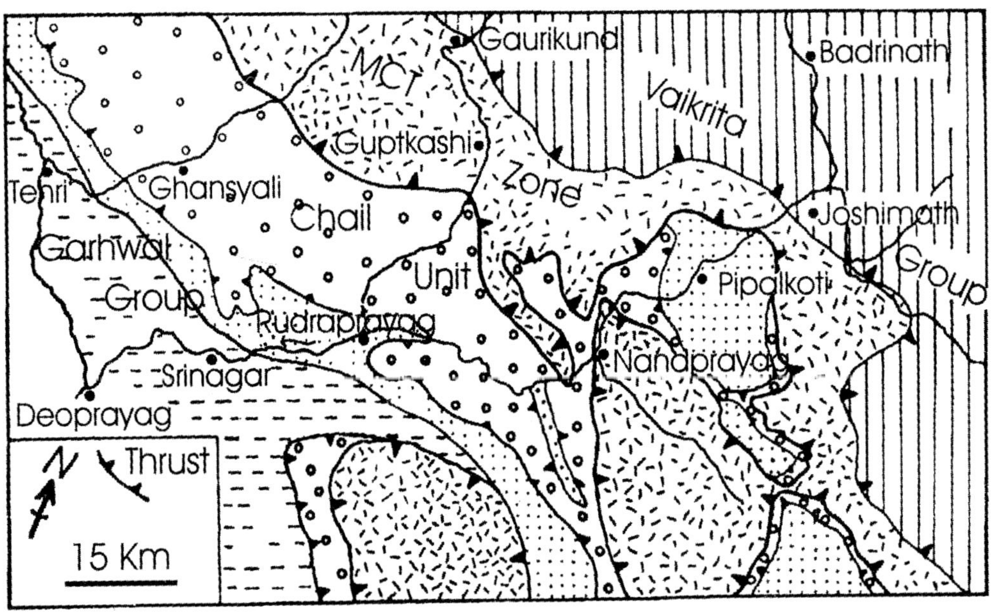

Figure 1: Regional geological map of Garhwal Himalaya, after Valdiya (1980); studies have been carried out along the Bhilangna and Alaknanda valleys.

interest. In the present paper an attempt has been made to deduce the structures, textures and P-T conditions of metamorphism of each episode of the Chail Unit and the MCT zone of Garhwal Himalaya.

REGIONAL GEOLOGICAL SETTING

Valdiya (1980) divided the Central Crystallines into three Groups viz;

i. **Bhatwari/Chail Group:** Porphyritic granitoids, gneisses, migmatites, quartzites, calc-silicates, amphibolites and low-grade metamorphics.

ii. **Munsiari Group:** Coarse-grained granitoids, gneisses, migmatites and medium grade metamorphics of pelitic, psammatic and carbonates.

iii. **Vaikrita Group:** Medium to high-grade metamorphics of pelitic, psammatic and calc-silicates, migmatites, gneisses, granitoids and intrusives of the Tertiary leuco-granitoids viz; Gangotri, Kedarnath, Badrinath, Malari etc.

The rocks of Munsiari Group have extensive shearing and are intermingled with various rocks, called as the "MCT Zone".

All the rocks from the studied area (Fig. 2) have a northward monotonous dip, which covers a part of the Chail unit and MCT zone, separated by the NE dipping MCT-II (MCT of Heim and Gansser, 1939).

Chail Unit

The intrusive granites are usually grey to dark-grey in colour, and porphyritic in nature. The mafic and felsic colour bands within porphyritic granite were also noticed at several places to the north of Budhakedar village (15 km north of Ghansyali), indicating partial melting and anatectie structures. The mafic bands consist of biotite, tourmaline, magnetite and hornblende, whereas felsic bands contain quartz, muscovite, feldspar and apatite. White to cream coloured, quartzofeldspathic schists (200-300 m thick) are also frequently present in southern part and are characterised by thin to thick zones of mylonites and cataclasites described as chlorite-muscovite-quartz and biotite-sericite-quartz schists. The Chail quartzite is usually massive white, rarely ferruginous and at places is characterised by a pebbly horizon (greenish blue, white, milky rounded to sub rounded pebbles). Numerous metabasic sills and dykes intruded in it and at places is mylonitised (Ghansyali). The porphyritic granite-gneisses or intrusive granites tectonically overlie the quartzite along the Chail shear within this Unit (Fig. 2).

MCT Zone

The MCT zone with exposures of the Central Crystallines (Vaikrita Group) of Higher Himalaya along MCT-I in the Garhwal region (Fig. 1), is about 10 to 20 km thick and is characterised by numerous shear planes. The MCT zone is directly thrust over the Chail unit (quartzite, phyllite, slate, carbonaceous slate, etc.; Singh et al. 1998a) along MCT-II and the southern limit of this zone is marked at different levels due to presence of the various imbrications delineated by sheared granite-gneisses. In the Alaknanda valley, the MCT zone

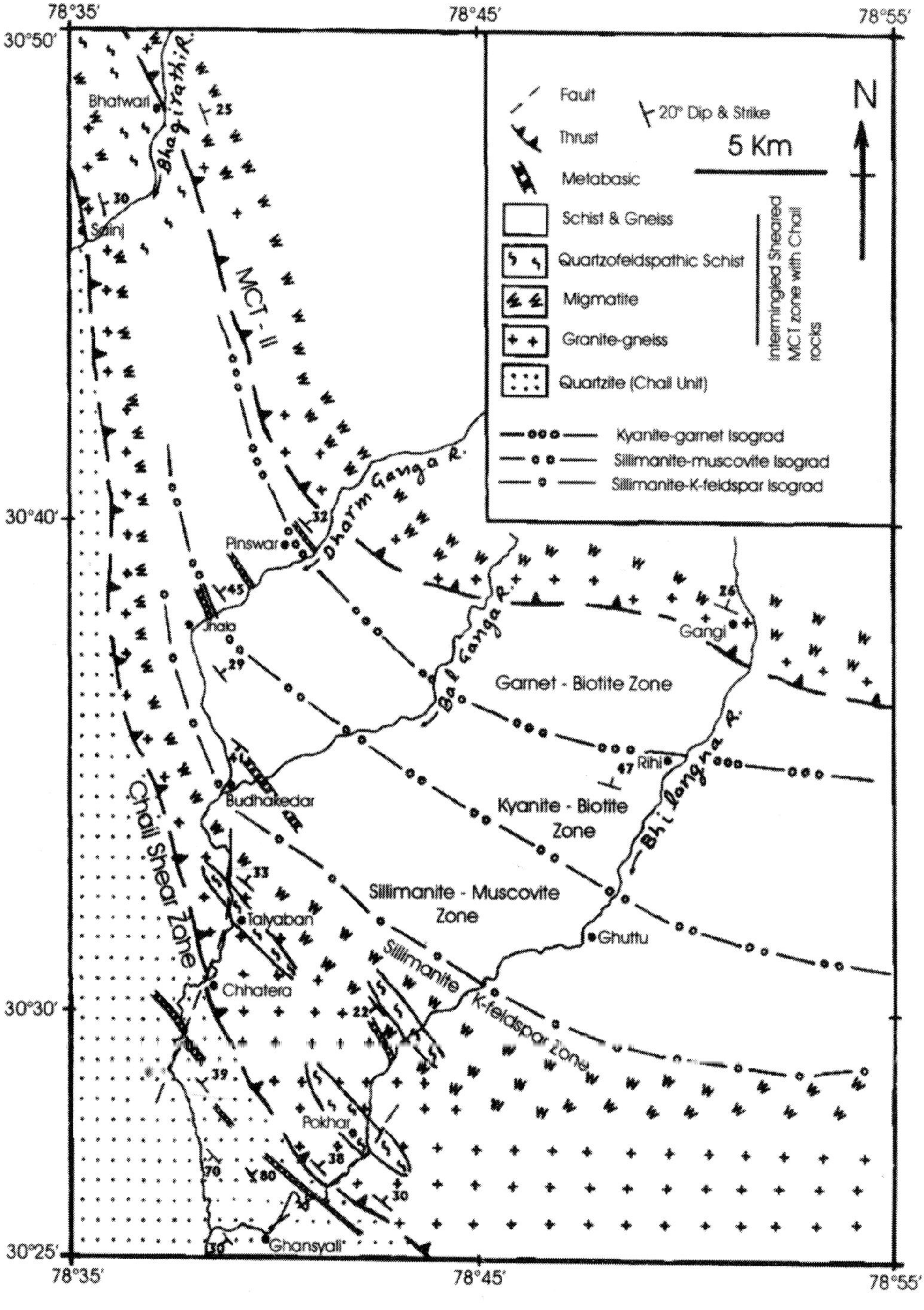

Figure 2: Geological map of the Bhilangna valley, Garhwal Himalaya, after Singh et al. (1996).

comprises of granite-gneisses with medium-grade metamorphics of pelitic, psammitic rocks and lenses of and calc-silicates. The granite-gneisses occur in the southern part and are highly sheared near the thrust where they are converted into chlorite-muscovite-sericite schists. Medium to coarse- grained metamorphic rocks also contain medium- grained leucocratic granite, and pegmatites (2-3 m).

DEFORMATION AND STRUCTURAL SETTING

The rocks of the studied area were affected by D1 to D5 deformation episodes, which were responsible for F1 to F4 phases of folding. The F1 folds are rare but are rootless. The tight and isoclinal F2 folds plunge at 25° towards ENE or WSW, and were formed during the D2 deformation (Fig. 3a). The F3 folds of D3 deformation are overturned and plunge 20-30° towards the NW. The intensity of deformational stresses during the late D3 phase resulted in thrusting. The F4 folds are of chevron type and plunge 10-15° towards the NE or SW and sometimes in N or S directions. During late D3 and D4 deformation, sliding took place in the MCT Zone along the S2 plane and retrograde metamorphic minerals were formed. D5 deformation was coaxial with D4 deformation and was accompanied by cataclasis, faulting and neotectonic movements.

The simple asymmetrical shears from the MCT zone and sedimentary units show a southward movement of shearing present during thrusting phenomenon (Singh et al., 1998b). Some normal faulting is dextral within the MCT zone, synchronous with crustal shortening. The small-scale structures and sense of shear indicate that at least two phases of deformation have taken place during development of the MCT zone. The presence of chlorite, biotite, andalusite development along the low angle shears can not be ruled out as a syn-tectonic thermal episode of low-grade metamorphic conditions which was superimposed on Precambrian regional metamorphism.

The palaeomagnetic studies also reveal that a secondary TRM with unblocking spectra between 200-350° C, indicating pyrrhotite of the last thermal event (cooling age ~20 Ma) in the Alaknanda valley (Schill et al., 1998, 2001 and 2002). The geodynamic exhumation of the MCT zone was initiated around 35 Ma (intense ductile deformation) and was active about 20 Ma (Searle et al., 1988) and was uplifted in different stages producing imbrications under brittle-ductile conditions. The late deformation affected the MCT zone towards eastern and western parts respectively along the Alaknanda river-valley.

GENERAL PETROGRAPHIC CHARACTERS

i. Chail Unit

a. **Chail Quartzite:** Medium to coarse quartz grains are the main components embedded in a ferruginous and argillaceous matrix. The elongated quartz grains and presence of minor amounts of chlorite and sericite define the foliation in quartzite. The quartzite contains a thick mylonitic zone near Ghansyali, where quartz grains are ribbon shaped (Fig. 3b).

Table 1: Tectonic setting, Structures and Metamorphism of the Chail Unit and MCT Zone, Garhwal Himalaya.

		Rock Type	Metamorphic Grade Grade	Structure
CCZ (VAIKRITA GROUP)		Gt-Bio-gneiss, Ky-Sill gneiss, migmatite, Ky-Mica schist, Gt-Mica schist, Mica schist and Quartzites.	High-grade (Ky, Sill Zone) of M1 metamor-phism	Rocks have prominent S_2 schistosity (during D_2 deformation).
------------- MAIN CENTRAL THRUST - I (VAIKRITA)---------				
MAIN CENTRAL THRUST ZONE		Chlorite schist, Mica schist, Chl-Mus Schist. Porphyritic gneiss, migmatite, quartzites. mylonites and cataclasites.	Low-grade metamorphics M_3 (super-imposed on granitoids, gneisses and schist). Recrystalli-sation along shear zone.	Mainly mylonites, ductile, brittle -ductile and brittle conditions, shear structures with, sheath folds, S-C and S-C' fabrics (D_3 & D_4 phase). The extensional strutures are also present (D_5 phase).
------------- MAIN CENTRAL THRUST - II (MUNSIARI) -------------				
LESSER HIMALAYAN ZONE (LHZ)	Chail Unit	Porphyritic gneiss, Augen gneiss, Streaky gneiss, intrusive granite, schists. Massive white quartzite and mylonitised quartz sericite schist	Low-grade green-Predominated by schist facies with S-C and S-C' fabrics amphibolites during D_3& D_4 episodes).	
	Garhwal Group	Phyllite, Slate, Shale, Carbonates, Quartzite	Very low-grade Rocks having prominent F_2 and F_3 foldings (D_2& D_3 phase).	

b. **Schist and Porphyritic Granite:** The - kyanite, + sillimanite (rare), andalusite, garnet, biotite, chlorite, muscovite and sericite have been recorded from schist and porphyritic granite which are highly sheared and retrograded (Green schist facies). The presence of various kinds of enclaves of pelitic, quartzofeldspathic and mafic materials within the schist and granites near the Budhakedar, Pokhar (and many other places) yield assemblages of higher-grade of metamorphism (amphibolite facies) (Fig. 3c).

ii. MCT Zone

The enclaves of quartzofeldspathic schists, metapelites, quartzites and amphibolites are important features noticed within these granite-gneisses of the MCT zone (Table 1). The Vaikrita Group with a NE plunging lineations and NW-SE folds indicates syn-thrusting deformation. The Sedimentary unit of Lesser Himalaya shows SSW to SW plunging fold axes of earlier origin (prior to thrusting) which have Aravalli trend to (Precambrian Indian Shield). S-C fabric studies reveal three distinct nature of the MCT movement: (i) shear surface developed parallel to earlier penetrative surface during ductile movement, (ii) shear surface at an angle of about $10\text{-}30^0$ with schistosity (low-grade development of metamorphic minerals, Fig. 3d), under brittle-ductile conditions and (iii) S-C' fabrics having dips of $50\text{-}65^0$ towards east under brittle conditions.

Garnet: The garnets from low-grade rocks of Chail Unit usually show inclusions of chlorite, magnetite and quartz. The inclusions are mostly spiral shaped and show syn- to post-tectonic texture. The rim of such garnet is usually free from inclusions and is idioblastic in nature. On the other hand, the garnet from the kyanite zone exposed near Pinswar is free of inclusions and is idioblastic in nature. The garnet is very rare in the gneisses and migmatites and is usually fine- to medium-grained, fractured and xenoblastic in nature.

Biotite: Light brown to dark brown pleochroic subidioblastic biotite of gneisses and metapelitic enclaves occur around K-feldspar, and occasionally show an intermingled relationship with sillimanite. The biotite associated with kyanite-sillimanite in a quartz matrix shows brown to light brown pleochroism. Some biotites associated with low- grade metamorphics are usually greenish brown to light green.

Muscovite: The prograde muscovite is very rare in the granite-gneisses as well as in pelitic enclaves associated with granitoids. The prograde relics of muscovite (often intermingled) with fibrolite, kyanite and biotite in migmatite are related to decomposition of white mica.

The medium to coarse subidioblastic muscovite from granite, and pegmatites is related to crystallisation of granitic melt and can be defined by the following reaction, as suggested by Thompson (1982) and Grant (1985).

Liquid = Biotite + Muscovite + Aluminosilicate + K-feldspar + Quartz ...(1)

On the other hand, the white micas from sheared zones of granite- gneisses are medium to coarse- grained and are usually associated with chlorite and greenish biotite (medium grained). It can be defined by the following hydration reactions:

K-feldspar + Biotite + H2O = Chlorite + Muscovite + Quartz ... (2)

and

Biotite + H2O = Chlorite + Sericite + Quartz ... (3)

Aluminosilicates: The kyanite and sillimanite are present in medium to coarse - grained prismatic crystals in the metapelite and in some enclaves of granite-gneisses. A bluish coloured rock exposed near Jhala (20 km north of Ghansyali), contains kyanite, sillimanite, and andalusite within a single thin section and shows textural relationship with magnetite, biotite and muscovite (Singh, 1993).

The presence of needle shaped sillimanite around coarse crystals of kyanite (Fig. 3c) suggest the following reaction texture:

Kyanite = Sillimanite ... (4)

The presence of coarse subidioblastic andalusite around sillimanite and rarely with kyanite suggests the low-pressure transformation of aluminosilicates in a subsequent event.

Sillimanite = Andalusite ... (5)

and

Kyanite = Andalusite (rare) ... (6)

In addition to reaction (4), Singh et al. (1996) also noticed the appearance of fibrolite/ sillimanite around a few subidioblastic muscovite in gneisses, which can be related to following reaction during prograde metamorphism.

Muscovite + Quartz = K-feldspar + Sillimanite + V ... (7)

K-feldspar: The sericitisation is very common in porphyroblastic K-feldspars of the granitoids, and suggests post-tectonic deformation. The secondary overgrowth textures i.e. feldspar crystallisation on sericitised K-feldspar in the granites and gneisses, suggests the recrystallization during ductile shearing. The perthitic feldspar is usually porphyroblastic in nature and is highly fractured, aligned in the S2 schistosity plane. Flame perthites are frequently found in ductile shears of granitoids.

Plagioclase: The high An content of plagioclase is common in mafic enclaves of the granite-gneisses. The plagioclase from sheared basic rocks is usually altered into epidote and sericite during the late retrogression. The albite rich plagioclase is also noticed from shear zones.

Quartz: Xenoblastic, medium to coarse grains are present in the ribbon, lensoidal and almond shaped grains in the rocks occurring near the thrust and the shear zones.

Chlorite: Medium to fine grained, retrograded chlorite occurs around biotite, feldspar, muscovite and garnet in the pelitic, quartzófeldspathic schists and granite-gneisses.

Chloritoid: It present in a schistose rock (chloritoid + magnetite + muscovite + quartz) exposed about 5 km north of Budhakedar along a shear zone. It is characterised by

Figure 3: **(a)** A mesoscopic tight isoclinal F1 fold, S2 and S3 foliation within MCT shear zone developed during D2 and D3 deofrmation (Bhagirathi valley).

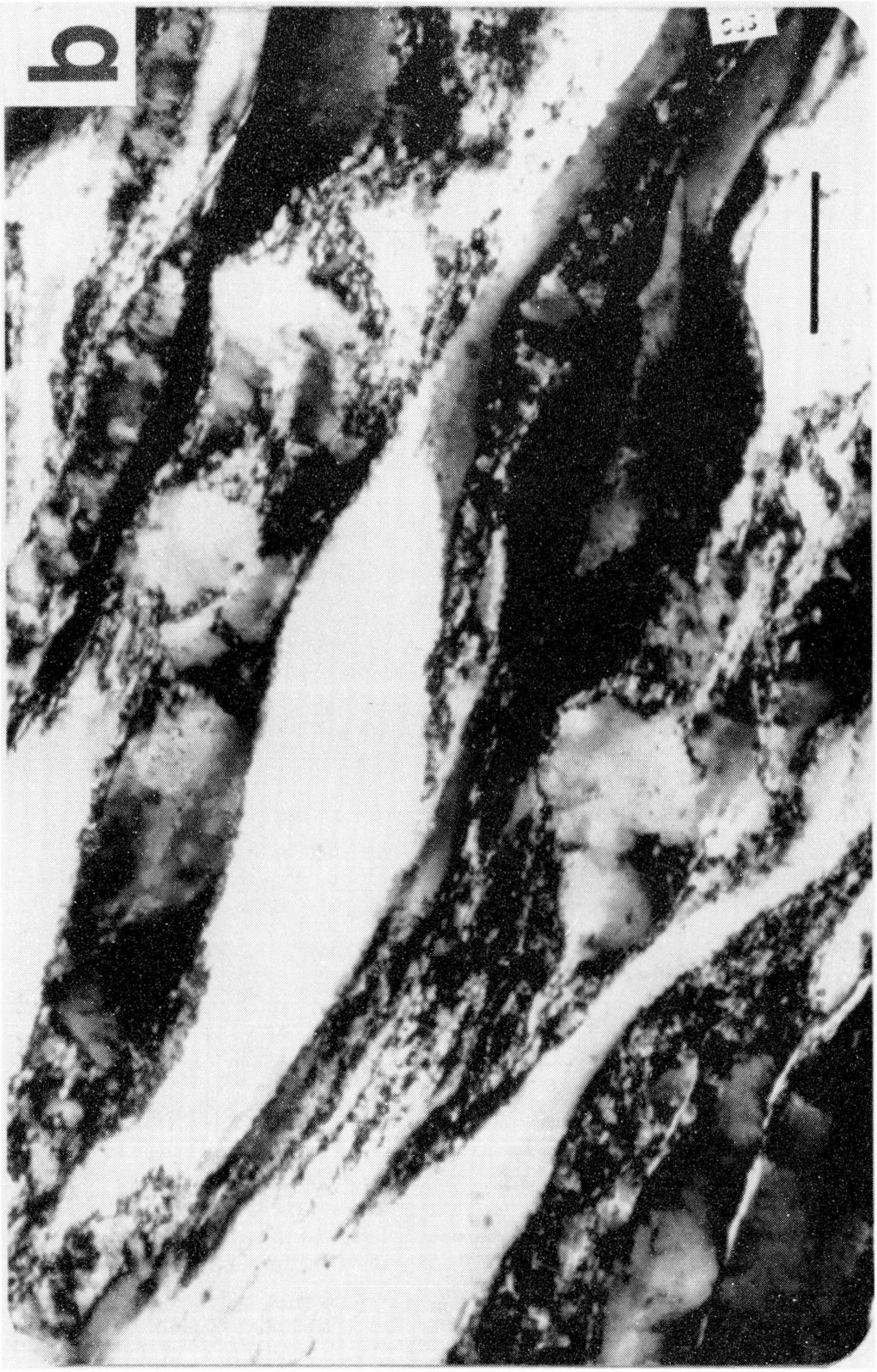

Figure 3: (b) Ribbon shaped quartz grains from mylonitised Chail Quartzite are from rocks exposed near Ghansyali, showing development of Boehm's lamellae along the main foliation. A few grains are mylonitised at the tapering ends during D3 and D4 deformation (scale bar = 500 m).

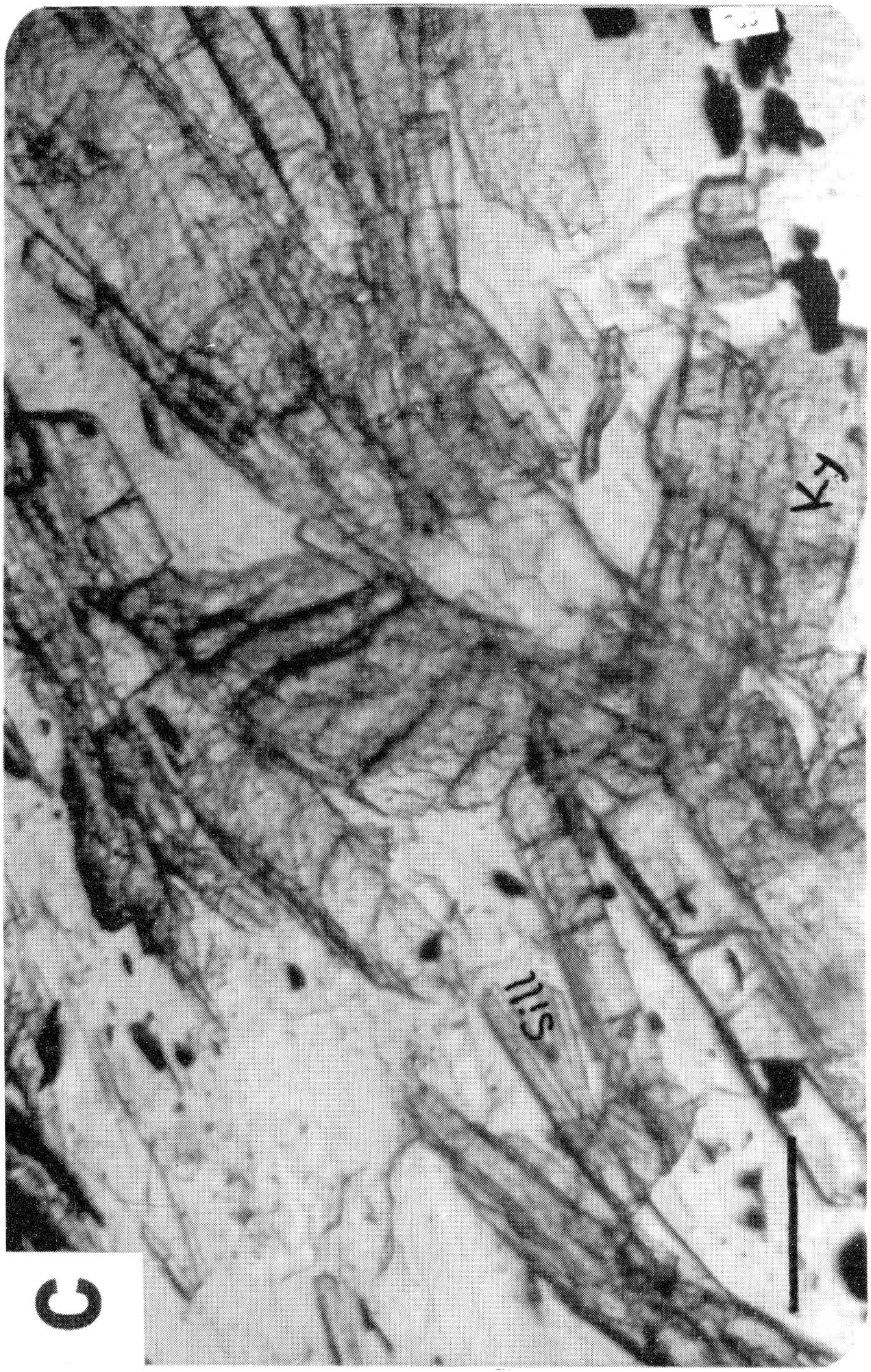

Figure 3: (c) The transformation of kyanite into sillimanite from sheared Chail schist exposed near Jhala showing prograde metamorphism in the kyanite-sillimanite schist (scale bar = 500 m).

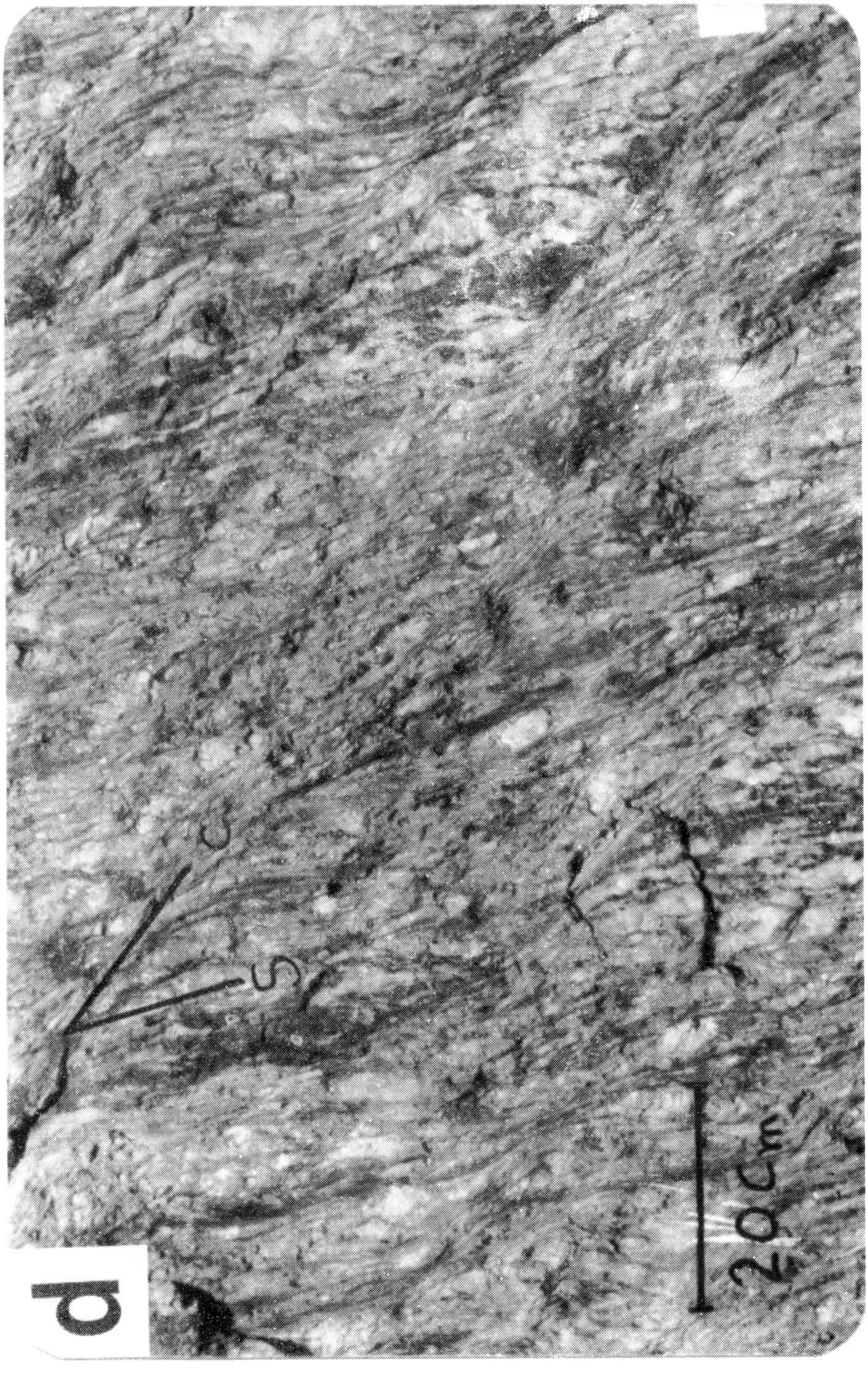

Figure 3: **(d)** Almond shaped S-C structures developed in granite-gneiss during late retrograde metamorphism (M2) and chlorite, sericite, biotite, muscovite minerals were formed along shear surface during D4 deformation, 500 m north of Budhakedar.

polysynthetic twinning. The textural relationship indicates post-tectonic crystallisation after MCT (D4-deformation).

Tourmaline: Medium to coarse- grained, strongly pleochroic with different shades, viz. dark yellow to green, brown to light yellow, green and brown. The tourmaline is stretched, and elongated along the S2 and S3 (=C) foliation.

Graphite: Medium to fine grained, defining the S2 schistosity, is present in some aluminosilicate schists.

METAMORPHIC EPISODE

The observed assemblages from various rocks of the area are shown by solid circles in AFM diagrams (Fig. 4). The different rock types with their varied mineral assemblages, textures, and grade of metamorphism are also given in Table 1. The textural study and mineral assemblages suggest that rocks of the Chail unit and MCT zone of Garhwal Himalaya are characterised by a Barrovian type metamorphism i.e., Ky-Bio to Sill-Mus zone, under medium pressure conditions (Singh et al., 1996 and 1997). Partial melting followed the sillimanite-K-feldspar zone, which led to the formation of migmatites associated with granite-gneisses. The kyanite-sillimanite and hornblende-quartz bearing pelitic and mafic enclaves, within the granite-gneisses (at many places), point out that possibly porphyritic granitoids were emplaced, at the late stage of metamorphism.

The textural relationship of the rocks of the Chail unit and MCT zone, discussed in the preceding section, suggests that the index minerals Ky-Mus+ Gt, Sill-Ky and Sill-Mus associated with different parageneses, were crystallised on S1 and S2 penetrative surfaces during the first episode of metamorphism (M1). The presence of all the reactant and product phases of the various reaction textures defining the different metamorphic zones along the S2 foliation point out that regional metamorphism was completed by D2 deformation (Singh et al., 1994 and 1996).

On the other hand, the crystallisation of biotite (greenish colour), chloritoid, epidote, muscovite (medium-grained) and sericite etc. were aligned on shear (-C) plane of granite and gneisses corresponding to the MCT/shear zone revealing superposed metamorphism (M2) on granite-gneisses, migmatites and metamorphites (of the M1 episode). The strongly rotated sigmoidal and porphyroclasts of feldspar indicate ductile to brittle-ductile conditions of deformation for the thrusting and shearing during D3 deformation. But the presence of spiral cores with idioblastic rims of garnet (very rarely present) and rotated cores of plagioclase mantled by idioblastic rims, suggest syn to post tectonic crystallisation during D3 deformation. The last phase of deformation (D4) is characterised by the crystallisation and alignment of chlorite, sericite, epidote, mica, apatite, tourmaline etc. in the new shear surface (C′). These fabrics cross the fabric developed during the D2 and D3 deformations (Fig. 3d).

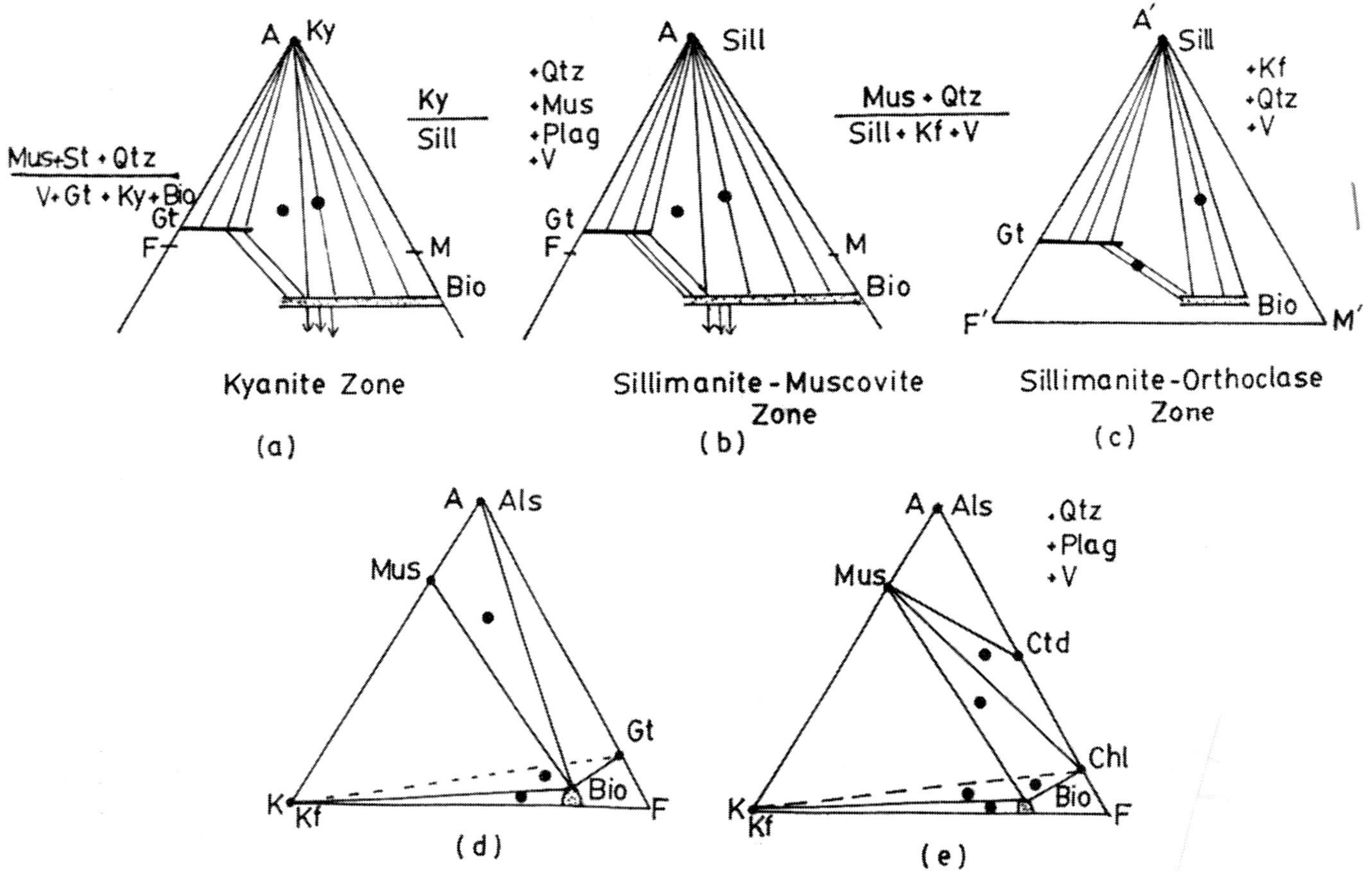

Figure 4: AFM and A'F'M' projections showing metapelites of amphibolite facies. The horizontal line represents the possible reaction isograd. AKF diagrams (d, e) represent the schematic changes of granite-gneisses into low-grade (M2) metamorphic assemblages during the late deformation stage. The solid circle represents the observed mineral assemblages.

Thus two phases of metamorphism and five phases of deformation can be suggested for the evolution of the Chail Unit and the MCT zone of Garhwal Himalaya (Fig. 5).

M1 Episode: The assemblages Ky+Gt-Mus-Bio-Kfs+Plag-Qtz, Sill+Ky-Mus-Kfs-Qtz observed from the MCT zone indicate the presence of the upper amphibolite metamorphic facies (shown in Fig. 4b & c and Table 1). The lower amphibolite facies rocks (staurolite zone) could not be traced out due to remote and difficult terrain. But it is reported from the MCT zone or equivalent rocks by several workers (Pande and Das, 1973; Sinha and Bisht, 1986; Islam and Thakur, 1988) from the adjoining areas of the Lesser Himalaya. The following reaction is proposed for the disappearance of staurolite at Devalikhal, Kumaon Himalaya (Pande and Das , 1973).

Staurolite + Muscovite + Quartz = Garnet + Kyanite + Biotite + V ...(8)

The presence of assemblages Bio + Ky-Mus-Qtz and Ky + Sill-Mus-Qtz and transformation of kyanite into sillimanite (Fig. 3c) from pelitic schist defines the beginning Sill-Mus zone by reaction (4) above.

However, the textural study points to the presence of several crystals of kyanite in the Sill-Mus zone, migmatites and gneisses. The stability of kyanite crystals at higher temperature may be due to: (i) reaction kinematics of aluminosilicate (Chatterjee, 1989) during the prograde sequence of regional metamorphism, (ii) related with sillimanite due to rapid heating during the course of granite emplacement at late stages of metamorphism followed by exhumation. The field study reveals that Sill-Mus and Sill-Kfs zones of the amphibolite facies are closely associated with anatexis and migmatisation (Fig. 2). The appearance of sillimanite/fibrolite along the cleavages of muscovite can be related to the discontinuous reaction (7) for the appearance of higher-grade metamorphic zones.

An experimental study (Chatterjee and Johannes 1974) yielded 650°C/5 Kbar for reaction (7) and marked the higher grade of metamorphism. The textural study reveals the presence of subidioblastic muscovite from migmatites and a few paleosomes of granite-gneisses. The presence of such muscovite at elevated temperatures may be due to the above reaction with addition of Al_2O_3 and FeO. Thompson (1982) pointed out that Tschermak-substituted muscovite can be stable and can co-exist with Sill + Kfs + Melt at higher temperature than the end-member muscovite. Therefore, this prograde subidioblastic muscovite may be considered as Tschermak muscovite.

M_2 Episode (Late Retrograde Stage)

Fine-grained chlorite, chloritoid, muscovite, sericite, epidote, tremolite and actinolite are widespread in the area, but sometimes coarse-grained flakes are also present along shear planes suggesting late retrograde metamorphism. This kind of metamorphism is characterised by low to medium grade assemblages viz. chlorite-muscovite-epidote-quartz-magnetite, chlorite-biotite-epidote-quartz, etc. (shown in Fig. 4e). These metamorphic assemblages are confined to shears of the MCT and are superimposed onto granite gneisses. The phase relationship is shown in the AKF diagram and the mineral paragenesis denoted by solid circles (Fig. 4d & e). The appearance of chloritoid, hornblende and andalusite confirms temperature of regional metamorphism.

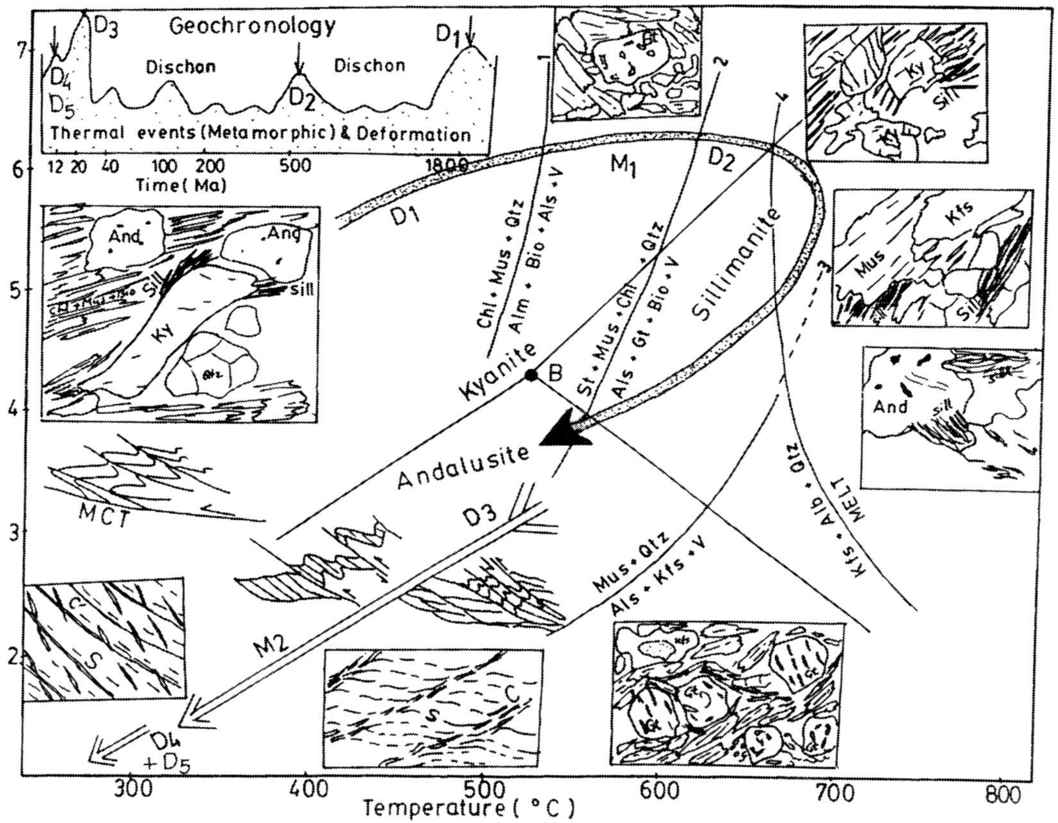

Figure 5: Tectonic and metamorphic evolution of the Chail unit and MCT Zone of Garhwal Himalaya. D1 to D5 represent the deformation espisodes, showing possible correlation with clustering of geochronological age (see text). The observed sketched photomicrographic textures in the inset. The experimentally obtained curves are (1) and (2) after Hoscheck (1969), (3) Chatterjee and Johannes (1974), (4) Ebadi and Johannes (1991). The dotted curve represents P-T path for metamorphism M1 from EPMA data (Saklani and Bahuguna, 1989). Al_2SiO_5 triple point is based on Bohlen et al. (1991).

Alteration and retrogression confined to thrusts, shears and fault zones can be related to the phenomenon of a late retrograde stage. Mylonitisation and cataclasis are well developed near the thrusts and fault contacts. The fracturing of porphyroblasts and ultra-mylonitisation are very common along the shear zones. The biotite and muscovite show alteration into chlorite and sericite. However, the late retrograde minerals also show S- C' foliation during D4 and D5 deformation (Fig. 5).

P-T-T PATH AND DISCUSSIONS

It is well established that the Himalaya evolved due to the collision of the Indian and Eurasian plates around 45 to 50 Ma ago (Fig. 6a). But pre-Tertiary metamorphism and granitoid emplacement within the rocks of the Himalaya had been reported by many workers (Sinha, 1989; Pognante and Lombardo, 1989; Ravishanker et al., 1995 and Singh et al., 1997).

Deformation in four phases (Valdiya, 1980 and 1986; Sinha, 1989 and Thakur, 1992) and in five phases (Schwan, 1980 and Singh et al., 1997) have been recorded from the Lesser and Higher Himalaya. Ray (1971) and Schwan (1980) have suggested that at least two phases of deformation of the CCZ in the Himalaya should be Precambrian in age. The clustering of geochronological data (ranging in age from 1850 Ma to 8 Ma for the Central Crystallines, see references in Singh et al., 1997) suggest many episodes of intrusive granitoids in the Himalaya viz. late-early Proterozoic (1800-2000 Ma), late Proterozoic to early Cambrian (500 + 50 Ma) and Tertiary (20 + 1 Ma).

Rb-Sr geochronological data obtained from different workers show a strong tectono-thermal event to have ocurred around 1800 Ma that may be related to the regional metamorphism, migmatisation, anatexis and granitization in the western Himalaya (Bhanot et al., 1979; Narayan Das et al., 1981; Trivedi et al., 1984; Singh et al., 1986; Chakraborty, 1993 and Ravishankar et al., 1995). The uninterrupted sequence of metamorphics associated with migmatites and granite-gneisses (MCT zone) of the Garhwal Himalaya, point that at least one of the metamorphic episodes was possibly Precambrian.

Vidal et al. (1982) recognized that the Manaslu leucogranite and metamorphism in Nepal Himalaya is of Tertiary age, but they suggested that Tertiary leucogranites have a high diversity in nature and might have been derived from older (Precambrian-Cambrian?) metamorphic and granitic terrains. This is also an indication of partial melting of older granitic and metamorphic masses of the pre-Tertiary sequences. Pognante and Lombardo (1989) noted that Central Crystallines contain the relics of pre-Himalayan tectonothermal and magmatic history apart from a Himalayan overprint. Due to this, the presence of chlorite, muscovite, chloritoid, andalusite, garnet and sericite from granitoids and from shears of the MCT zone metamorphics can be considered to be either (i) due to retrograde metamorphism, or (ii) related to Tertiary prograde metamorphism overprinted on M1.

In the polyphase metamorphic and deformation history of the rocks associated with Chail unit and MCT zone (Fig. 5 & 6), the presence of kyanite-sillimanite-andalusite-bearing schists points to a unique tectonic history. These textural studies show the transformation of (i) kyanite to sillimanite (Fig. 3c), (ii) sillimanite to andalusite and rarely kyanite to andalusite (reactions 4, 5, and 6). The transformation of kyanite into sillimanite may be related to the process of regional metamorphism, or may be related to rapid heating. But the appearance of andalusite at the expense of sillimanite (frequently), and kyanite (rarely), is important and it may be related to (i) decompression processes subsequent to regional metamorphism (M1) and emplacement of porphyritic granitoids (1800 Ma), or (ii) to Tertiary low pressure metamorphism, or (iii) combination of both the processes.

Most workers believe that the metamorphism in the Himalaya is only the consequence of a Tertiary orogeny, and they describe two episodes of metamorphism: (i) Prograde regional metamorphism (M1 - medium pressure- temperature) followed by, (ii) A prograde (low-pressure; medium to high temperature) metamorphism M2 (e.g., Hodges et al., 1988). However, there is a considerable disagreement about the age of the metamorphism (Singh et al., 1997). Some workers considered that both events are of Tertiary age while others ascribed Precambrian events followed by Tertiary events.

i. If both episodes (M1 and M2) are Tertiary as discussed by some workers, then M2 did not increase in temperature subsequent to M1.

ii. If the M1 metamorphism is Precambrian then presumably the metamorphic gradient of M1 relaxed in much lower values after this earlier orogeny (M1 metamorphism). In this case, Tertiary metamorphism (M2) should require an increase in temperature from the relaxed geothermal gradient.

The presence of prograde chlorite, biotite, andalusite, epidote, chloritoid, actinolite, hornblende and garnet minerals developed along shear planes and also overprinted on older schist, gneiss and granitoids from the Lesser Himalayan zone point to a low-grade regional metamorphism. Thus, the present study reveals that Tertiary metamorphism affected the early recrystallization to a significant depth. The presence of green schist facies assemblages along the thick shear planes in granite-gneisses is a strong evidence for a low-grade superimposed metamorphism (M2). The MCT zone comprises low- to medium- grade metamorphism of pelitic, psammitic and basic rocks. These older granitoids, calc-silicates and quartzites occur as lensoidal bodies within the MCT zone, which were transported towards south over the rocks of sedimentary unit from deeper levels (Fig. 6b). The folded metamorphic isograds within the MCT zone indicate their existence prior to thrusting.

CONCLUSIONS

The rocks of the MCT zone rmingled with Central Crystallines ramped up to various stages along imbricated MCT under brittle-ductile conditions from a depth of 14 km by (Jain and Anand, 1988; Saklani et al., 1991 and Saklani, 1993) during the Tertiary (20+1 Ma). However, the exact place and path of the MCT is still controversial.

Wide distribution of older ages around 1800-2000 Ma of granite-gneisses of the southern part of the Central Crystallines by different techniques indicate a strong tectono-thermal event which can be by correlated for granite, migmatite and the first regional metamorphism (M1) during D1 and D2 deformations.

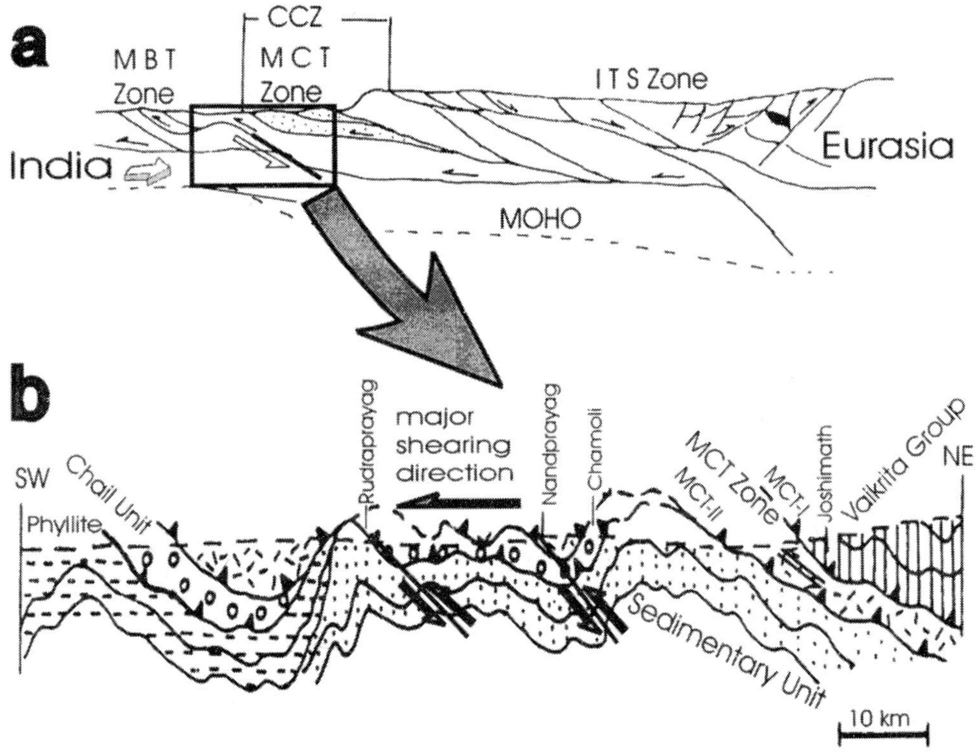

Figure 6: (a) General model of tectonic evolution of the Himalaya, after Hodges et al. (1988).

(b) Cross section along the Alaknanda valley, the Main Central Thrust zone shows folding and transportation towards south over the rocks of sedimentary unit from deeper level during late D3 deformation. High angle shears also developed during late D4 and D5 deformation under brittle condition.

The transformation of kyanite to sillimanite from metapelitic enclaves occurring within the granite-gneisses, and a sequence of prograde metamorphism without tectonic interruption, suggest that crystallines were subjected to 600-650° C /5-7 Kbar P-T conditions.

The prograde history of different metamorphic assemblages especially obtained from shear zones, suggests a Tertiary thermal event. The S-C' fabrics from low-grade assemblages of the M2 episode further suggest a subsequent deformation (D4 and D5) at shallower depths. Thus, a Precambrian prograde regional metamorphism can be inferred in the Garhwal Himalaya.

ACKNOWLEDGEMENTS

The author is grateful to Prof. P.S. Saklani for his supervision and guidance through out the work. I am also thankful to Prof. A.B. Thompson ETH, Zurich for his critical comments on manuscript and to Prof. S.P. Singh for assisting in the field. The financial grant provided by CSIR, New Delhi is gratefully acknowledged.

REFERENCES

Bhanot, V.B., Bhandari, A., Singh, V.P. and Kansal, A.K. (1979): Geochronological studies and geological studies on a granite of Higher Himalaya NE of Manikaran Himanchal Pradesh. J. Geol. Soc. India, V. 20: pp. 90-94.

Bhattacharya, A.K., Bhatnagar, G.S., Narayan Das, G.R., Gupta, J.N., Chabria, T. and Bhalla, N.S. (1984): Rb/Sr dating and geological interpretation of sheared granite-gneiss of Brijranigad, Ingedinala, Bhilangna Valley, Tehri Dist., U.P. Him. Geol., V.12: pp. 212-224.

Bohlen, S.R., Montana, A. and Kerrick, D.M. (1991): Precise determination of the equilibrium Kyanite - sillimanite, and kyanite -andalusite and a revised triple point for Al_2SiO_5 polymorphs. American Mineralogist, V.77: pp. 677-680.

Chakraborty, B.K. (1993): The Proterozoic orogeny and evolution of the granitic rocks of the Himalaya. Proc. Nat. Acad. Sci., V. 63 (a): pp. 253-280.

Chatterjee, N.D. (1989): Thermodynamics of minerals. Springer Verlag, New York.

Chatterjee, N.D. and Johannes, W. (1974): Thermal stability and standard thermodynamics properties of synthetic 2M1-muscovite, $KAl_2AlSi_3O_{10}(OH)_2$. Contrib. Mineral. Petrol., V.48: pp. 89-114.

Das, B.K. and Pande, I.C. (1973): Zone of progressive regional metamorphism in Dudhatoli syncline, Garhwal Himalaya. Him. Geol., V.3: pp. 190-208.

Ebadi, A. and Johannes, W. (1991): Beginning of melting and composition of first melts in the system Qz-Ab-Or-H_2O-CO_2. Contrib. Mineral. Petrol., V.106: pp. 286-295.

Grant, J.A. (1985): Phase Equilibria in Partial melting of pelitic rocks. In: J.R. Ashworth (ed:) Migmatites; pp. 86-144.

Heim, A. and Gansser, A. (1939): Central Himalaya, Geological observations of Swiss expedition 1936. Mem. Soc. Helv. Sci. Nat., V.73: 245p.

Hodges, K. V. and Silverberg, D.S. (1988): Thermal evolution of the greater Himalaya, Garhwal Himalaya. Tectonics, V.7: pp. 583-600.

Hodges, K. V., Hubbard, M.S. and Silverberg, D.S. (1988): Metamorphic constrains on the thermal evolution of central Himalayan orogen. Phil. Trans. Roy. Soc. Lond., V.A326: pp. 257-280.

Hoscheck, G. (1969): The stability of staurolite and chloritoid and their significance in metamorphism of pelitic rocks. Contrib. Mineral. Petrol., V.22: pp. 208-232.

Islam, R. and Thakur, V.C. (1988): Metamorphic history of Bhilangna valley, Garhwal Himalaya. Pub. Cent. Adv. Stud. Geol. P.U. Chandigarh, V.3: pp. 127-138.

Jain, A.K. and Anand, A. (1988): Deformational and strain patterns of an intracontinental collision ductile shear zone-an example from the Higher Garhwal Himalaya. J. Struct. Geol., V.10: pp. 717-734.

Metcalfe, R.P. (1993): Pressure, temperature and time constraints on metamorphism across the Main Central Thrust zone and High Himalayan Slab in the Garhwal Himalaya. In: P.J. Treloar and M.P. Searle (eds:) Himalayan Tectonics. Geol. Soc. Sp. Pub. No.74: pp. 485-509.

Narayan Das, G.R., Bhatnagar, G.S., Bhattacharya, A.K. and Sharma, Y.C. (1981): Uranium in metamorphites around Brijranigad, Raunsal, Tehri Garhwal (U.P.). In: Saklani, P.S. (ed:) Metamorphic Tectonites of the Himalaya; pp. 111-121.

Pognante, U. and Lombordo B. (1989): Metamorphic evolution of the High Himalaya crystallines in SE Zanskar India. J. Meta. Pet., V.7: pp. 9-17.

Ravishanker, Kumar, G. and Singh, G. (1995): Sequence stratigraphy and major geological events of Himalayas. Spl. paper Geol. Surv. India.

Ray, S. (1971): Petrogeny in Himalaya - A brief study of it's granites. Him. Geol., V.1: pp. 564-573.

Saklani, P.S. (1993): Geology of Lower Himalaya (Garhwal). International Periodicals & Publications, New Delhi; 245p.

Saklani, P.S. and Bahuguna, V.K. (1989): Geothermometry of Central Crystallines of the Bhagirathi valley, Garhwal Himalaya. In: Saklani, P.S. (ed:) Metamorphism, Ophiolites and Orogenic Belts; pp 1-16.

Saklani, P.S., Nainwal, D.C. and Singh, V.K., (1991): Geometry of the composite Main Central Thrust (MCT) in Yamuna valley, Garhwal Himalaya, India. N. Jb. Geol. Palaont. Mh., V. 1991(6): pp. 364-380.

Schill, E., Appel, E., Gautam, P. and Singh, V.K., (1998): Preliminary Palaeomagnetic Results of Medium grade Metacarbonates from the Lesser Himalaya. J. Nepal Geol. Soc., V.18: pp. 205-215.

Schill, E., Appel, E., O. Zeh, Singh, V.K. and Gautam, P. (2001): Coupling of late-orogenic tectonics and secondary pyrrhotite remanents: towards a separation of different rotation processes and quantification of rotational underthrusting in the western Himalaya (northern India). Tectonophysics, V.337: pp. 1-21.

Schill, E., Crouzet, C., Gautam, P., Singh, V.K. and Appel, E. (2002): Where did rotational shortening occur in the Himalayas? Inferences from Palaeomagnetic analysis of remagnetisations. EPSL (in press).

Schwan, W. (1980): Shortening structures in Eastern and Northwestern Himalayan rocks. (ed: P.S. Saklani), New Delhi, 62p.

Searl, M.P., Cooper, D.J.W. and Rex, A.J., (1988): Collision tectonics of the Ladakh-Zansker Himalaya. Phil. Trans. R. Soc. Lond., V.A 326: pp. 117-150.

Singh, R.P., Singh, V.P., Bhanot, V.B. and Mehta, P.K. (1986): Rb-Sr ages of the gneissic rocks of Rihee- Gangi, Bhatwari, Hanumanchatti and Naitwar areas of the Central Crystalline zone of Kumaon Himalaya (U.P.). Ind. J. Earth Sci., V.13: pp. 197-208.

Singh, V.K. (1993): Geology of Budhakedar area, district Tehri Garhwal, Lesser Himalaya, U.P. Unpublished Ph.D. Thesis, University of Delhi, 135p.

Singh, V.K., Singh, S.P. and Saklani, P.S. (1994): Kyanite- sillimanite -andalusite - bearing schist from Central Crystallines of Budhakedar area Garhwal Himalaya. Bull. Ind. Geo. Assoc. Chandigarh, V.27: pp. 143-153.

Singh, V.K., Singh, S.P. and Saklani, P. S. (1996): Metamorphic and geodynamic evolution of the Central Crystallines of the Garhwal Himalaya, India. N. Jb. Geol. Palaont. Abh., V.199: pp. 89-109.

Singh, S.P., Singh, V.K. and Saklani, P.S. (1997): Metamorphic evolution of the Central Crystallines of Higher Himalaya along Dhauli Ganga valley, Garhwal Himalaya. N. Jb. Geol. Palaont. Abh., V.206: pp. 249-275.

Singh, S.P., Singh. V.K. and Saklani, P.S. (1998a): Metamorphism and Thrust Tectonics in the Munsiari Group of Central Crystalline Zone of Garhwal Himalaya, India. J. Nepal Geol. Soc., V.18: pp. 85-95.

Singh, V.K., Singh, S.P., Saklani, P.S. and Dubey, C.S., (1998b): Mesostructures and deformational history of the Central Crystallines: an example from Garhwal Himalaya, India. J. Nepal Geol. Soc., V.17: pp. 59-69.

Sinha, A.K. (1989): Geology of the Higher Central Himalaya. John Wily & Sons. 219p.

Sinha, A.K. and Bist, K.S. (1986): K-Ar dating of the crystalline and basics rocks from Garhwal, Central Himalaya. Ind. Jour. Earth. Sci., V 13: pp. 209-215.

Sorkhabi, R.B. and Stump, E. (1993): Rise of the Himalaya: A Geochronologic approach. GSA Today, V.3: pp. 87-92.

Thakur, V.C. (1992): Geology of western Himalaya, (Pergamon Press Ltd) London, 355p.

Thompson, A.B. (1982): Dehydration melting of pelitic rocks and the generation of H_2O under-saturated granitic liquid. Am. J. Sci., V.276: pp. 425-454.

Trivedi, J.R., Gopalan, K. and Valdiya, K.S. (1984): Rb-Sr age of granitic rocks within the lesser Himalayan Nappe, Kumaun, India, J. Geol. Soc. Ind., V.25: pp. 641-654.

Valdiya, K.S. (1980): Geology of Kumaun Lesser Himalaya. Wadia Institute of Himalayan Geology, Dehradun; 219p.

Valdiya, K.S. (1986): Phanerozoic Plate Motions and Orogenesis: Central Himalayan Sector. The Indian Lithosphere, INSA, pp.73-82; New Delhi.

Vidal, P.H., Cocherie, A. and Le Fort, P. (1982): Geochemical investigation of the origin of Manaslu leucogranite (Himalaya Nepal). Geochem. Cosm. Acta., V.46: pp. 2279-2292.

A CONTRIBUTION TO METAMORPHIC HISTORY OF AN AREA OF CHAMOLI GARHWAL

B.C. Joshi

Department of Geology, Bundelkhand University, Jhansi - 284128, (UP)

ABSTRACT

Mesozonal as well as epizonal metamorphites characterize the northern proximity of Chamoli Garhwal. These metamorphites have been classified as medium- to high-grade gneisses, garnetiferous biotite schist with bands of amphibolites and quartzite, while medium- to low-grade metamorphites are constituted by pyritiferous carbonaceous schist, quartzite-phyllite-migmatite-marble with lenses of magnesites, amphibolites chlorite schist. These metamorphosed polytectonites represent the rocks of the Main Central Crystalline Axial Zone of the Himalaya, which came into its present phase by a process of cyclic uplift due to activation of "deep seated " or "mantle" faults which varied in space and time.

The vertical uplift of the Main Central Crystalline zone in stages caused folding, dislocation and igneous action in different phases of the rise of the Himalaya. This resulted in the development of overturned folds and their tectonic transportation towards south under gravitational force. This caused thinning and finally shearing of the inverted limb of the folded crystalline mass. The plane of separation formed the floor thrust. The gliding of Central Crystalline sheet along the floor thrust was accompanied by migmatisation, folding and faulting. During the course of tectonic translation of the over riding crystalline mass the body had been under great tensional stress in N-S direction. Due to being under tension a number of subsidiary low angle faults were developed from the floor thrust in form of roof thrust and branch line thrust. In the area under investigation the floor thrust has been formed by the Nagolgad thrust plane while Kalsir thrust forms the roof thrust. Hafla dislocation has been recognized as branch line dislocations.

INTRODUCTION

The Chamoli Garhwal is characterized by the Central Crystallines MCCA belt (Pande 1987) as well as less metamorphosed rocks of the Inner sub-zones of the Lesser Himalaya (Pande 1981), or "Inner schistose series" (Middlemiss 1887). Pande (1981) described similar metamorphites under the "Tectonic Crystalline Nappe Zone" of the Lesser Himalaya. These tectonometamorphic rocks form an integral part of the Garhwal Group (Auden 1939). Saklani. (1993) and Singh et al. (1998) reported the rocks of Garhwal Himalaya occurring as a schuppen zone formed by MCT-I, MCT-II and MCT-III.

The cyclic evolution of Himalayan polymountain system is evident from the occurrence of parallel to sub parallel structural belts from north to south (1) Trans Himalayan Zone (2) Main Central Crystalline Axial Zone (3) Lesser Himalayan Zone and (4) Tertiary Foot Hill Zone.

The axial zone became a positive landmass at the end of Caledonian tectonism. Since then the axial zone has been refolded and uplifted from time to time during the different phases of Tertiary uplift. The rocks of axial zone were subjected to repeated metamorphism, migmatisation and deformation. Earlier studies revealed that this region is constituted by polymetamorphosed polytectonites but the mechanism responsible for the origin of tectonostratigraphic and tectonometamorphic belt is yet to be solved.

Table 1: Showing various litho-units of Chamoli area.

Central Crystallines	1.	Migmatites
	2.	Garnetiferous-biotite schist
	3.	Mica schist
	4.	Interbanded schist and flaggy quartzite
	5.	quartzite
	--------------------- Kalsir -Nagthat Dislocation-------------------	
	(M.C.T. Roof Thrust)	
Medium -High Grade Metasedimentaries	Interbanded feldspathic biotite schist and flaggy quartzite	
	--------------------------Hafla Dislocation-----------------------	
	(Branch Thrust)	
Epizonal Metasedimentaries	1.	Quartzite
	2.	Biotite schist
	3.	Carbonates
	4.	Phyllites
	5.	Slates
	-------------------------Nagol dislocation-----------------------	
	(Floor Thrust)	
	Low grade Metasedimentaries of the Pokheri Group	

GEOLOGICAL SETTING

The present study forms a part of the Chamoli and Gopeshwar hills and falls between the latitude 30°22' to 30°30' N and longitude 79°15' to 79°30' E. The metasedimentries of the area belong to the Chail series of the Shimla Himalaya (Pilgrim and West, 1928; Ashgirie et al., 1977; Pande 1991 and Heim and Gansser, 1939). Heim and Gansser, (1939) observed in and adjoining areas of Shri Badrinath that the high grade rocks (Chail-Jutogh Group) are thrust over the less metamorphosed ones of the lesser Himalayan belt. They designated the tectonic junction plane as the "Main Central Thrust" (cf. Main Central Fault, Pande, 1987). In the Helang Mandal area the junction planes between the Central Crystallines, consist of mesozonal metamorphites and less metamorphosed Pokheri rocks forming a Schuppen Zone (Duplex Structure Fig. 3). The area evidenced three phases of deformation designated as D1, D2 and D3.The whole region witnessed four systems of folds i.e.F1, F2, F3 and F4. The MCT delineates the metasedimentaries from the gneisses of Chamoli region (Figs. 1 and 2). The lithounits of the Chamoli are summarized in Table 1.

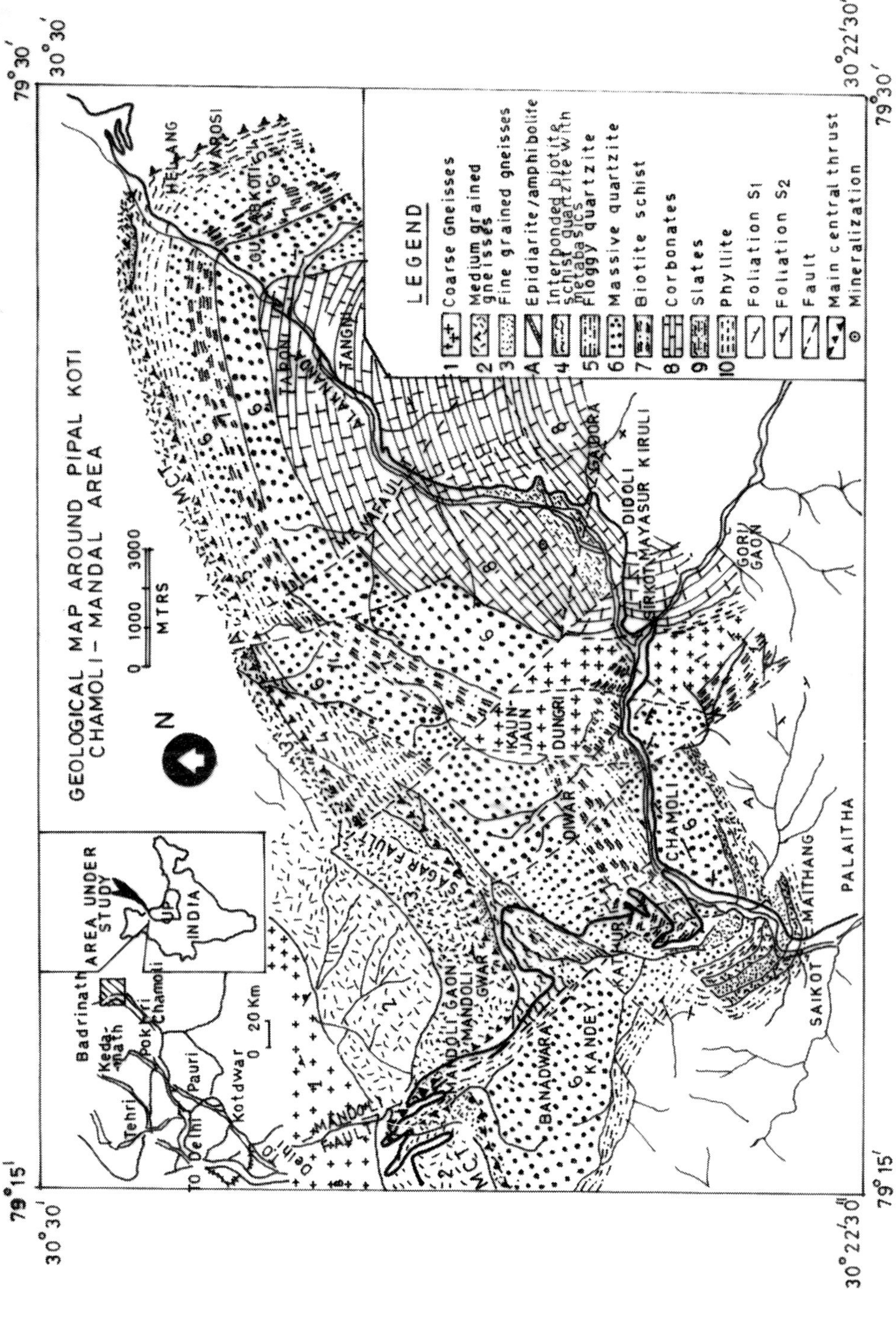

Figure 1: Geological map around Pipalkoti-Chamoli-Mandal area.

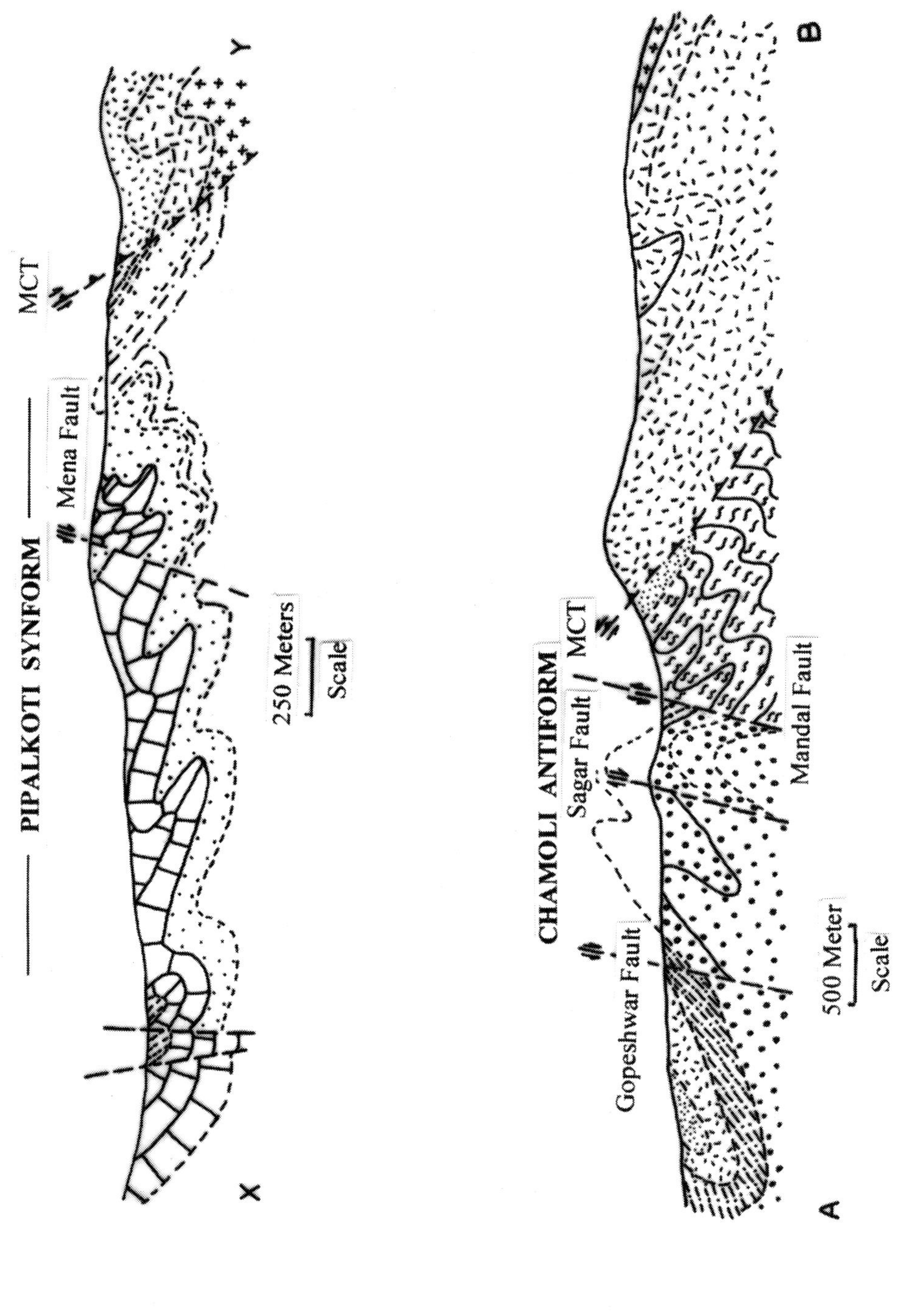

Figure 2: Geological cross section along X-Y and A-B lines.

The area is characterized by intricate folding and faulting and is represented by tectonometamorphic zones (Table 2). The geological history of the map area is closely related to the evolution of the "Main Central Crystalline Axial Belt" (Pande 1987,1991).

The original sediments of the MCCA basin measuring about 3000 m in thickness suffered under "Plutonic" metamorphism of the basinal history, during the Caledonian orogenesis. At the close of these earth movements the metasedimentaries with in the basin were subjected to deformative stresses with the commencement of the "orogenic stage" of the Basinal history. The event marked the upliftment of the MCCA belt (Table 2). At the end of orogenesis the MCCA belt (Fig. 1) became a positive land form i.e. origin of Proto Himalaya.

The Central Crystallines (Proto-Himalaya) suffered phased upliftment and related regional metamorphism during the successive orogenesis i.e. the Hercynian (Episode II), the Karakoram (Episode III) and the Tertiary (Episode IV, Table 2).

The metamorphites of Chamoli Garhwal area constitute a part of the "Lesser Himalayan zone " (Middlemiss, 1887); or form a segment of the "Garhwal series" or the "Garhwal Group" of Auden (1939). Mehdi et al. (1972) have described these lithounits as the "Chamoli Formation". The tectono-metamorphic rocks of the area, formed by medium to low-grade metamorphites fall under the "Tectono Crystalline Nappe Zone" (Pande, 1981). The metasedimentaries constituting the lithology of the map area partly belong to the "Inner sub-zone" of Lesser Himalaya. A geological traverse from north to south of the Chamoli Himalaya reveals the occurrence of the following lithounits (Table 3).

On the basis of mineral composition, texture, grade of metamorphism and tectonic setting, the lithounits of the area have been recognized to represent the medium grade metasedimentaries of the MCCA belt, in the northern part of the area. Kalsir-Nagthat Thrust (Fig. 3) tectonically delineates the metamorphites of the migmatite zones.

The thrust sheet forms a duplex structure. The schuppen zone consists of (a) the migmatite zone in the north, (b) the Helang Chamoli sub-group, delineated by Kalsir Nagthat dislocation (a) and tectonically rest over the southern part of nappe sheet. It has been designated as the Pipalkoti formation.

The rock units of the area represent varied rock suites. These range from gneisses, metapelites, metasemipelites, metapsammites, crystalline dolomitic limestone and metabasics.

Gneisses: These gneisses are products of pyrometasomatic derivatives representing mesozonal metamorphites of the MCCA belt. Quartz, feldspar (orthoclase, microcline,) are the major mineral constituents of these migmatites while biotite, muscovite and some chlorite form minor minerals. Apatite, epidote, zircon, tourmaline are the accessory minerals (Table 3).

Metasedimentaries: These rock units are (1) metapelites, (2) metapsammites and (3) Crystalline dolomitic limestones which are intricately folded and faulted.

Black slate is fine grained and splits into thin slabs. Quartz, chlorite and sericite form main components while black opaque minerals and iron oxides represent minor minerals.

Table 2: Showing metamorphic derivatives form the original minerals sediments laid in MCCA basin in the Pre-Caledonian time.

Possible Origin	Detritus quartz I, Sericite I, Chlorite I, Clay minerals, Oxides of Al, Fe, K, Na, and other minor elements		
Episode	Orogeny	Zones and Types	Minerals developed
I	Caledonian Orogeny (1050-390) a. Main stage of MCCA basin Chandra valley granite (512 Ma) Kawa diorite (414 Ma)	Load metamorphism M1 (M1a) Plastic viscous flow zones and migmatisation (M1b) Plastic flow zone (M1c) Plastic flow zone and Fracture zone	Muscovite II, Quartz, Biotite I, Garnet I,Sillimanite, Kyanite, Andalucite. Staurolite, Cordierite and Plagioclase I. Quartz II, Biotite I, Garnet I, Chlorite II, Kyanite (Rare). Sericite II, Chlorite II, Quartz II, Green Biotite, Iron oxides,
	b. Orogenic stage Kawa diorite (414 Ma)	Retrogression of high grade minerals, migmatisation II	Sericite II, Chlorite II and III, K -feldspar I with relict core of clouded plagioclase I, Garnet II, Biotite II with Pleochroic haloes.
II	Hercynian (520-240 Ma) Barkot Pandoh granite (325Ma)	Regional metamorphism M2 (M2a) Progressive migmatisation III	Quartz III, Biotite II, Garnet II, Plagioclase II, Orthoclase I, Muscovite II, Chlorite II.
		(M2b) Retrogressive stage	Chlorite III & IIIa, Fracturing of Garnet and Quartz, Biotite II degenerated to Chlorite IIIa with the formation of ilmenite and sphene.
III	Karakoram (280-60 Ma)	Regional metamorphism (M3a) progressive stage of migmatisation IV	Biotite III, Garnet III, K felspar II, Quartz IV, Plagioclase III
		(M3b) Retrogressive stage	Chlorite IV, Plagioclase III (clouded) K-feldspar highly sericitised.
IV	Tertiary (195-2 Ma Alpine orogeny) The MCCA belt witnessed repeated movements responsible for final uplift of the Himalayan polymountain system. Basic TH (90-130 Ma) Basic LH (50-40Ma) Amphibolites MCCAB (22-35 Ma), Ranibag Granite (10-25 Ma)	Regional metamorphism (M4a) progressive stage migmatisation Va Vb Vc Vd Ve	The metasedimentaries were migmatised during Cretaceous-Eocene time and suffered migmatisation (75-55 Ma). The high-grade minerals biotite III, Muscovite, Garnet III, Albite I, Plagioclase IV, were formed. The rocks again suffered migmatisation (50-40 Ma). The rocks were migmatised again (40-25 Ma). At final stage of the uplift of the orogen the rocks suffered fourth phase of migmatisation (25-10 Ma) The end stage of the tectogenesis is marked by appearance of chlorite V showing the late cataclasis and final uplift of the Himalaya.

Table 3: Showing lithology, mineral assemblage and metamorphic zones of the studied area.

Zone	Lithology	Mneral Assemblage
1. Migmatite Zone Mesozonal metamorphites of the MCCA belt (Pande 1987)	a. Coarse grained Biotite migmatite	Feldspars, augen measure 2.5x1.7 cm. Biotite I, Muscovite, Quartz II
	b. Coarse grained streaky augen migmatite	Biotite-II and Quartz-II form interbanding. Quartz-II and feldspars form augen, 1-2 cm in length.
	c. Medium to fine grained migmatite	Feldspar crystals measure about 0.7 cm in length, Biotite, Muscovite and minor minerals.
	d. Fine grained chlorite migmatite	Chlorite and Biotite wisps give rise to schilleren structures.
Kalsir Nagthat Thrust (MCF) Roof Thrust		
2. Metapelites Helong Chamoli sub group forming thrust sheet over LH belt	a. Inter banded felspathic biotite schist and flaggy quartzite	QuartzII, Biotite II, Feldspar and minor minerals. Feldspars are clouded and Quartz forms lenses
	b. Massive Quartzite with bands of mica schist strings	QuartzII, Biotite II, Chlorite sericite, few grains of clouded plagioclase
Hafla Dislocation (MCF2 branch thrust)		
3. Metapelites Pipalkoti Formation	a. Carbonate Unit	Crystalline dolomatic limestone with siliceous bands and lenses of magnesite
	b. Slates	Quartz I, Sericite, Chlorite, Calcite and iron oxides
	c. Phyllites with schistose Amphibolites	QuartzII, Chlorite, Sericite, Hornblende, and iron oxides,
	d. Migmatite	Quartz, Feldspar, Biotite II, Chlorite and minor minerals
Nagol gad Dislocation (MCT3 floor thrust) Pokheri Group		

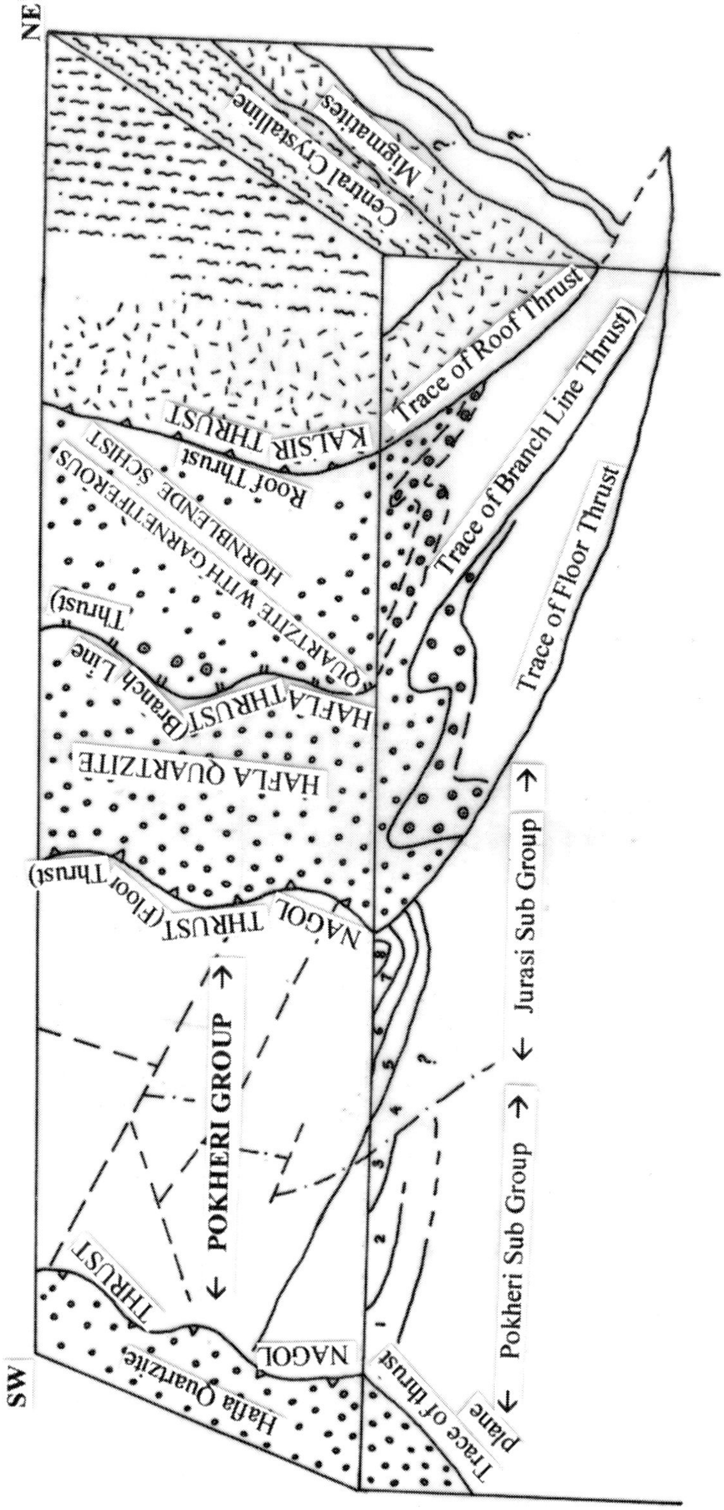

Figure 3: Diagrammatic Model of Pokheri-Mandal area showing Duplex structure (after Pande, 1991). Legend: 1: Pyritiferous Graphite Schist, Quartzite & Flaggy Schist, 2: Interbanded Schist & Quartzite, 3: Interbanded Quartzite-Phyllite, 4: Interbanded Chlorite Schist & Quartzite, 5: Phyllite, Quartzite, Marble with Lenses of Magnesite, 6: Amphibolite, 7: Talc Schist, 8: Metabasics.

Greyish to silvery grey or greenish grey phyllite is a fine to medium grained rock. Quartz, biotite muscovite, chlorite with some feldspars are the main mineral constituents while sphene, hematite and dusty opaque minerals form the accessories.

Bluish green to silvery grey or copper brown schist, with three sets of cleavages are constituted by quartz, biotite, muscovite, and chlorite, some feldspars along with zircon, sphene and iron oxides are accessory minerals.

The geologic attributes preserved in these rocks have been minutely studied. The results of the studies reveal that the litho-belts of the Helang Mandal are the products of a long and phased tectonic metamorphic history.

TECTONOMETAMORPHIC HISTORY

The metasedimentaries of the Chamoli-Garhwal are represented by the metamorphites, ranging from mesozonal to epizonal rocks and reactivation the Main Central Crystalline Axial belt (MCCA Belt), where as the low grade metamorphites, constituting the southern part of the map area, belong to the inner sub-zone of the Lesser Himalayan Belt (LH Belt; Pande, 1987).

These metamorphites are the products of phased metamorphic and geo-tectonic "Episodes". These episodes are closely related to rejuvenation of old Achaean basement (1350-1700 ma). Activation of mantle faults in response to igneous activity, formation of different arena of sedimentation, development of metasedimentary/sedimentary prism and their phased upliftment, varied in space and time (Joshi, 1986; Joshi and Pande, 1989; Pande, 1983, 1984 and 1987). The geological imprints in these metasedimentaries point towards a complex metamorphic history related to phased evolution of the Himalaya. Four major episodes have been recognized for the evolution and development of the Himalaya. The tectonometamorphic events are related to different orogenic cycles which commenced with the dawn of the Caledonian orogeny and then followed by the Hercynian and Karakoram tectonic movements; and finally terminated with the earth movements of the Tertiary era (Table 2).

Episode-I: This tectonometamorphic event is related with the formation of MCCA depositional basin, with the commencement of Caledonian orogeny as a result of mobilization of basement rock, caused activation of mantle faults, thereby marked by slow subsidence of the faulted blocks and deposition of sediments, as referred previously by Pande (1991).

The rocks of the MCCA belt representing the oldest sediments were subjected to varied P.T conditions, according to the depth of burial. This resulted in plutonic metamorphism of Proterozoic to Ordovician sediments (550-320 Ma, Pande 1991).

The imprints of the load metamorphism in the rocks of the area are poorly preserved. However, the occurrence of migmatites in the northern part of the area reveals that these metasedimentaries belong to the zone of "plastic flow". Biotite, garnet with some kyanite in these rocks support the mesozonal conditions of metamorphism under P.T.conditions of 4.5-6 Kb and 450-600 degree centigrade. During this event, the rocks acquired bedding schistosity

S1 (S0=S1), marking the "main stage "of basinal history. The "orogenic stage " is marked by upheaval of the metasedimentary prism with the development of isoclinal folds (Fd1). This stage also marks the degeneration of high-grade minerals. These rocks were refolded and regionally metamorphosed during successive periods of orogenesis.

Episode II: This is related to Hercynian orogenesis (520-240Ma). During this event the MCCA belt was further uplifted and the rocks of the belt suffered regional metamorphism. During the progressive stage development of biotite II, muscovite and garnet II took place. The Garnet II is marked by quartz inclusions, which define 'S' or 'Z' arrangement developed under stressed conditions. The platy minerals are the products of recrystalization and define axial plane cleavage (S2a) in these metamorphites with quartz grains running parallel to the schistosity (Pande and Das, 1973). The retrogressive stage is marked by appearance of Chlorite III with development of tight isoclinal folds (Fd1a) and chlorite III. This event is marked by intrusion of quartz diorite (414 Ma) and Chandra valley granite (510 Ma) into the MCCA belt.

Episode III: During Karakorum orogenic movements, the streses migrated to the north and south of the MCCA belt. This caused development of new amphitheatre of sedimentation i.e. Trans Himalayan and Lesser Himalayan sedimentary basins, in response to igneous action (Alkali granite, 235 Ma; Porphyritic granite, 190 Ma; Basics and ultrabasics, 130-80 Ma).

During this episode the rocks were further uplifted and suffered regional metamorphism. Appearance of reddish brown biotite III, quartz III, garnet III, and chlorite III took place in response to the reaction of the invading magma, which resulted in migmatisation of the country rock, with the development of feldspars.

The retrogressive stage is marked by degeneration of high grade minerals and appearance of chlorite IV.The rocks were refolded as open folds nearly coaxial with Fd1. The rocks also acquired slaty cleavage S2a and crenulation-cleavage-S3 (Pande, 1981).

Episode IV: With the incoming of Tertiary orogeny the MCCA belt along with T-H and L-H belts, was further uplifted with intrusion of igneous rocks (Trans Himalayan granites 130-80 Ma; Khardung volcanism 555 Ma; Mandi granite 37-24 Ma and Badrinath tourmaline granite 10-19 Ma,

The tectono-metamorphic history related with the episode IV is rather complicated. This period is marked by phased uplift of the MCCA belt along with and LH belts and development of southern most depositional basin (Tertiary outer foot hill zone). The axial structural belt might have reached a critical height by then. The MCCA belt during post Eocene period suffered southward gravitational translation to form the crystalline nappe over the LH belt (Pande, 1991). This sheet apparently rode over the epizonal metamorphites of the area Fig 1.

During the Tertiary tectogenesis the rocks of MCCA belt were refolded (Fd3 and Fd4). During the end stage of upliftment the rocks acquired flexure folds (Fd4) and fracture cleavage (S4). The final upheaval of the Himalaya is marked by the occurrence of Chlorite IV associated with late cataclasis.

CONCLUSIONS

The rocks of the region suffered repeated metamorphism, folding and faulting. The occurrence of varied types of gneisses and migmatites around Mandal and Gwar confirms the view that these rocks are products of both load and regional metamorphism. Minor isoclinal folds, striations, graded bedding and current bedding throw light on the lithostratigraphic history of the area. The minerals of different generations are evidences of the phased regional metamorphism that took place during different periods of upheavals.

Intracrustal fracture MCF is a major structure of the area, which demarcates the boundary between the MCCA belt and LH belt. Further the nappe, apparently resting over the rocks of the LH belt, is characterized by a Duplex structure.

ACKNOWLEDGEMENT

The author is highly thankful to Prof. I. C. Pande, FNA for his constant encouragement and improving the manuscript during the preparation of this paper. Financial assistance from DST (Govt. of India) is also acknowledged.

REFERENCES

Ashigirie, G.D; Sinha, A.K; Pande, I.C. and Malik, B.C. 1997 A contribution to the geology and geochronology and history of regional metamorphism of Himachal Himalaya, Him. Geol; Vol.7, pp. 102-117,

Auden, J.B. Director's annual report for 1939,Record Geol.Surv.India, 78, pp.70-78, 1940.

Heim, A and Gansser, A.1939 Central Himalayas, Geological observation of the Swiss expedition, 1936, Mem. Soc. Helv. Sci. Nat. V.13, pp.1-245, .

Joshi, B.C. 1986 Geology of the Pokheri area, District Chamoli, Garhwal, Unpublished PhD thesis, Garhwal University, Srinagar Garhwal, 132 p,

Joshi, B.C and Pande, I.C. 1989 Geology of the Helang Mandal area, Distt.Chamoli (Garhwal). Metamorphism, Ophiolites and Orogenic belts, In Saklani ; P.S. (Ed.) Today and Tomorrow's Printers and Publishers, 223-234, .

Joshi, B.C and Pande, I.C. 1992 Structural studies: A case history from Pokheri area, District Chamoli Garhwal, Uttar Pradesh. Jour. Him. Geol 3 (I),102-110, .

Mehdi, S, H., Kumar, G and Prakash, G. 1972 Tectonic evolution of Eastern Kumaun Himalaya: A new approach, Him Geol (2), 481-501, .

Middlemiss, C. S. 1887 Crystalline and Metamorphic rocks of the lower Himalayas, Garhwal and Kumaun, Sec. (1), Rec. Geol. Surv. India 20, (4) 134-143,

Pande, I.C. 1991 Tectonic and Metamorphic Investigations of Kumaun- Garhwal Himachal Lesser Himalaya: In: Saklani, P.S (Ed), Today and Tomorrow's Printers and Publishers, New Delhi 216p,

Pande, I.C.1983 Shifting of geosynclines vis- a -vis Himalaya.Abstr. Proc. Ind. Geol. Assoc. 4th Session, B.H.U; Varanasi,

Pande , I. C. 1984 Deep fracture tectonics vis-à-vis NW Himalayan orogen. Abstr. Proc. Nat. Sem. Geol; Delhi University; Delhi, p.3,

Pande I. C. 1987 Deep-seated faults and tectogenesis. Proc. Nat. Sem. Ter.Orog; B.H.U; Varanasi.pp.77-89,

Pilgrim, G.E and West, W.D. 1928 The structures and correlation of the Simla rocks. Mem. Geol. Surv. Ind. 53, 140 p.

Saklani, P.S., 1993. Geology of Lower Himalaya (Garhwal). International Periodicals & Publications, New Delhi, 245p.

Singh, S.P., Singh, V.K. and Saklani, P.S. 1998 Metamorphism and thrust tectonics in the Munsiari Group of Central Crystalline Zone of Garhwal Himalaya. J. Nepal Geol Surv., 18, 85-95.

Pande, I. C. and Das B.K. 1973, Zone of Progressive regional metamorphism in Dudatoli syncline, Garhwal Himalaya. Him, Geol., 3, 190-208.

NEOTECTONIC AND MORPHOTECTONIC EVOLUTION OF A PART OF NORTH EASTERN KUMAUN HIMALAYA

Rameshwar Bali

Department of Geology, Lucknow University.

ABSTRACT

A part of the Northeastern Kumaun Himalaya in the Dharchula-Tawaghat sector has been studied from the viewpoints of slope instability including catastrophic landslides that continue to damage life and civic constructions. The washout of the Sobala microhydel power station and other similar occurrences are common in the area. The functioning of several microhydel projects as well as the upcoming Dhauliganga Hydroelectric project is getting hampered due to constant problem of hill slope instability. The satellite data and field studies show that the entire stretch from Tawaghat up to Sobala is characterised by the occurrence of simple and compound debris cones resting at an angle repose of around 45°. These unconsolidated, noncohesive and geotechnically weak fans constitute the sites prone to hill slope instability in the area. An attempt has been made to relate the morphotectonic evolution of these agradational landforms to the thrust movement during the Himalayan orogeny. The neotectonic activities, together with creep movement and anthropogenic activity on the reactivation of the debris cones are discussed.

INTRODUCTION

In tectonically active regions, the tectonic processes control largely the geomorphological set-up of a region as well as its geomorphic processes and evolution. The surface of the earth acts as a limiting surface between tectonic processes that displace the crust, and geomorphic processes that redistribute material across its surface. The Himalayan mountain belt was evolved due to the collision of the Indian and the Asian plates. The youthful topography of the Himalayan region was evolved in response to continuous tectonic adjustments taking place due to interaction of the two plates.

In the light of a detailed morphotectonic study carried out in the area around Tawaghat and Sobala of Northeastern Kumaun Himalaya, it has been observed that the present day geomorphological set-up of the region is characterised by numerous debris cones, terraces, deep gorges, ingrown meanders etc.

A number of rivers such as the Kali and Dhauliganga drain through the area. Due to the availability of water, a number of hydroelectric projects (e.g. the Dhauliganga hydroelectric project, Kanchuti, Sobala, Chhirkila and Khet microhydel projects etc.) have come up in recent years. Because of ground instability, on a general basis, there is always a possibility of

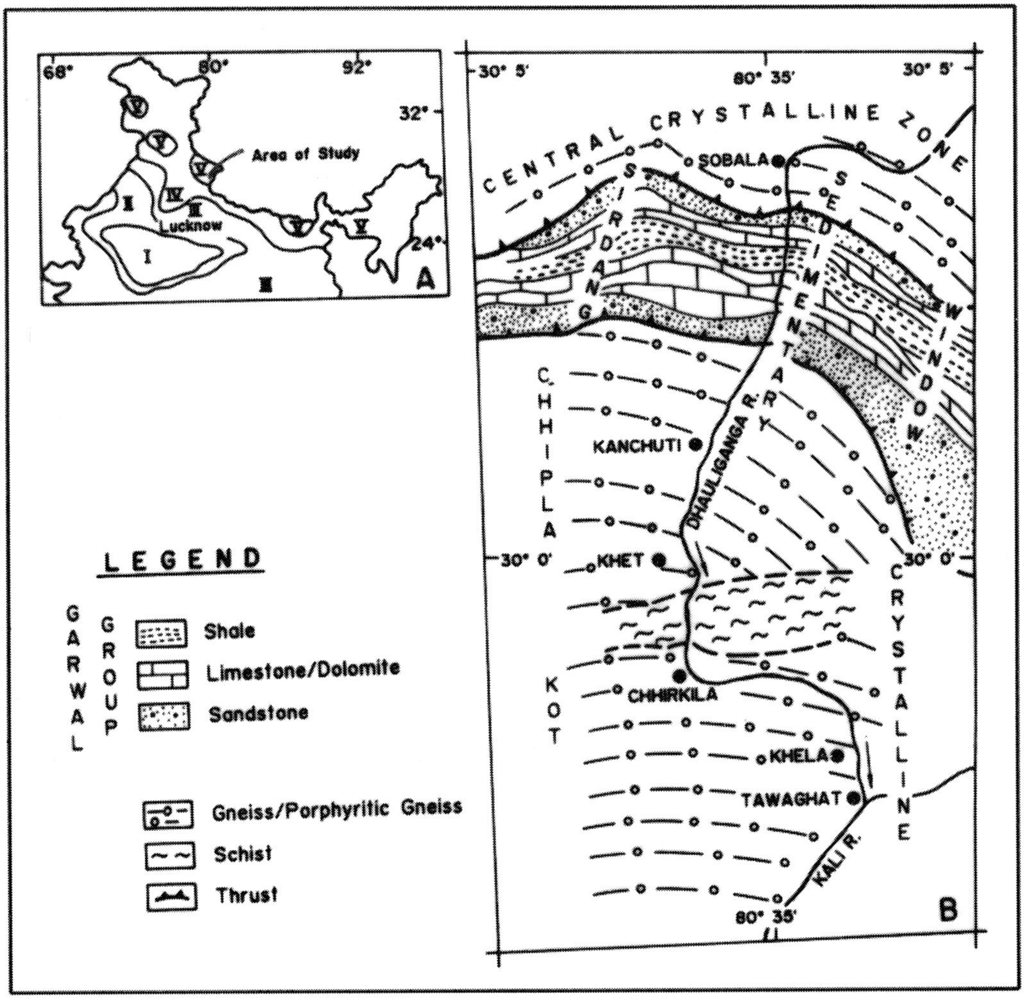

Figure 1 A: Location of the area in the Seismic Zonation map of India. The area falls in Zone V of the map.

their instability. The paper presents a study of the morphotectonic features of the region with special reference to the nature of instability of the ground.

GEOLOGICAL SETTING OF THE AREA

The area of investigation forms part of the Northeastern Kumaun Himalaya and extends from Tawaghat to Sobala for nearly 30 kms. north of Dharchula township (Fig.1). The area is demarcated by the rocks of the Chhiplakot Crystalline mass in the south and Central Crystalline Zone in the north. The intervening area is occupied by the Lesser Himalayan sedimentary rocks exposed in the form of a tectonic window commonly referred to as the Sirdang Sedimentary Window (Sinha, 1989). The Chipplakot Crystalline Mass is dominantly constituted of gneisses, porphyritic gneisses and schists. The porphyritic gneissic rocks have foliation planes dipping 40° - 45° in NNE to NE direction and are made up of medium to large augen of quartz and feldspar wrapped within the mica layers. The rocks in general appear to be highly competent and weather- resistant. The rocks of the Inner Sedimentary Belt (= Garhwal Group) having tectonic contacts towards its north and south are constituted by limestone, dolomite, sandstone and shales. These rocks at places, show the effects of shear metamorphism on a local scale as a result of which they resemble marble, quartzites and slates respectively.

TECTONIC AND STRUCTURAL DESIGN

The present-day morphological set-up of the area seems to be controlled by the tectonic evolution of the Himalayan mountain belt during the Himalayan orogeny. The northward movement of the Indian plate resulted in large-scale south directed tectonic transport of thrust sheets. The thrust sheets with their root zone in the Central Crystallines had once covered the entire Lesser Himalayan Sedimentary Belt (Agarwal, 1994). Subsequently, part of these thrust sheets have been eroded resulting in the exposure of the underlying Lesser Himalayan rocks (Singh and Agarwal, 1996) in the form of a tectonic window. The Sirdang Sedimentary Window is bound by Chhiplakot thrust (fault) in the south and the Munsiari thrust and the Main Central Thrust in the north (Valdiya, 1980). The sub-aerial translation of the crystalline thrust sheets caused brittle deformation of the underlying sedimentary rocks as well as the base of overriding crystalline sheets (Searle et.al.,1987). The Chhiplakot crystalline mass represents a crystalline thrust sheet in the area. During this phase of tectonic transport of the crystalline sheet, the underlying sedimentary rocks were subjected to shear metamorphism.

The Northeastern Kumaun Himalaya in general is highly prone to seismotectonic activities (Verma et al., 1977). The study area falls in Zone V of the Seismic Zoning map of India (National Atlas of India, 1996). Apart from regular micro-seismic events, this region shows the highest frequency of 5 and above magnitude earthquake in the entire Himalayan belt (Valdiya, 1986), the strongest being the M 7.5 Dharchula - Bajang (Nepal) earthquake of 1916. The 1979 Seraghat earthquake with a magnitude of 6.5 affected nearly 700 sq kms of the area around the epicentre. Dharchula has been the epicentre of two earthquakes, viz. 1980 (M 6.1) and 1987 (M 6.2) earthquake. A number of rockfalls, slips and slides have been initiated in the entire region due to these movements thereby causing extensive damage to various masonry structures and civic constructions (Narula and Shome, 1992). A number of smaller faults run along and across the Dhauliganga river (Gupta and Agarwal, 1994). Three major faults are present in the vicinity of the area viz. the E-W trending dextral strike-slip faults called the Chhiplakot fault to the south, the Baram fault to the south-west and the NNW-SSE trending fault to the west. The three faults intersect near Seraghat and movement along one or all of these resulted in the ground instability of the area (Paul,1986).

GEOMORPHOLOGICAL FEATURES OF THE AREA

The Tawaghat- Sobala area has a typical geomorphological set-up that gradually evolved in response to the prevalent tectonic, structural and other related geological processes.

The entire stretch from Tawaghat to Sobala is typically marked by the presence of a thin veneer of fan shaped deposits or Debris cones on the hard and compact rocks bordering the Dhauliganga river. These cones mostly occur, at places, as compound cones due to the merger of the individual fans (Fig.2). Such features are observed commonly along the right bank of the river from Tawaghat up to Kokalkhet village and around the Khet and Chhirkila areas. The debris material in most of the cones is highly heterogeneous and is constituted of angular boulders of gneiss embedded in a matrix of sand and silt size particles (Fig.3 B,C). The material as such is highly non-cohesive and unstable. A number of cones have stabilised at an angle of repose of around 45^0, however, others have been reactivated due to toe erosion during flash floods as well as due to anthropogenic activities. It has been observed that such intense activities of fan formation in similar geological conditions are absent in nearby areas.

The area seems to be affected by neotectonic activity at a number of places. Both the Dhauliganga and Kali rivers flow along structurally controlled valleys, the former flowing along a north-south trending fault from Sobala to Tawaghat (Fig. 3 A). This fault has even offset the east-west trending MCT at around Nyu (Gupta and Agarwal, 1994). This valley, in turn, is further offset by two later east-west trending transverse faults, one at Sobala and the other around Chhirkila. Both the rivers are entrenched and cut deep into the valley floor. Detailed study of the IRS- PAN satellite data shows the presence of three to four levels of well developed ingrown meanders within the Kali valley around Dharchula (Fig. 4). It is observed that around such meanders, the cliff faces are developed on the outer side of meander lobes

Well-developed terraces, constituted of heterogeneous debris material, are conspicuously noticed above the present-day Dhauliganga river channel (Fig.3 D). Persistent downcutting resulted in the formation and preservation of these geomorphic units. At least two to three levels of such terraces, which are mostly unpaired in nature, are prominently present between Sobala to Tawaghat. Although the terraces are developed on both the banks of the river, those around Nyu, Kanchuti and Khet villages are very prominent.

Soil creep constitutes another significant mode of slope modification in the area. This has resulted in the development of prominent tensional cracks in the slopes, bending of

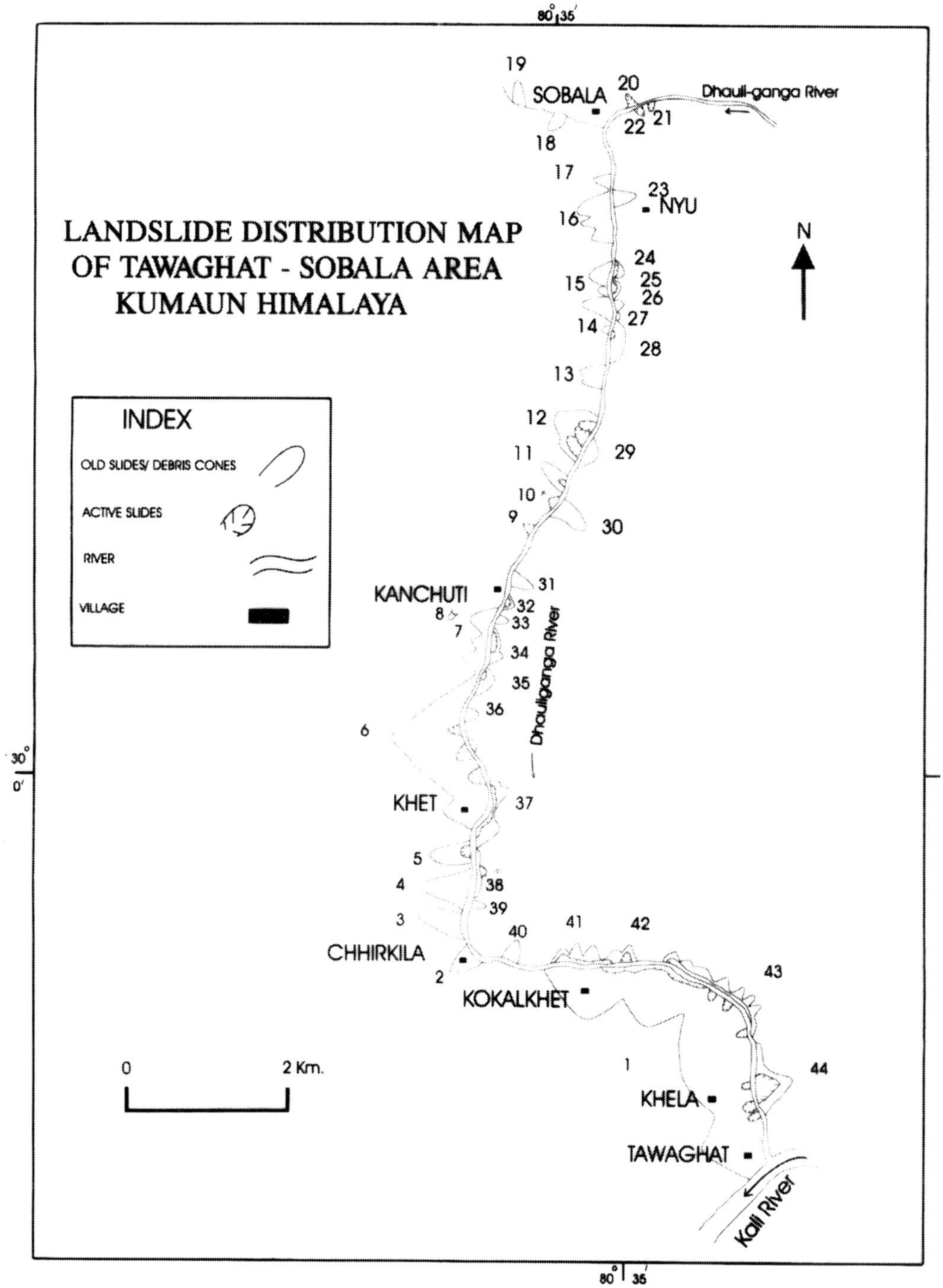

Figuer 2: Landslide distribution map of the area prepared using IRS I C PAN data . The distribution of simple and compound debris cones has been shown.

Figure 3: **A.** Dhauliganga river flowing along a faulted and entrenched valley upstream of Tawaghat. The area is marked by the presence of a series of failed debris cones present along the streams.

B. Road constructed along a stable debris cone downstream of Kanchuti. The slope has started showing signs of failure

C. Flash flood led toe erosion has reactivated movement along the debris cone south of Nyu.

D. River terrace indicating neotectonic movement along the Dhauliganga valley.

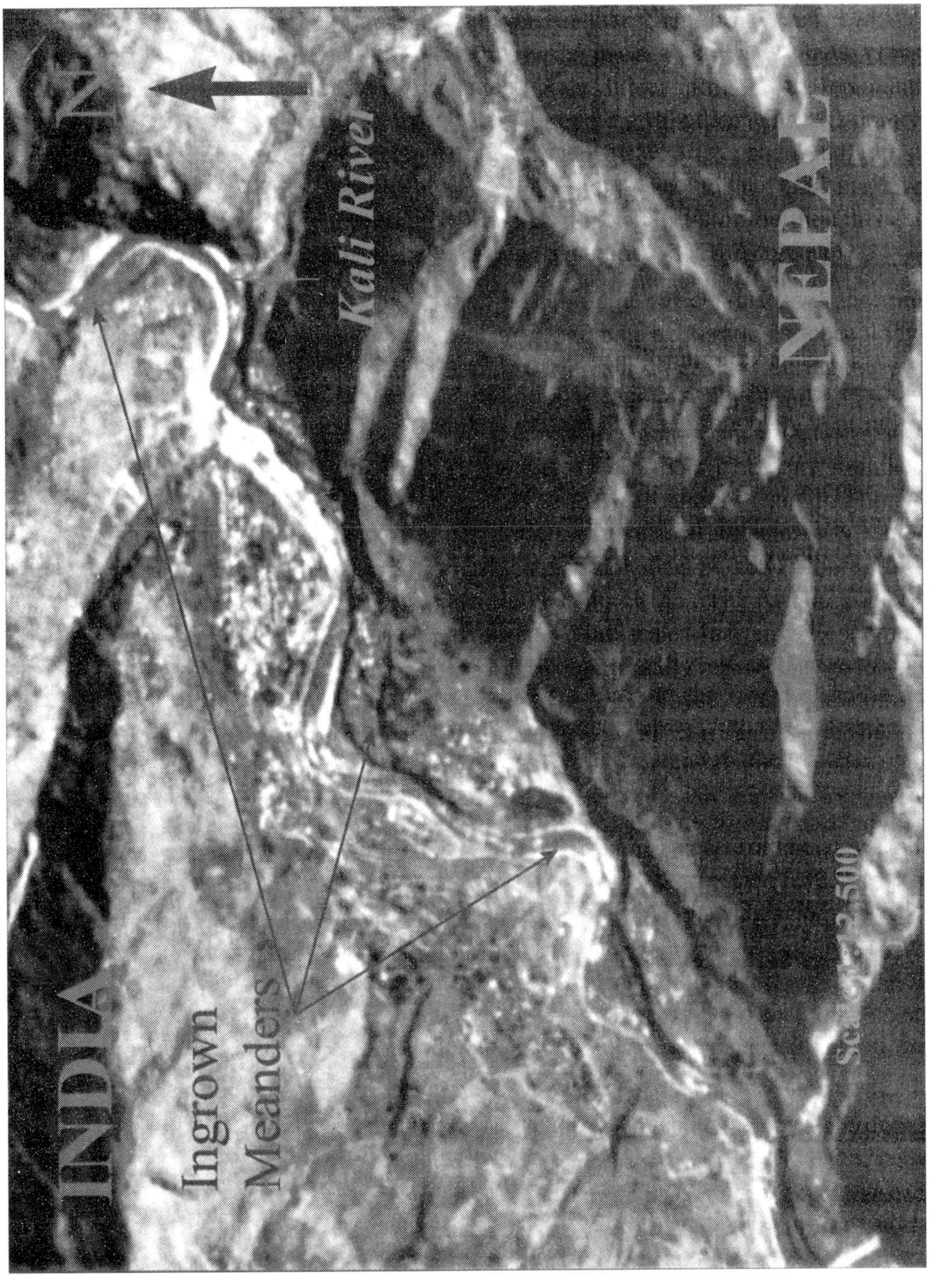

Figure 4: Ingrown meanders along the Kali river as seen in the IRS I-C PAN data which indicate neotectonic activity in the area..

the Chhirkila penstock pipes, breaking of the Kanchuti power channel , opening of joints, etc. Constant creep movement along some of the slopes is also responsible for the destabilisation/ subsidence of the Tawaghat -Sobala motor road.

EVOLUTION OF DEBRIS CONES

An attempt has been made to understand the intense fan formation activity in this part of the Himalaya. The supply of the debris material and the growth of the fans seem to be related to the tectonic evolution of the area.

During some of the later phases of the Himalayan orogeny, crystalline thrust sheets moved from the Central Crystalline Zone along the Main Central Thrust and covered the Lesser Himalayan Sedimentary sequence (Agarwal, 1994). During the process of sub-aerial transport of the crystalline thrust sheets near surface conditions, the rock masses on both the sides of the thrust plane underwent intense brittle / cataclastic deformation. This possibly resulted in fracturing, breaking and pulverisation of the rock mass thereby rendering them weak (Fig.5).

The process of erosion that followed after the thrusting events, gradually exposed the brittle deformed rocks adjoining the thrust planes to active denudational processes. This process of disintegration were possibly enhanced due to impact of excessive precipitation experienced by the area. Further, the area lies in close proximity of snow clad peaks of the Higher Himalaya and might have been covered with thick pile of snow in the past also. Frost action, on the cataclastically deformed rocks further resulted in their disintegration. Besides the area also falls in zone of high seismicity and increased neotectonic activity. Thus all these factors, i.e., tectonic, climatic and neotectonic, have contributed to the higher rate of erosion and subsequent transportation of the heterogeneous unconsolidated rock material to be finally deposited in the form of debris cones.

CONCLUSIONS

Detailed field studies and IRS-IC PAN and LISS III satellite data have been effectively used to understand the geomorphological framework of the area. The Tawaghat-Sobala area, due to its typical geological setting, is actively prone to denudational processes. The area around Tawaghat shows numerous evidences of neotectonic activity including the recurrent seismotectonic activities, that seem to have played a key role in modifying the geomorphological set up of the area. The presence of successive levels of ingrown meanders further confirms that they evolved as a result of structural changes (uplift) of the valley floor rather than any changes in the stream.

The area from Tawaghat to Sobala is highly prone to landslides. The landslide distribution map (Fig. 2, Bali et al., 2002) shows, that out of the 44 landslides present in the area, 21 are still active. The entire stretch in general, and a few sectors in particular, are prone to land instability conditions throughout the year. These sectors are almost completely covered by thin veneer of debris cones. The simple and compound cones, typically present on either side of the river gradually evolved through geological time and presently rest against

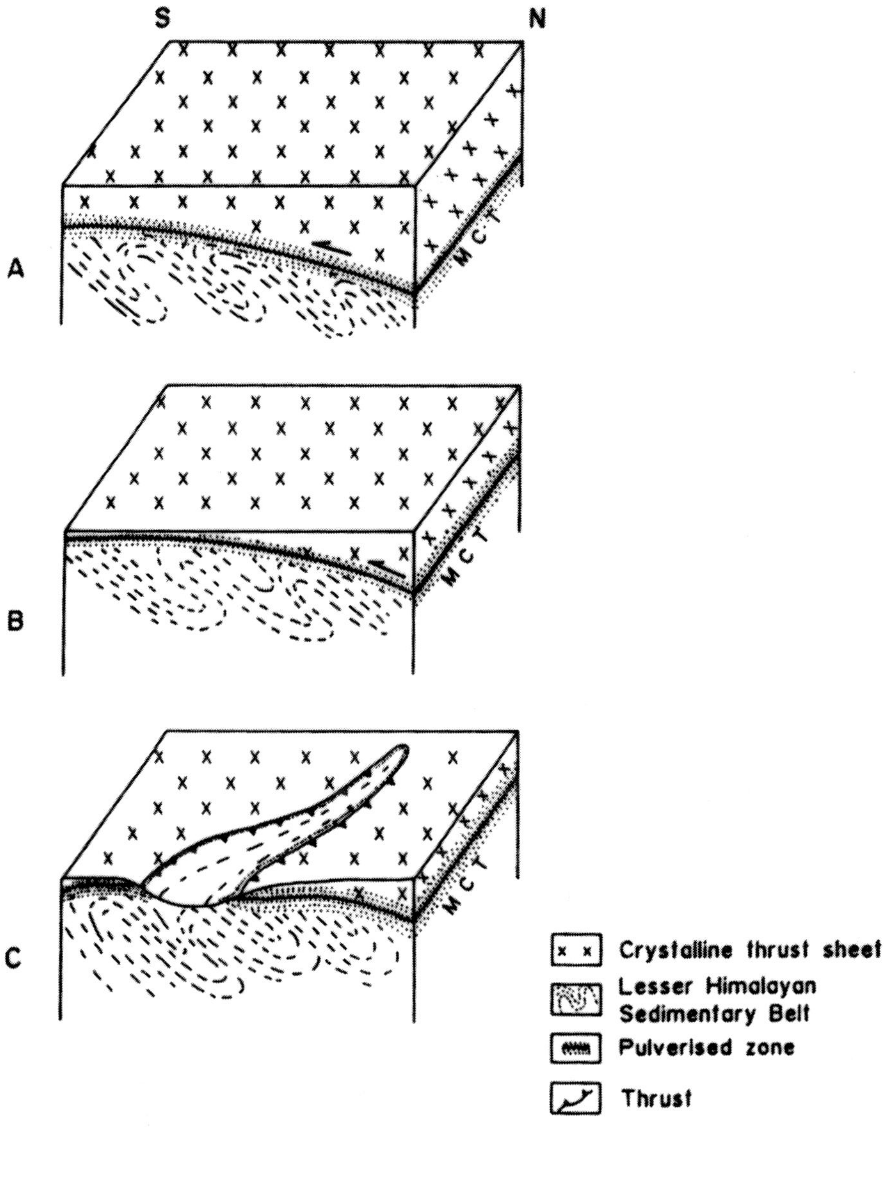

Figure 5: Schematic diagram showing the development and localisation of pulverised zone due to thrust sheet movement.

A. Initial position of thrust sheet over the Lesser Himalayan Sedimentary Belt due to southward translation.

B. Partial erosion of the overriding Crystalline thrust sheet.

C. Present day disposition of the Sirdang Sedimentary Windows due to the removal of a part of the overriding thrust sheet. The erosion has led to the exposure of the pulverised zone along the thrust contact.

their angle of repose. Due to a number of causative factors such as the intense precipitation, toe erosion associated with flash floods and neotectonic activity; the debris material was persistently subjected to creep process (Fig. 3). Numerous shallow rotational slip planes, developed in these debris cones, are at times daylighted especially during the flash floods. The problem of failure of the debris cones further increased due to recent anthropogenic activities of a number of microhydel schemes including the Dhauliganga Hydroelectric project. It is relevant note that the Sobala microhydel project was washed away due to such slope failure a few years ago.

ACKNOWLEDGEMENT

The author is thankful to the Head Department of Geology, Lucknow University for providing working facilities. Discussions with Prof. A.R.Bhattacharya and Dr. K.K.Agarwal of Department of Geology, Lucknow Uniersity are thankfully acknowledged. Financial support for the work has been provided by the Department of Science and Technology, New Delhi under the project No.ES/11/275/97.

REFERENCES

Agarwal, K.K., 1994. Tectonic evolution of the Almora crystalline zone, Kumaun Lesser Himalaya: An interpretation. Jour. Geol. Soc. Ind. vol.43 (1), pp. 5-14.

Bali, R., 1998. Evidences of hill slope instability around microhydel projects in part of Pithoragarh district, Kumaun Himalaya. In Proceedings 13th Himalaya-Karakoram-Tibet workshop, Peshawar, Pakistan p. 25-27.

Bali, R., Yudhbir, Sharma, A.K. and Ranhotra, P.S. 2002. Landslide Hazard Microzonation of Sobala-Tawaghat area, Kumaun Himalaya. In Aspects of Geology and Environment of Himalaya. Ed. C.C. Pant and A.K. Sharma. Gyanodaya Prakashan. India. pp. 199-218.

Gupta, G.K. and Agarwal, R.K., 1994. Geothermal studies on the Darma Valley thermal spring belt, Pithoragarh district, Uttar Pradesh. GSI records. vol. 127 (8). pp. 240-242.

Narula, P.L. and Shome, S.K.(1992). Macroseismic studies of recent earthquakes in northwest Himalaya- A review. In special Issue on Seismology in India-An overview. Current science v.62.(1/2), pp.24-33.

National Atlas of India, (1996). Seismic Zoning Map of India, Plate 60, National Atlas and Thematic Mapping Organisation, Department of Science and Technology, Govt. of India

Paul, S.K. (1986). Seismic condition in Northeastern Kumaun Himalaya and the adjoining area in far most western Nepal. In Proc. Int. Symp. on Neotectonics in South Asia . Survey of India . Dehradun pp. 396-406.

Searle, M.P., Windley, B.F., Coward, M.P., Cooper, D.J.W., Rex, A.J., Ting Dong, Li., Xuchang, Xiao., Jan, M.Q., Thakur, V.C., and Kumar, S. 1987. The closing of the Tethys and the tectonics of Himalaya. Bull. Geol. Soc. Amer. 98. pp. 678-701.

Singh, I.B. and Agarwal, K.K., 1996. Foot wall block deformation patterns in the Krol belt Nainital, Uttar Pradesh. Proc. Symp. NW Himalaya and Foredeep. GSI Spl. Pub. vol.21(1). pp.87-91.

Sinha, A.K. 1989. Geology of the Higher Central Himalaya, John Wiley & Sons Ltd. pp.219.

Srivastava, H.N.,1973. The crustal seismicity and the nature of faulting near India-Nepal-Tibet trijunction. Him. Geol. vol. 3.pp.381-393.

Valdiya, K.S.(1980). Geology of Kumaun Lesser Himalaya. Wadia Institute of Himalayan Geology. Dehradun, p. 291.

Valdiya, K.S., 1986. Neotectonic activities in the Himalayan belt. Proc. Int. Symp. on Neotectonics in south Asia, Survey of India, Dehradun, pp. 241-267.

Verma, R.K., Mukhopadhyay, M. and Roy, B.N. 1977. Seismotectonics of the Himalaya, and the continental plate convergence. Tectonophysics, v. 42, pp. 319-335.

SEISMOTECTONIC STUDIES IN MANIPUR AND ADJOINING AREAS

Arun Kumar and Manichandra Sanoujam

Department of Earth Sciences, Manipur University, Imphal 795 003.

ABSTRACT

It is well established that a major part of seismicity of the earth is a consequence of interaction of tectonic plate boundaries. The present study deals with the seismotectonic investigations along the eastern margin of Indian Plate and Burmese arc using the seismological data. The region displays an interesting results of the seismic data analysis based on Harvard Centroid Moment Tensor solutions (1996-2001). The existing knowledge of tectono-stratigraphy is collected from available literatures. Seismotectonics studies in Manipur and its adjoining areas are interesting as it falls in the eastern margin of the Indian Plate and the Burmese area which constitutes an active subduction zone. The present study addresses the complex tectonics of this zone using the Harvard Centroid Moment Tensor data, a technique accepted today, replacing the conventional fault plane solutions.

INTRODUCTION

It is now well established that a major part of the seismicity is due to the interaction of tectonic plate boundaries. Measurement of magnetic anomalies provides a measure of the rate of motion, whereas the earthquake source mechanism supplements it with the direction of relative motion. Presently, the source mechanism studies are widely used in understanding the active tectonic processes, which are far more complex on a small scale rather than the global tectonics would suggest.

The eastern margin of the Indian Plate and Burmese Arc constitute an active subduction zone. The geotectonic evidences suggest an eastward subduction of Indian Plate beneath the Burmese plate. Gravity and seismicity data confirmed the presence of a subducted and eastward dipping Indian lithospheric slab. The most significant geological feature in this region is N-S trending Indo-Burman Ranges, considered to be formed around Mid-Cretaceous as a consequence of collision process between the oceanic extension of Indian Plate with its neighbouring Burmese Arc (Brunnschweiler, 1966). The bulk of sediments of these ranges consists of thick turbidite sequence of Cretaceous to Upper Eocene shales and sandstone (Disang Group), which progressively were thrust on the Indian plate, along with the slices of underlying oceanic crust, and ophiolites during subduction (Mitchell and McKerrow, 1975). The presence of blue schist in ophiolite mélange (Acharyya et al 1990), and volcanic arc composed of andesitic rocks with local basalt and rhyolite mostly of Miocene age, confirm the subduction process. The subduction of this region appears to be still active as evidenced by sea floor ophiolites and sediments scrapped off due to the under thrusting of Indian plate (Curray et al., 1979).

Indo Burman Ranges have longitudinal tectonic provinces which are characterized by distinct lithotectonic assemblages and thrusts demarcating various zones from NNW trend in south at latitude 19 N to NNE in north of 24 latitude. They display varying components of movement and are characterized by a change from wrench to thrust type along the fault trace (Fig 1). The prominent dislocations/ thrusts are Kadi-Disang, Hail-Hakalula-belt of Schuppen consisting of Naga-Disang thrust and Eastern Boundary Thrust.

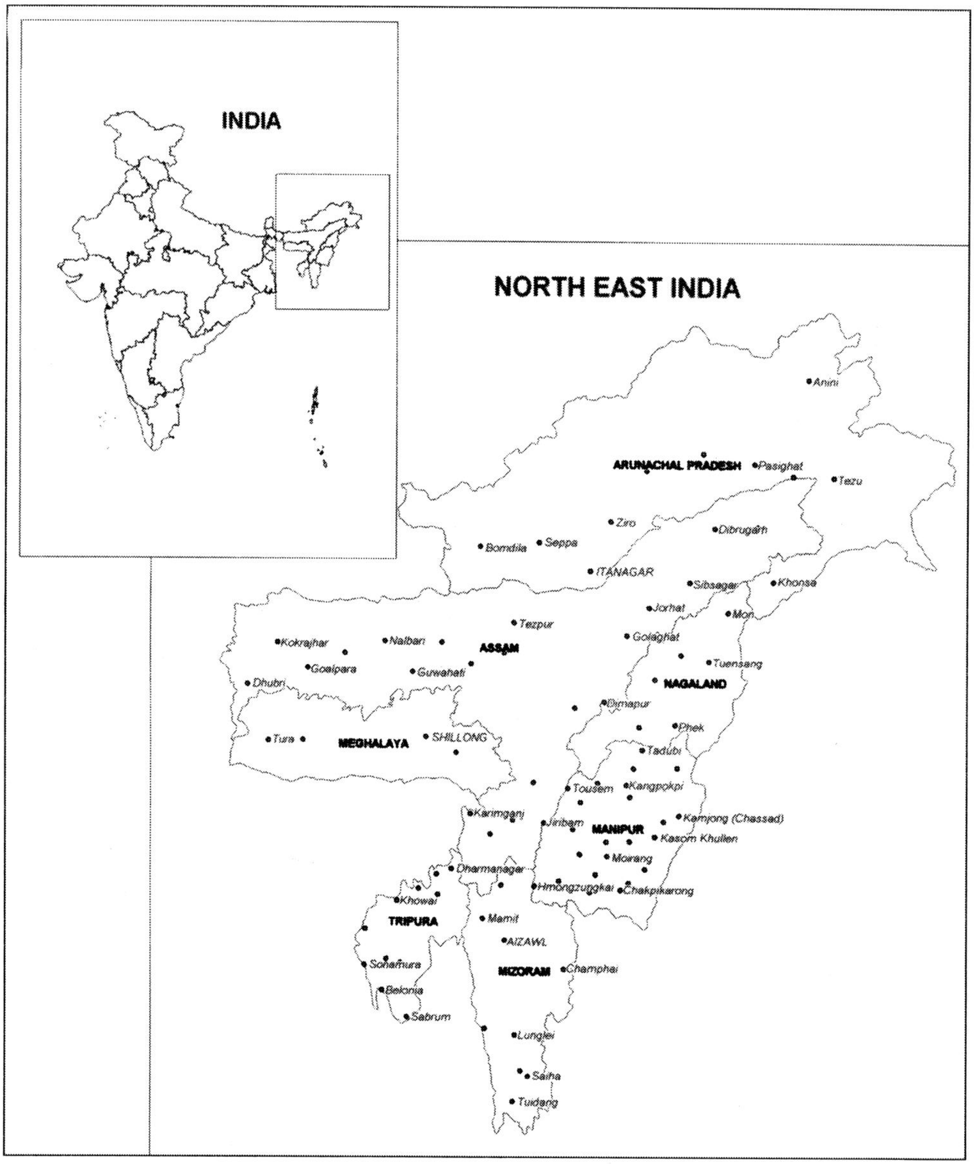

Figure 1: Location map of NE India.

The occurrences of earthquakes in Burmese arc region are extended down to intermediate depth. Seismological studies reveal the distribution pattern along an inclined trend inferred as the Indian lithospheric slab, dipping east at about 45 beneath the Indo-Burman ranges (Mukhopadhyay and Dasgupta, 1988; LeDain et al., 1984). Based on the analysis of seismic, gravity and other geological data, Gupta and Bhatia (1986) suggest that earthquakes in this region are associated with remnants of already subducted, but not totally assimilated, Tethys oceanic lithosphere beneath the Burmese plate. Based on source mechanism data, the principal compressive stress direction in this region is in the NNE rather than the subduction direction of Indian plate. It has been suggested that the eastward subduction either continues relatively aseismically or has stopped recently (LeDain et al., 1984; Ni et al., 1989; Chen and Molnar 1990). Recent studies made by (Kumar et. al., 2000) also support the shallow to deep seismic activities along the Indian plate and subduction region respectively.The present study addresses the complex tectonics of the region using Harvard Moment Tensor data for focal mechanism of earthquakes.

REGIONAL TECTONIC SETUP

In order to understand the regional tectonic setup of Manipur and adjoining areas, attempts have been made to compile the geological data from various workers. The geographic extent of the subduction process encompasses a large geographic area. Therefore, the scope of the present study extends beyond Manipur. The Indo-Burman Ranges form a folded mountain system with slight sigmoidal shape. IBR abuts against Shillong Plateau near N25 latitude whereas in NE with Mishmi hills at 97 E longitude. It Shows about to 300 km width around Tripura at meridian 24 and further south it tapers to 15-20 km near Cape Negaris. Dasgupta (1977) linked the tectonics of IBR with Indonesian arc through Andaman and Nicobar Archipelago. The tectonics of IBR, Bay of Bengal and Arakan coast is deemed to be the result of collision since Miocene (Bender, 1983). It is believed that the collision process started between oceanic extensions of Indian Plate with its neighbouring plate around Mid-Cretaceous. The sediments of shelf area and collision zone were many kilometers apart. The gap was decreased during Late Eocene between the shelf and peripheral tectonic basins. The configuration of colliding Indian plate and Myanmar continental mass started towards the north and progressively extended southward. South of Kohima, the continental blocks remained still sufficiently apart for the Disang to plunge down to shrinking gulf and to form the tectonic frame for Kohima Synclinorium which opens into a wide expanse, now represented by Cachar-Mizoram-Tripura fold belt and Surma valley cum residual Bengal basin. Towards the east and south, all the Late Eocene to Pleistocene sedimentation was thus confined to an already existing shelf zone (Shillong-Mikir Plateaux edge and submerged basement spur), deep marginal basins of sedimentation (occupied by Naga-SchuppenZone) and Kohima synclinorium and the receeding Indo-Myanmar gulf (occupied by Surma valley-Chittagong hills and residual Sylhet and Bengal basins).

Mathur and Evans (1964) and Das Gupta (1977) recognized three major structural zones in Assam-Arakan belt: (i) an eastern zone of argillites with varying degree of metamorphism and ultrabasic intrusive, (ii) central zone of Disang and Barail sediments and (iii) a belt of Schuppen in northwest with imbricate thrust. The northern Assam-Arakan bulge around latitude 24 N is due to eastward movement of Shillong plateau for about 250 km along Dauki fault having strong vertical component.

The Indo-Myanmar ophiolites broadly represent marginal ocean basin crust mainly based on geochemical studies of the volcanics in the ophiolites suites (Sengupta et al., 1989; Vohra et al., 1989). The 100 km wide leading edge of overriding Myanmar plate was an oceanic lithosphere flooring the forearc marginal basin. Mid-Jurassic-Early Cretaceous rifting and splitting of eastern Myanmar continental margin (and older arc) and migration of the western part (that included slivers of pre Jurassic rocks) away from Eastern half due to diffusion or irregular spreading (Hamilton, 1988) generated a thin oceanic crust which constituted the leading edge of the overriding Myanmar plate that rammed on to the thick accretionary prism of the subducting Indian plate (Early Mid Eocene). The rifted and transported fragments of the Myanmar continental arc crust (Naga Metamorphics, Kanpetlet schist, Triassic schist etc.) and the ophiolites form a part of the accreted terrains within the accretionary prism (Das Gupta and Nandy 1995). Thus, the Indo-Myanmar ophiolites are not part of the underplate oceanic crust of the leading edge of the Indian plate below the forearc trough; The continued subduction of Indian plate below the central Myanmar basin upturned the outer ridge sediments during late Oligocene along the Eastern Boundary Thrust (EBT).

It is, however, possible that the time of generation, accretion and thrusting of the ophiolites were along the strike of the arc. Ophiolites are likely to vary along the strike of the arc. Ophiolites of Indo-Myanmar, Andaman and also of western Sunda arc represent accreted forearc basin crust. It is known from geological account and tectonic history that the collusion process existed in the geological past now depicted from its present status. The geological data support the evidences for wrench fault and thrust, associated with vertical component along the faults besides the strike slip component. It is difficult to determine the prominent tectonic history due to the lack of direct field evidences to recapitulate the entire process of subduction. However, the presence of major right lateral strike slip faults in the east, running parallel to Burmese arc can be related to the oblique nature of convergence (Fitch 1972; Maung, 1987). Based on the gravity and seismic studies, various workers (Verma et al., 1976; Mukhopadhyay, 1984; Mukhopadhyay and Das Gupta 1988; Ni et al., 1989; Gupta 1990) confirmed the presence of subducted slab. They further believed that the Indian plate could be dragged towards east dipping lithospheric slab, of the Indo-Burman Ranges, through the asthenosphere. This work appears to be inadequate to describe the complex tectonics of subduction region.

EARTHQUAKE SOURCE MECHANISM

The occurrence of earthquake in the plate boundary is the manifestation of the plate motion. Therefore, a study of nature of the faulting and its slippage along the fault plane is of great importance for knowing the tectonics of a region (Fig 2). Since 1960, with the help of digital simulations using computers, several methods for the numerical determination of fault-plane solution were proposed. The first formulation was that of Knopoff (1961), reformulated by Kasahara (1963), and applied in the computer program developed by Wickens and Hodgson (1967). Now a days, the conventional fault plane solutions are being replaced by the Moment Tensor Solutions, to represent an accurate seismic source. It is now a practice to invert the waveforms recorded at seismological observatories for obtaining the moment tensor components and to derive the best double couple solution and the P, B and T axes orientations.

Data of moment tensors for thousands of earthquake provide a more objective and

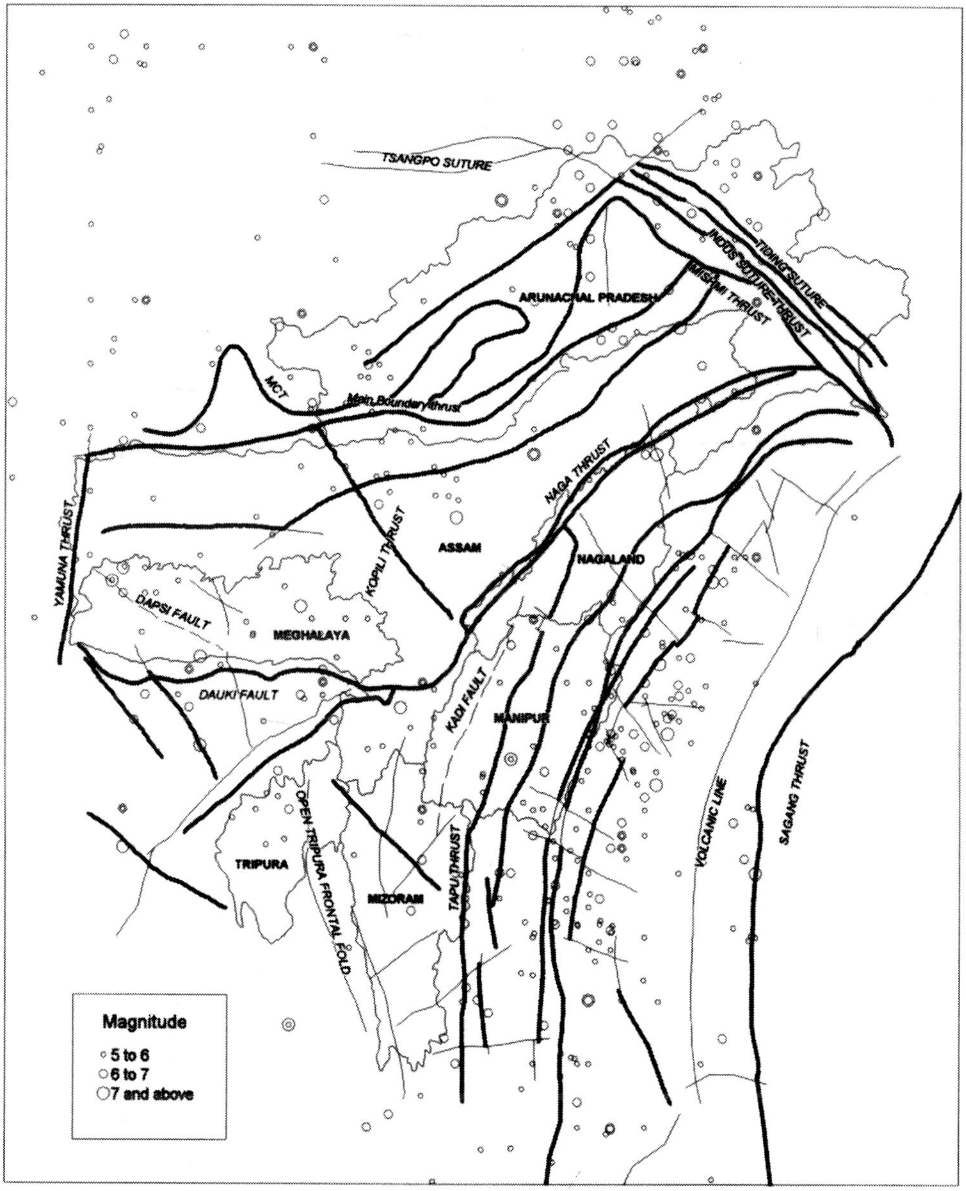

Figure 2: Seismotectonic map of Manipur and adjoining areas

complete description of the earthquake source, and also provide valuable source for tectonic analysis. The use of seismic moment and the moment tensor to describe earthquake size and geometry is prefered to the more traditional magnitude and focal mechanism (Frohlich et al., 1992). The reasons are: conventional beach-ball focal mechanisms and meaningful magnitudes can be obtained directly from earthquake moment tensors. There is relationship between earthquake moment tensors and the physical conditions near the focus of earthquake, such as fault area, amount of slip, and strain released by earthquake. Now global data of moment tensors for thousands of earthquakes are available for routine analysis of the digital seismograms. Since 1977, the data of earthquake are complete and less influenced by subjective interpretation.

TECHNIQUES USED TO ANALYSE THE HARVARD CMT DATA

The concept of Mean Slip Angle (MSA), (Rao et al., 1994) is used to classify the focal mechanisms from the CMT data instead of the earlier method of Frohlish and Apperson (1992), The total seismic energy in the study area is partitioned into thrust, normal and strike-slip components using the moment magnitude (Mw).

Technique

Using the Mean Slip Angle (MSA) technique the focal mechanism can be classified into thrust, strike-slip and normal components. The slip angles on the auxiliary planes of a focal mechanism are used as inputs. A slip angle of -90 and 90 represents pure normal and thrust respectively. In case of pure strike- slip the slip angle is 0 or 180 . For uniformity, the slip angle is transformed into modified slip angle (mSA), so that the pure strike slip is always 0 . The mSA and true slip angle(s) are given below :

MSA = s When |s| = 90

MSA = 180 x sign(s) - s When |s| => 90

In this relation, the pure normal, thrust and strike-slip can be defined by mSA as -90 ,0 and 90 respectively, as mSA ranges from -90 to 90 .

Due to the ambiguity of fault plane, a unique choice of a fault plane for any focal mechanism is difficult, therefore, an average of the mSA values of the two auxiliary planes are used for the classification. The average of the mSA termed as Mean Slip Angle (MSA) can be used for the classification of the focal mechanisms as follows:

-90		MSA		-45	=	Normal fault
-45	<	MSA	<	45	=	Strike-slip fault
45		MSA		90	=	Thrust

The above concept of the classification using MSA is shown in the Fig. 3. The justification for average mSA values is due to the fact that with the degree of fault plane ambiguity the Slip Discrepancy (Sd = | mSA1 - mSA2 |) is generally low for majority of the earthquakes. Hence, the error in averaging the mSA values is usually very less. Fig 2 a.b. shows a slip discrepancy function forming a decaying curve in which most of the events fall in low slip discrepancy (Sd.). This behaviour is confirmed by the global data comprising about 10,600 CMT solutions. Therefore, by omitting events with high Sd, the problem due to fault plane ambiguity is reduced. Thus, the classification enables a quick and systematic examination of large data sets of focal mechanisms, especially in the absence of the actual plane of faulting (Kumar et al., 1996).

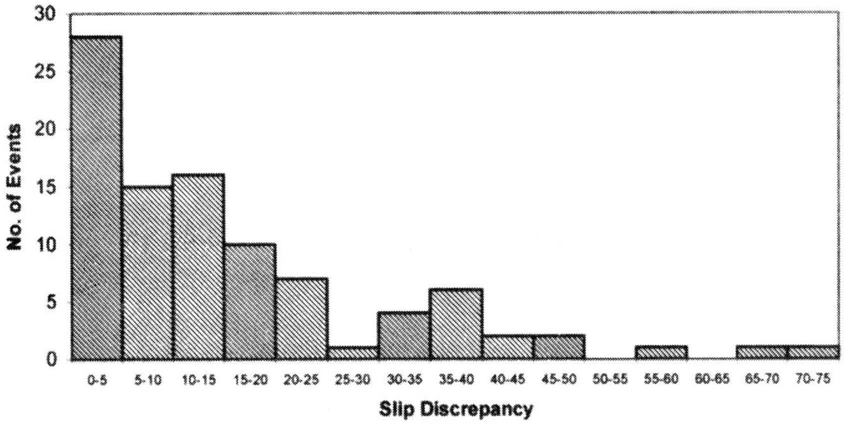

Fig. 2a : Slip discrepancy (Sd) histogram for the Centroid Moment Tensor solutions of the Indo-Burma region

In this study, 94 focal mechanism solutions in the Indo-Myanmar region (20 -28 N, 90 -100 E) are chosen. The spatial distribution for each of the fault type is plotted using different symbols. The epicentral map is superposed on the tectonic map of the region, and it can be seen that some of the epicenters can be well correlated with the known tectonic features.

The predominantly thrust, normal and strike slip mechanism in the whole study region is listed in Table 1a, b. It is seen that, the dominant mechanism is strike-slip followed by thrust and normal mechanisms. As a different type of mechanism does not reflect the actual mode of energy released in the study area, the seismic energy has been partitioned in different categories.

Table 1a: Distribution of earthquakes and their respective energy in the Indo-Myanmar region. (values are in %).

Type	Burmese a	Arc b	Indian Slab b
Strike-slip	55.30	54.50	52.53
Thrust	32.89	33.33	36.84
Normal	11.78	12.12	10.62

a - Ravikumar (1996)

b - Present study

Table 1: b

Focal mechanisms data from Harvard Centroid Moment Tensor catalogue for earthquakes in the Indo-Myanmar region (1976-2001). (Stk1, Dip1, Slip1) and (Stk2, Dip2, Slip2) are the strike, Dip and Slip of the two planes respectively. Mom x 10Exp is the scalar seismic moment. (PPlg and PAzi) and (TPlg and TAzi) are the plunge and azimuth of the P and T principal axes respectively. Mw is the moment magnitude.

Date	Lati	Long	Depth	Stk1	Dip1	Slip1	Stk2	Dip2	Slip2	Mom	Exp	PPlg	PAzi	TPlg	T_Azi	Mw
05/29/76	24.39	98.65	15.0	323	80	-172	232	82	-10	1.22	26	13	187	1	278	6.7
05/29/76	24.29	98.58	15.0	242	88	0	152	90	178	1.12	26	1	197	1	107	6.6
05/31/76	24.26	98.60	15.0	342	72	-169	249	80	-18	1.63	25	20	205	5	297	6.1
07/21/76	24.74	98.57	15.0	338	88	-178	248	88	-2	1.98	25	3	203	1	293	6.1
08/12/76	26.55	97.12	15.0	148	45	93	323	45	87	7.99	24	0	55	88	154	5.9
05/12/77	21.60	92.77	40.0	216	72	3	125	87	162	9.91	24	10	172	15	79	5.9
10/01/77	23.27	93.16	60.8	145	41	-171	48	84	-49	1.62	24	37	354	27	107	5.4
02/23/78	23.16	94.93	122.0	331	31	44	201	69	113	5.37	23	21	274	59	144	5.1
01/01/79	20.51	93.59	95.0	93	32	62	305	62	106	1.12	24	16	23	69	248	5.3
05/29/79	24.92	95.05	109.0	109	30	134	241	69	68	1.31	24	21	348	60	120	5.3
07/13/79	25.13	95.59	117.0	44	28	99	215	62	85	3.88	23	17	308	72	114	5.0
11/02/79	25.21	96.32	10.0	357	79	-175	266	85	-11	9.16	23	11	221	5	312	5.2
12/02/79	26.61	97.78	10.0	119	42	72	323	51	106	1.13	24	4	42	77	292	5.3
11/01/80	27.42	89.05	44.1	209	51	-2	301	89	-141	2.98	25	28	172	25	68	6.z
11/02/80	22.69	94.49	20.0	208	19	131	345	76	78	7.79	23	30	85	57	239	5.2
04/25/81	24.99	95.52	153.0	135	44	80	329	47	99	3.89	24	2	52	83	308	5.7
05/01/81	23.32	94.64	99.1	247	22	21	137	82	111	3.34	23	34	210	49	70	5.0
06/30/81	23.00	95.45	10.0	7	40	142	128	67	57	2.76	23	15	242	55	355	4.9
08/14/81	25.61	98.51	10.0	153	46	-141	33	63	-52	9.00	23	55	352	10	97	5.2
08/16/81	25.53	97.02	37.0	298	66	-7	30	83	-156	4.95	23	22	257	12	162	5.1
04/17/83	21.86	94.19	112.0	236	26	-22	345	81	-115	4.53	23	49	229	31	95	5.0
08/23/83	24.48	94.69	148.0	297	44	58	158	54	118	6.76	23	5	229	67	127	5.2
08/30/83	25.34	94.90	68.9	161	65	159	260	71	26	4.24	24	4	29	32	122	5.7
10/21/83	21.87	94.37	92.0	109	5	151	350	87	-85	8.6	23	47	265	42	76	5.2
11/01/83	26.58	96.38	144.0	177	58	147	286	62	37	1.35	24	3	51	44	144	5.4
03/05/84	24.85	95.49	95.8	114	21	144	238	78	72	1.73	24	30	343	54	127	5.4
05/06/84	24.33	93.42	61.3	157	69	161	254	73	22	1.19	25	2	25	28	116	6.0
11/02/84	26.52	96.96	16.0	311	49	17	210	77	138	3.78	24	18	267	38	162	5.7
12/03/84	24.75	92.99	102.0	352	46	123	129	53	61	1.33	25	4	239	67	339	6.0
04/24/85	25.83	95.97	56.4	324	38	58	182	58	113	8.75	23	10	256	68	139	5.2
09/05/85	25.29	98.23	60.7	14	65	177	105	87	25	2.15	24	16	237	19	333	5.5
02/08/86	23.79	93.09	33.0	224	62	-15	321	77	-152	1.46	24	29	186	10	90	5.4
02/19/86	24.89	91.18	18.0	340	50	180	70	90	40	1.08	24	27	197	27	302	5.3
04/26/86	23.31	94.92	129.0	87	30	136	217	69	67	4.75	23	21	324	59	95	5.1
07/26/86	23.86	94.19	64.9	130	72	15	36	76	162	1.74	24	2	83	23	352	5.4
11/01/86	25.53	96.91	69.8	28	45	132	157	58	56	6.96	23	7	270	61	13	5.2
05/18/87	24.58	93.94	75.3	67	68	-14	163	77	-158	2.76	25	25	27	6	294	6.2

Contd......

Date	Lati	Long	Depth	Stk1	Dip1	Slip1	Stk2	Dip2	Slip2	Mom	Exp	PPlg	PAzi	TPlg	T_Azi	Mw
08/24/87	23.04	94.53	127.0	135	42	144	254	67	54	9.76	23	14	9	54	119	5.3
02/06/88	24.05	91.66	31.0	239	76	9	147	82	166	6.7	24	4	194	16	103	5.8
07/03/88	22.22	94.36	86.0	133	18	-104	327	72	-85	4.45	23	63	244	27	54	5.0
08/06/88	25.19	94.89	101.0	284	45	55	148	54	120	8.9	26	5	217	65	117	7.2
08/13/88	24.94	95.24	126.0	307	35	69	152	58	104	4.56	23	12	232	73	98	5.0
08/21/88	24.94	95.89	93.8	67	49	28	318	69	136	6.78	23	12	16	46	274	5.2
10/02/88	20.74	94.67	45.6	168	27	-58	313	68	-105	5.18	23	64	198	21	54	5.1
02/12/89	26.18	96.83	33.0	295	60	24	193	69	148	9.53	23	6	246	37	151	5.3
03/08/90	25.11	96.61	56.5	32	73	173	124	84	18	7.68	23	8	256	17	349	5.2
01/05/91	23.61	96.18	20.9	2	8	66	97	7	3	3.11	26	6	228	25	321	6.9
05/11/91	23.42	93.25	73.6	159	77	171	251	81	13	1.43	24	3	25	16	116	5.4
12/04/91	24.19	93.83	70.0	245	68	11	151	79	157	2.00	24	8	200	23	106	5.5
12/02/91	24.47	93.08	101.0	258	54	30	150	66	140	1.19	24	7	206	45	109	5.3
03/25/92	24.18	95.20	120.0	272	30	43	143	70	113	6.53	23	22	216	59	85	5.1
03/27/92	21.12	94.52	86.2	159	12	-85	334	78	-91	2.76	24	57	242	33	65	5.6
04/15/92	23.96	94.57	143.0	281	45	56	145	54	120	4.09	24	4	214	66	114	5.7
04/23/92	22.54	98.97	15.0	354	77	-168	262	79	-13	1.58	25	17	218	1	308	6.1
04/23/92	22.40	98.85	15.0	345	68	-173	252	84	-22	1.93	25	20	207	11	301	6.1
06/15/92	23.95	96.03	22.9	8	69	-173	275	83	-21	3.14	25	20	229	10	323	6.3
07/08/92	21.07	93.51	79.0	191	33	-75	353	58	-100	8.00	23	75	237	13	90	5.2
07/09/92	20.96	90.20	30.0	79	59	13	342	79	148	1.11	24	14	34	30	296	5.3
11/02/92	20.43	94.51	71.4	187	29	-43	317	71	-112	7.61	23	58	197	23	63	5.2
04/01/93	23.55	94.00	107.0	99	44	-169	1	83	-46	5.68	23	37	309	25	59	5.1
01/11/94	24.80	97.40	15.0	106	25	-85	281	65	-92	1.6	25	70	186	20	12	6.1
04/06/94	25.62	97.05	15.0	96	57	26	352	68	144	7.15	24	7	46	40	310	5.8
05/29/94	20.45	94.17	49.1	220	32	-8	316	86	-122	6.5	25	41	197	33	73	6.5
08/03/94	21.61	93.88	64.5	216	73	0	306	90	-163	4.57	24	12	173	12	80	5.7
08/08/94	24.76	94.97	146.0	111	34	120	257	61	71	1.83	25	14	0	68	129	6.1
11/02/94	25.37	96.83	42.7	34	76	179	124	89	14	9.25	24	10	258	10	350	5.9
02/17/95	27.48	92.62	35.0	322	46	-172	226	84	-44	1.83	24	34	174	25	282	5.4
05/06/95	24.83	95.02	148.0	278	39	60	135	57	112	4.84	25	9	209	69	94	6.4
05/09/95	25.00	94.94	124.0	142	33	118	289	62	73	6.84	23	15	31	69	164	5.2
07/26/96	24.68	96.53	67.6	93	70	11	359	80	159	1.21	24	7	48	22	315	5.3
07/27/96	20.80	94.85	129.0	204	60	33	96	62	146	7.8	23	1	150	43	59	5.2
11/01/96	24.05	93.38	53.0	102	34	59	317	61	109	1.3	24	14	34	68	265	5.3
04/14/97	22.55	94.18	110.0	150	61	165	247	77	30	8.66	23	11	15	30	112	5.2
05/08/97	24.51	92.36	35.0	78	68	4	347	86	158	8.57	24	12	35	18	300	5.9
07/11/97	21.41	94.64	150.0	208	37	24	98	76	124	1.21	24	23	162	48	44	5.3
07/31/97	23.80	93.43	41.8	330	16	40	201	80	103	9.06	23	34	280	54	126	5.2

Contd......

Date	Lati	Long	Depth	Stk1	Dip1	Slip1	Stk2	Dip2	Slip2	Mom	Exp	PPlg	PAzi	TPlg	T_Azi	Mw
11/02/97	22.21	92.70	54.4	163	37	168	263	83	54	1.55	25	28	21	41	139	6.1
12/03/97	25.01	96.52	54.4	122	84	5	32	85	174	5.14	24	1	77	7	347	5.7
05/02/98	24.84	95.09	127.0	132	68	156	232	68	24	1.96	24	0	2	32	92	5.5
09/26/98	27.87	93.60	33.0	233	26	118	22	67	77	3.6	23	21	122	65	270	5.0
10/01/98	23.82	94.74	112.0	86	12	180	177	90	78	1.12	24	44	278	44	75	5.3
02/22/99	23.15	93.99	51.0	47	35	-103	242	56	-81	4.22	23	77	180	11	326	5.0
04/05/99	24.50	93.96	65.1	86	62	16	348	76	151	2.51	24	9	39	30	304	5.5
10/05/99	25.88	91.89	33.0	244	68	12	149	79	158	6.74	23	7	198	23	105	5.2
06/07/00	26.70	97.15	37.0	290	38	41	166	66	120	3.74	25	16	234	57	118	6.3
06/08/00	26.43	97.41	18.8	126	33	67	333	60	104	5.77	23	14	53	71	276	5.1
07/02/00	24.45	94.67	103.0	133	37	155	244	75	56	7.74	23	22	359	48	117	5.2
10/06/00	24.18	97.13	72.7	144	77	166	237	77	14	1.64	24	0	11	19	101	5.4
10/01/00	23.58	94.63	122.0	343	14	98	155	77	88	2.41	24	31	246	58	62	5.5
11/01/00	21.94	93.04	74.0	216	83	-5	307	85	-173	1.71	24	8	171	1	81	5.4
03/03/01	23.99	93.43	64.1	140	40	-141	19	66	-57	7.00	23	56	332	15	85	5.2
04/10/01	25.35	94.91	149.0	8	34	28	254	75	121	9.14	23	23	321	51	199	5.2
04/12/01	24.77	98.81	15.0	57	63	-28	161	65	-150	3.21	24	38	19	1	288	5.6
08/12/01	24.43	94.99	142.0	285	40	48	155	61	119	4.41	23	12	224	62	112	5.0

Seismic Energy

The moment magnitude provides a uniform scale for comparison and can be derived from the scalar seismic moment (Mo) of an earthquake as

$$M_w = \frac{2}{3}\log_{10} M_o - 10.7$$

(Hank and Kanamori, 1979)

To estimate the different seismic energy contributed to the strike-slip, thrust and normal components, the slip vector is resolved into two orthogonal components viz., strike component and normal or thrust component depending on the sign of the slip angle. The resulting components are scaled with the moment magnitude Mw. The corresponding components of the two planes are averaged to obtain the two strike components and the normal or thrust component for each of the events.

It can be seen from the above table that, 55% of the total energy was released by strike-slip mechanism only. The thrust and normal mechanisms released about 33% and 12% of energy respectively.

The P axis azimuths are plotted and superposed on the tectonic map of the region (Fig.3.). The P-axes of some of the events can be correlated with the structures of the region, but other trends are not discernible, due to the complexity of the Indian subducted slab below the Burmese arc. It is also inferred from the Harvard Centroid Moment Tensor solution that the present subduction was from NNW-SSE to NNE-SSW. It may be remarked

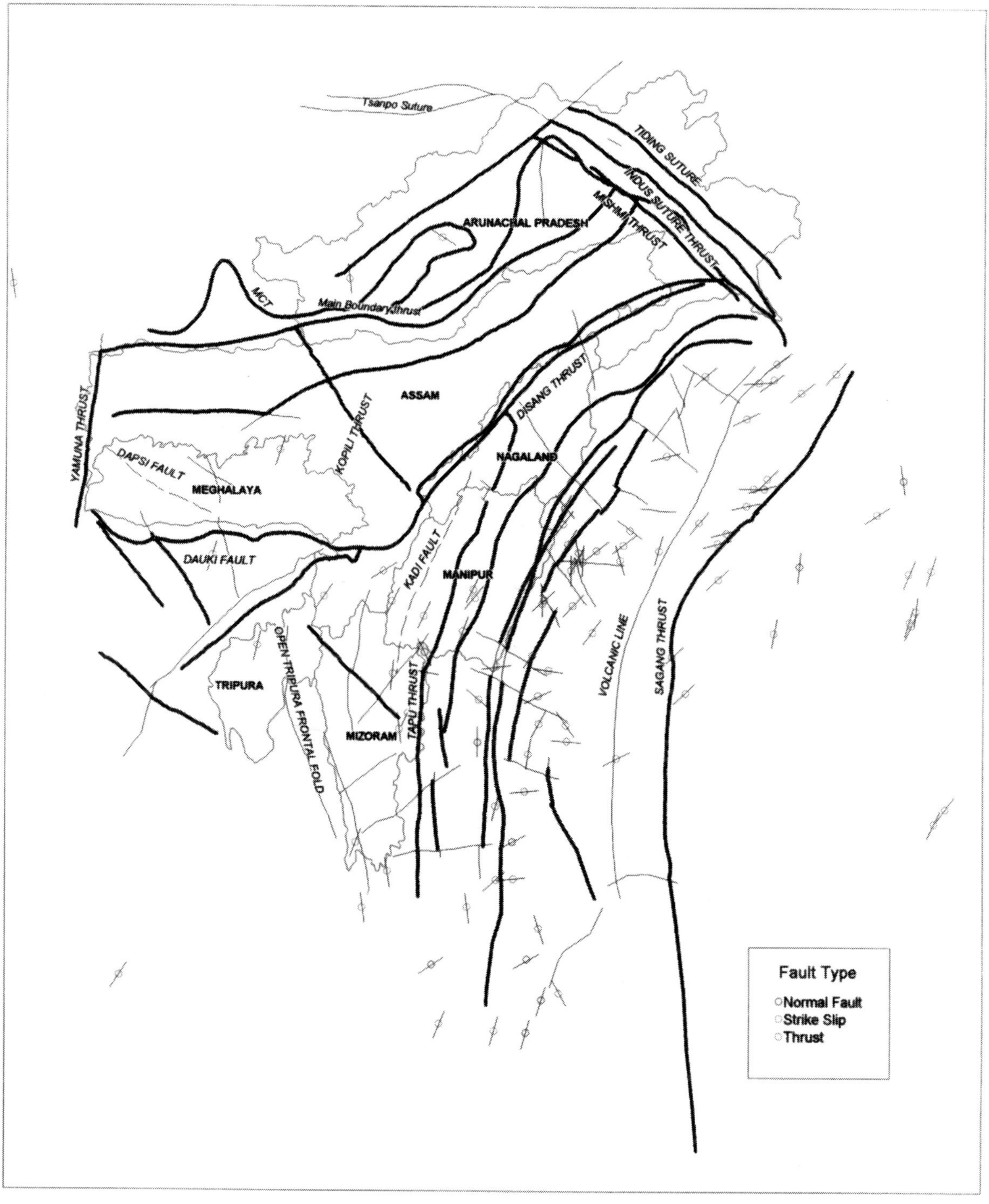

Figure 3: Focal mechanism (along with P-axi azimuth) along the major tectonic trends in NE India.

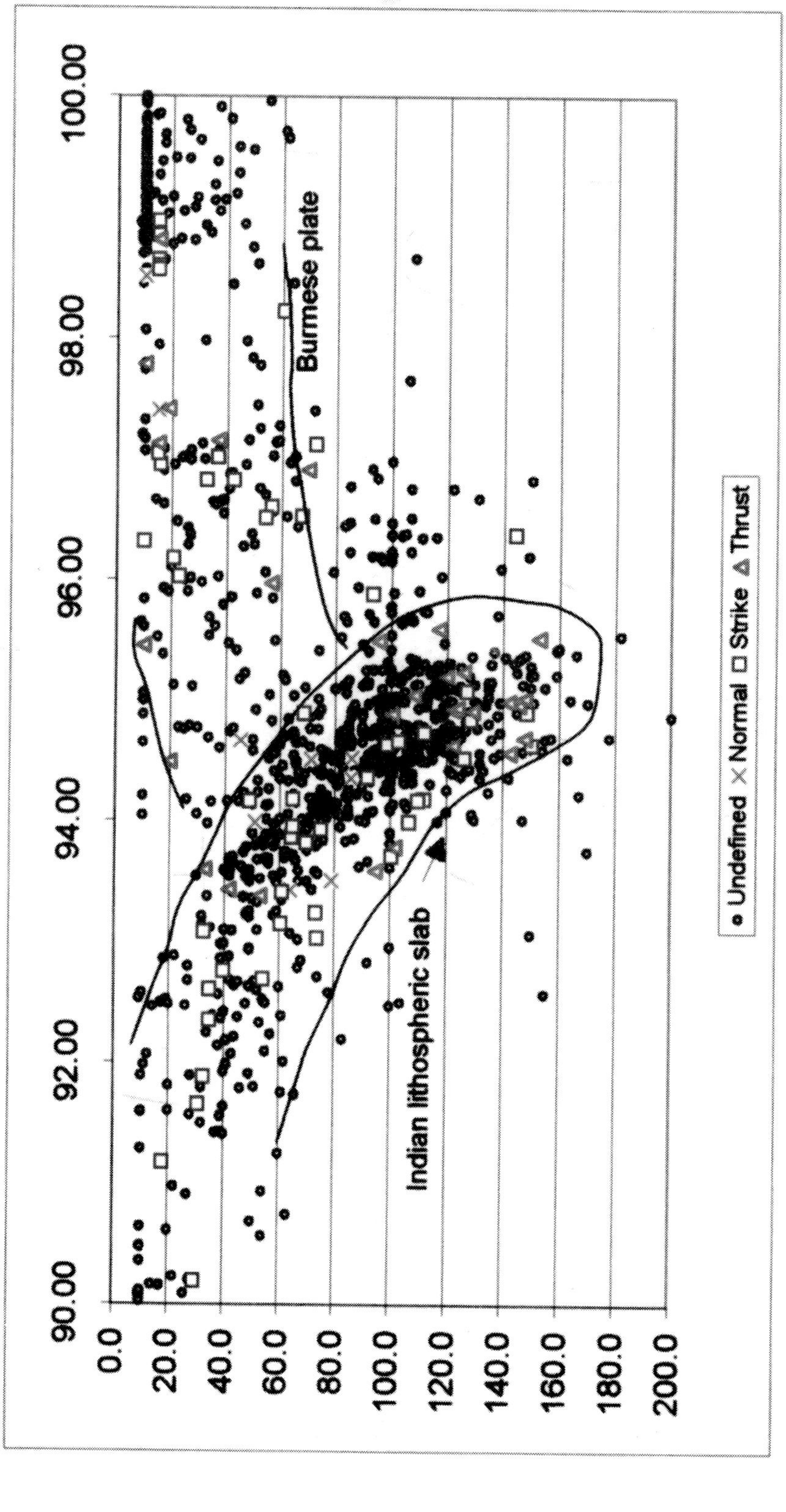

Figure 4: Section perpendicular to Burmese-Arc.

that, the subduction process ceased or its a E-W trending component was almost absent. LeDain et al., (1984) and Chen and Molnar (1990) opine that, eastward subduction was stopped recently or the Indian slab became aseismic. Present investigations reveal the high seismicity of Indian slab below the Moho (60-65 km) with strike slip mechanism as well as thrust mechanism (90 to 120 km.) and is still continuing (Fig.4). The various type sections, thrust and strike slip movements are comparatively more than the normal faulting.

The seismicity in the Indian plate is presently at intermediate depth whereas it is shallow along the Burmese arc. Based on the P velocity model of this region (Kumar and Manichandra, 2002), the Depth of Moho lies for about 60-65 Km. which is due to the past subduction process. It is possible that the subduction process started during Late Cretaceous with a regional E-W trend and is still continuing with a northerly trend (NNE-SSW to NNW-SSE). Based on the available HCMT data along the Shillong plateau, the NNE-SSW trend is dominant. The eastern part of the Indian slab is presently being dragged in a northerly direction below the Burmese arc in the east as well as below the Himalayan region towards in the north.

CONCLUSION

It is observed from a Harvard Tensor Centroid Moment analysis that the complex focal mechanism is due strike slip solutions which trend from NNW-SSE to NNE-SSW. However, the E-W trend is not observed. Based on the previous studies, it is suggested that the eastward subduction either continued relatively aseismically or stopped recently (LeDain et al., 1984; Ni et al., 1989; Chen and Molnar, 1990). Based on the present studies, it is confirmed that no P-axis was oriented in E-W direction within the Indian plate. The NNW-SSE to NNE-SSW dragging of the Indian plate was confirmed from the focal mechanism of Harward Tensoid Movement analysis below the Moho, at a depth ranging from 60 to 65kms, which is based on the Crustal Velocity Structure Model (Kumar and Manichandra,2002). The shallow seismicity within the Indian slab is not significant in the subduction process and related to the intraplate seismicity. The deep focal depth in the strike slip mechanism is from 62 to 140 km. within the Indian plate whereas in Burmese arc the strike slip mechanism is dominant with average shallow focal depth (40-45 km) and the strike slip mechanism trends NNE-SSW. The upper part of Indian lithospheric slab is seismically less active to aseismic and the subduction is active much below the Moho, while the shallow seismicity led the subduction process in the Burmese arc.

ACKNOWLEDGEMENT

The authors are grateful to the Seismology Division, Department of Science & Technology, Government of India, for providing the financial support to carry out the present investigations.

REFERENCES

Acharyya, S.K., Ray K.K., and Sengupta, 1990. Tectonics of the ophiolite belt from Naga Hills and Andaman Island, India. Proc. Indian Acad. Sc. (Earth Planet. Sc.)99, 187-199.

Bender, F. 1983. Geology of Burma : Gebruder Brontraeger Berlin, Stuttgart.

Brunnschweiler, R.O., 1966. On the Geology of Indoburman ranges, Geol. Soc. Australia Jour.13,127-194.

Chen, W.P. and Molnar, P., 1990. Source parameters of earthquakes and intraplate deformation beneath the Shillong Plateau and the northern Indo Burman Range. J. Geophys. Res. 95:12527-12552.

Curray, J.R., Moore, D.G., Lawyer, L., Emmel, F.J., Raitt, R.W., Henry, M. and Kieckhefer, R. 1979. Tectonics of Andaman sea and Burma. In : J.S. Watkins, L. Montadert and P. Dickerson (Editors), Geological and Geophysical investigations of Continental margins AAPG Memoir, 29, 189-198.

Das Gupta, A.B., 1977. Geology of Assam Arakan region : Quart. Jour. Geol. Min. Met. Soc. India, Vol. 49, No. 122, 1-53.

Das Gupta, Sujit and Nandy, D.R., 1995. Geological Framework of the Indo-Burmese convergent margin with special reference to ophiolite emplacement : Indian Jour.Geol.,, V.67, No. 2, 120-125.

Frohlich, C., and Denise Appeason, K. 1992. Earthquake Focal Mechanisms, Moment Tensors, and the consistency of seismic activity near Palte Boundaries. Tectonics, V. 11, 2:279-296.

Gangu, J.L., and Khar, B.M. 1983. Landsat and photogeological studies in Naga Hills between east of Merapani-Nazira and Indo-Burma border, KDMIPE(unpublished report).

Gubbins, D., 1990. Seismology and plate tectonics, Cambridge University Press, Cambridge.

Hamilton, W., 1977. Subduction in the Indonesian region in Island Arcs. Deep Sea Trenches and Back- Arc Basins, Maurice Euring Ser., vol.1, edited by M-Talwani and W.C. Pitmass III, pp. 15-31. AGU. Washington, D.C.

Hank and Kanamori, 1979. A moment magnitude scale. J. Geophys. Res., 84:2348-2350.

Knopoff, K. 1964. Q. Reviews of Geophysics, 2:625-660.

Kumar, M.R., Rao, P.N., Chalam, S.V. 1996. A seismotectonic study of the Burma and Andaman arc region using Centroid moment tensor data. Tectonophysics, 253:155-165.

Kumar, A., Manichandra,S. and Roy Someshwar. 2000. Seismological investigations in and around Manipur Res. Highlihgts in Earth System Sciences DST spl vol 2 on Seismicity 85-96

Kumar A., Sanoujam,M, 2001. Determination of crustal P velocity in Manipur and adjoining areas Proc., International Conference on Seismic Hazards with particular reference to Bhuj Earthquakeof January26, 2001, 3-5 Oct 2001 pp305-308

LeDain, A.Y., Topponier, P. and Molnar, P. 1984. Active faulting and tectonics of Burma and surrounding regions. J. Geophysics. Res. 89;453-472.

Mathur, L.P. and Evans, P. 1964. Oil in India : Internat. Geol. Cong. India. 22nd Session. Proc., 1-85.

Mukhopadhyay, M. and Das Gupta, S., 1988. Deep structures and tectonics of Burmese arc : Constraints from earthquake and gravity data : Tectonophysics, Vol. 149, 299-322.

Ni, J.F., Speziable, M.G., Bevis, M. Holt, W.E., Wallace, T.C. and Seager, W.R., 1989. Accretionary tectonics of Burma and three dimensional geometry of the Burma subduction. Geology, 17:68-71.1989)

Rao. N.P., Kumar, M.R., and Chalam, S.V., 1994. The Mean Slip Angle and its utility in the classification and analysis of focal mechanism solution. Proc. Of 30th Annual convention and seminar on Space Application in Earth System Sciences, 21-23 December 1993, IGU.

Sengupta, S., Acharyya, S.K., Van Den Hul, H.J., and Chattopadhyay, B., 1989. Geochemistry of Volcanic rocks from from the Naga hills ophiolites, Northeast India and their inferred tectonic setting : Jour.Geol.Soc. London, V.146, 491-498.

Vohra, C.P., Halder, D. and Ghosh Roy, A.K., 1989. The Andaman-Nicobar ophiolite complex and associated mineral resources : Current appraisal, in N.C. Ghost (ed), Phanerozoic Ophiolites of India, 281-315.

STRIKE-SLIP-RELATED STRUCTURAL PATTERNS AND OROGENIC GEOKINEMATICS OF THE MEDITERRANEAN-CENTRAL ASIAN OROGENIC BELT

M. Kopp

Geological Institute RAS, Pyzhevsky per.7.109017, Moscow, Russia

ABSTRACT

The method of structural pattern analysis allows us to recognize not only the kinematic type of structures (wrench fault, thrust, etc.) but also the concerned movements of the whole set of blocks.This paper is an attempt to develop the technique of the structural pattern analysis for finding the movements and deformations specially for the foldbelts, such as longitudinal, transverse and rotational displacements in relation to the general structural trend, different shapes of curving on a horizontal plane (bending, buckling), lateral squeezing-out and supercharge of deformed material. On the basis of the structural pattern analysis and with the use of the palaeomagnetic data the geokinematic map (scale 1:5 000 000 in original) of the Mediterranean-Central Asian orogenic belt was compiled by the author. Geokinematic analyses of the Mediterranean-Central Asian orogenic belt according to structural patterns make it possible to confirm independently that its structure was formed as a result of indentation of the Gondwana microcontinents (Apulian, Arabian and Hindustanian) into Eurasia. Moreover, additional emplacement centers (Caucasus, Kopetdagh, Pamirs, ets.) were established and block rotational movements were discovered.

INTRODUCTION

The orogenic geokinematics of the Mediterranean-Central Asian collision belt has not received an adequate attention. Research groups which adhered the mobilistic concept, concentrated their efforts on the pre-collision history of horizontal movements. The main subjects of discussions were nature of ophiolites, location and age of sutures as well as of the terrain. At the same time complicated horizontal movements occurring after the collision were not studied thoroughly. Till now, problems of collision in tectonics are a prerogative of a very limited circle of researchers. This is evidently reflected in compilation of graphic material: a set of perfect palaeogeodynamic maps for the preorogenic stage was resulted from the collective work of the "Tethys" Program [Atlas Tethys, 1993], whereas geokinematic maps for the orogenic stage were mainly compiled by a few scientists [Bazhenov, Burtman, 1990; Burke, Sengör, 1986; Burtman, 1984, 1989; Cobbold, Davi, 1988, Kopp, 1991; Tapponnier, 1977; Tapponnier, Molnar, 1976; Tapponnier et al., 1986].

However, the horizontal movements of the orogenic stage infect, determine the present appearance of folded belts. The orogenic structure of the Mediterranean- Central

Asian belt at the orogenic stage differs greatly from that of the preorogenic stage: folded zones and intermountain depressions overlie the Mesozoic sutures and intersect them. Frequently pronounced principal incompatibility of the actual geological structure with the global kinematics of rigid plates can be considered as one of the most complicated problems of mobilisation.,

The method of structural pattern analysis allows us to recognize not only a kinematic type of structure (wrench fault, thrust etc.) but also the concerned movements of the whole set of blocks. Luk'ianov (1991) defines a structural pattern as "stable and regulated in space combination of simultaneous parageneses of structures". This article is an attempt to develop the technique of the structural pattern analysis to find out the movements and deformations typically of the foldbelts, such as longitudinal, transverse and rotational displacements relative to the general structural trend, different kinematic shapes of flexure in horizontal plane (bending, buckling), lateral squeezing out and supercharge of deformed material. A number of researchers have done investigations in this direction e.g., Caire (1979), Chinnery (1966), Cloos (1928), Carey (1955), Freund (1974), Luk'ianov (1991), Woodcock (Woodcock, Fischer, 1986) and others.

There emerges a problem of space for volume of rock masses lengthening in the horizontal plane when wrench fault kinematics is analyzed. Two alternative variants are possible: (a) axial lengthening of the foldbelt and (b) its curving in the horizontal plane.

Two kinematic types of strike-slip faults

Cloos (1928) noticed that there are two varieties of normal faults: those that increase dip of a layer and those that reduce it. He named them synthetic and antithetic faults, respectively. These terms have been widely used to designate faults with contrasted rotation of the walls: In one case, rotation is concurrent with displacement, though it reduces the amplitude of the fault; in the other, it increases the amplitude. These terms have been applied not only to normal faults but also to reverse faults and thrusts. That is why there is apparently no reason to disregard this terminology in the case of strike-slips (Fig.1). In this case the elements displaced are not layers but axial planes of folds and other elements of subvertical structures. This was already done by Wilcox et al. (1973) although in a slightly different context: syn- and antithetic strike-slip faults were compared with R and R1 shears of simple shear zones, respectively. In the first approximation, this does not seem to contradict the accepted interpretation since these two types of shears differ also in the direction of the rotation of the walls. Yet, real strike-slips are not always elements of major simple shear zones, and hence the original," understanding of these terms (characterizing solely the direction of rotation of faults' walls) by H. Cloos is concrete and comprehensive.

To analyze deformation of a large volume incorporating a whole group of strike slips, it is observed that it differs considerably in these two cases. Synthetic strike-slips (Fig.1B) tend to broaden the folded zone by curving it, whereas antithetic strike-slips extend longitudinally towards the strike (see Fig.1A). These respective types of deformation of a

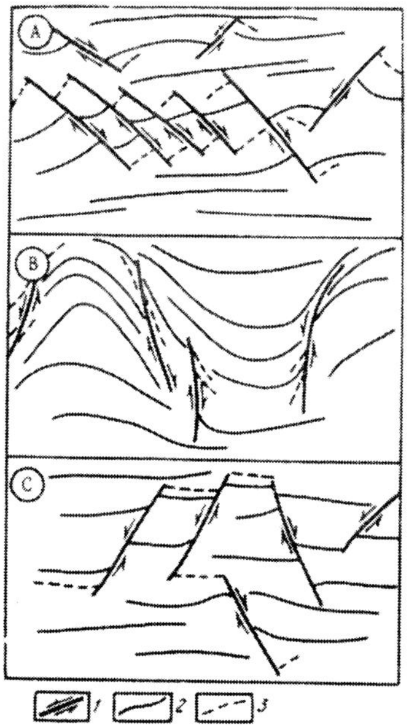

Figure 1: Two kinematic types of wrench deformations that differ in the direction of mass movement relative to orientation of the foldbelt. A - longitudinal lengthening (accompanied by antitethic flank rotations along oblique strike-slips), B - transverse advancement (accompanied by syntitethic flank rotations along oblique strike-slips), C - A combination of of transverse advancement and longitudinal lengthening (zero flank rotation). Key: 1, strike-slip faults; 2, axes of pre-strike-slip faulted folds and subvertical structural elements; 3, lines showing the trend of strike-slip-related structures of compressional origin

folded zone are referred to as "transverse advancement" and "longitudinal lengthening". There are many diagnostic criteria which must be considered for the comparison of structures of transverse advancement with the structures of longitudinal lengthening (Table 1).

The structures of longitudinal lengthening

General characteristics. Longitudinal lengthening of a foldbelt is best seen in compressed zones which generally lie in front of the blocks that act as indenters. Mechanically, it is a pure shear deformation, whereby lines diagonal to the orientation of compression are rotated towards the axis of maximum extension. The type of structural pattern is largely dependent on the type of deformation: when the rock is affected by shearing with attendant formation of diagonal strike-slips, the result is antithetic turn of structural units and of rotation (Fig. 2-4). When elongation is due to the transportation along the planes parallel to the extension axis, apparent signs of rotation are absent. The rotation of strike-slips was first described by Cloos (1957). later by Freund (1974). The respective paleomagnetic data were reported by Bazhenov, (1987) Carter et al., (1987) and Ron et al., (1984).

Table1: The comparison of the dingnostical criteria of the structures of the longitudinal lengthenin and transverse advancement forward arisen under the condition of compression of the folded belt.

#	Diagnostical criterium	Types of structures			
		Structures of lengthwise lengthening		Structures of transverse advancement	
1	Shape in plan of corresponding section of folded zone.	Approximately straightforward		Curved	
2	Manner of slewing primarily lengthwise elements and strike-slip wings.	Antithetic (slew increases amplitude of shear)		Synthetic (slew decreases amplitude of shear)	
3	Direction of slewing of: fault lines refer to main axis of stress and their resulting shape in plan	Lines of faults turn to axis of extension		Lines of faults turn to axis of compression in the course of deformation	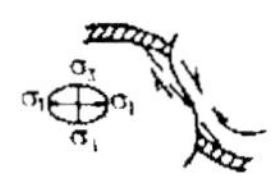
4	Presence or absence of compression transverse of shears	Locally present	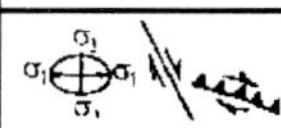	Everywhere present	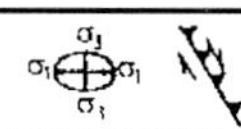
5	Morphology of shear zone (in connection with type of destruction.)	More often - brittle breaks, expressed by narrow, clear traced lines		Usually-viscous breaks, expressed by broad zones and able to change by flexures and zones of echelonic folds	
6	Surface variation of amplitude of compression structures, compensating strike-slip displacement	Value of horizontal shortening has maximum close to strike-slip		Value of horizontal shortening increases as we move father and father from strike-slip	
7	Surface variation of amplitude of extension structures, compensating shear displacement	Extension structures open to strike-slip		Extension structures open from strike-slip	
8	Orientation and position of strike-slip-related folds refer to line of strike-slip	Folds of transverse-diagonal orientation locate at ends of strike-slip		Folds have close to one of strike-slip orientation and locates in its zone	
9	Orientation (refer to general structures plan)of structures compensating deformation as a whole	Transverse		Lengthwise	

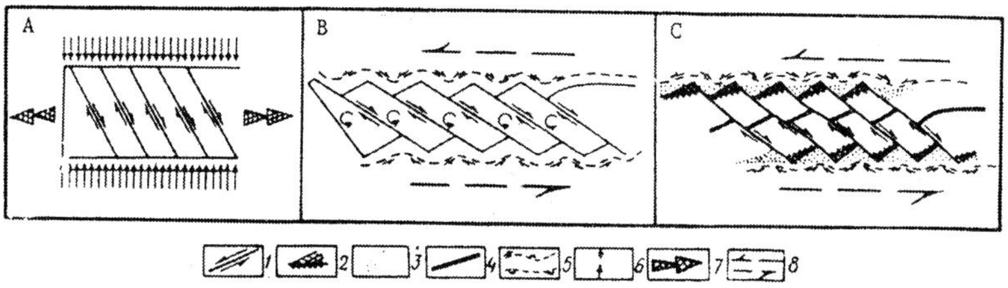

Figure 2: Structural pattern that results from coordinate unidirectional rotation of slice-like blocks (in the presence of one set of strike-slip predominating - the case of "domino" mechanism): a general diagram. A - before deformation, B - after deformation, C - types of structural pattern. Key: 1, strike-slips; 2, thrusts and other structures of compression (the width corresponds to the relative value of horizontal constriction); 3, areas of local decrease of horizontal compression or areas of horizontal extension at the rear of displaced fault flanks; 4, the displaced structural element, initially of longitudinal strike; 5, imaginary material lines after deformation (arrows denote the directions of block movements associated with local salients and gaps); 6, direction of common horizontal compression of the whole deformated volume; 7 - direction of lateral squeezing-out of the deformed material; 8, direction of the simple shear accompanying the deformation.

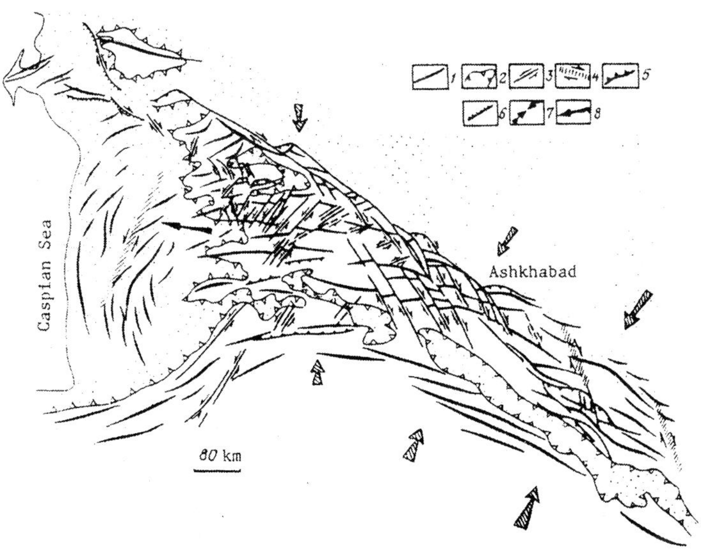

Figure 3: Kopetdagh: example of domains experienced by longitudinal elongation resulting from anticlockwise rotation of one set of slice-like blocks which was displaced along dextral strike-slips. Key: 1, fold axes; 2, molasse basins; 3, strike-slip faults; 4, simple shear zones; 5, thrust and reverse faults; 6, normal faults; 7, orientation of the horizontal compression; 7, direction of the lateral squeezing-out of the deformed material

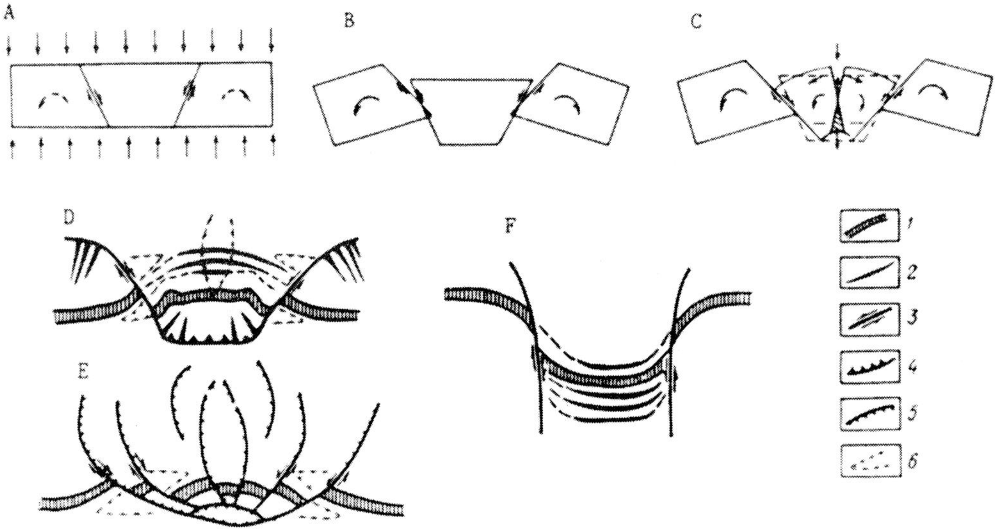

Figure 4: Structural patterns resulting from variably directed rotations of blocks (in the presence of two conjugated sets of strike-slips). A-C - general diagram: A - before deformation, B,C - after deformation B - without plastic deformation of blocks, C - with plastic deformation of blocks); D, E - structural patterns formed in the setting of: D - transpression, E - transtension; F - the structural pattern of an orocline formed under transverse advancement of a segment of the folded zone (for comparison). Key: 1, the displaced structural element, initially of longitudinal strike; 2, strike-slip-related folds; 3, strike-slips; 4, thrust and reverse faults; 5, normal faults and rift-type structures; 6, areas of local decrease of horizontal compression or areas of horizontal extension at the rear of displaced fault flanks.

Rotation of blocks with only one system of strike-slips present. Cloos and Freund considered a case when a set of strike-slips of the same sign is formed. The blocks then rotate like a pack of books, losing none of their intrinsic rigidity (so-called "the domino mechanism").

Such deformation can be expected in the Kopetdaghian meganticlinorium, as already suggested by Bazhenov, (1987) and (Kopp, (1991). The Kopetdagh system was folded in the Neogene due to the compression of shelf and slope sediments in the south of the Turan block. The resulted large northwest-trending linear folds (Fig.3), became more compressed in the east, and flattened out in the west. The eastern boundary of the Kopetdagh is much more abrupt, and is situated on the strike of a major Aden-Mangyshlak dextral shear zone towards east of the Lut block. All this seems to tie in with the assumed anticlockwise rotation of the East Iran Block. In the west, at the conjuction with Alborz, the southern Kopetdaghian folds turn southwest forming a north-advancing arc.

The Kopetdagh was cut into narrow slice-like blocks by a set of densely spaced strike-slips which were dextral in the east and sinistral in the west. Such symmetrical arrangement of contrasted strike-slips at the sides of the Kopetdagh suggests that it was curved in plan due to transverse advancement. This view was proposed by Ognev [1937]. However, the situation here is not that simple. Because of the contrasted strike-slips of different age the dextral ones formed at the end of the Miocene were reactivated in the Late Pliocene-Quaternary, whereas the sinistral strike-slips were covered by Pliocene sediments. However, both strike-slip sets might have coexisted for some time (Miocene-Pliocene boundary). In addition, sinistral strike-slips had smaller amplitude than the dextral ones (1-2 and 7-10 km., respectively). Finally, both sets of strike-slips differed in the type of compensation: The dextral strike-slips were conjugate in the north with thrusts, whereas the sinistral ones turned into normal faults or fanned with minor strike-slips. If the structural arcs were bent due to a gradual northeastward movement, both sets of strike-slips would be limited in the north only characteriesed by structures of compression.

The above facts suggest that the observed situation can be explained by allowing the deformation of longitudinal elongation with blocks rotated around a vertical axis. There are other arguments in support of this interpretation: 1) an anomalous big angle between left- and right- lateral strike-slips at their conjunction. (Theoretically, the double shear angle is known to be 90^0. However, because of internal friction in rocks it was actually much smaller for about $60\text{-}70^0$) (Höeppener et al., 1969). In the present case, the double shear angle is 90^0 and more). There was a respective reduction in the angle formed by the direction of the strike-slips and the compression structures; 2) antithetic rotations of fold axes can be especially noticed southwest of Ashkhabad; 3) the amplitude of thrusts was highest at the conjunction point with right-lateral strike-slips and waned with distance from it; and 4): if the Kopetdaghian arc emerged only as a result of northward movement, its frontal part should have huge thrusts and nappes with amplitudes ranging to several dozens of kilometers. Actually, Krymus [1966] and Rastsvetayev (1966, 1973) described that the front of the Kopetdagh was dominated by right-lateral strike-slips. On the other hand, although anomalous compression was recorded towards west of the Kopetdagh, however, it was oriented latitudinally and that it was likely most associated with the longitudinal lengthening with transverse advancement [Kopp, 1979).

The hypothesis of Cloos-Freund can account for the grouping of similarly signed strike-slips in separate domains as well as for some features of their mutual arrangement. However, Freund himself noted (1974) that it was unacceptable for the deformed fractures by two, affected by sets of conjugate strike-slips.

Rotation of blocks in presence of two conjugated sets of strike-slips of equal significance. A rectangular bar (Fig.4A-C) representing a subvertical tectonic unit can break up under compression into three blocks by conjugate strike-slips with opposite signs. Rotation can form some free space between the extreme blocks where the central trapezoid portion of the bar subsides and gets bent in the process. The resulting lens- and amygdaloid-shaped blocks (Fig.4D, E) gradually become longer and thinner under the continuing deformation. The convex side of the lens that overlapped the former boundary of the blocks experienced an additional horizontal compression with formation of a local structural arc; the concave side experienced an extension which created a basin that later was filled with sediments (see Fig.4C-E). The stresses also changed along strike of the bent segment: its lower ("southern") part developed an imposed "sublatitudinal" compression while in the "north", the "sublatitudinal" extension was concentrated.

The resulting specific structural pattern largely depends on the properties of the deformation field. Under transpression (Fig.4D), a trapezoid block bounded by strike-slips was pressed between the adjacent blocks. Pushing through, it exerted pressure on the walls, bending them outward. The resulting structural arc was complicated for lack of space by marginal minor arcs ("pitches"). With the wedge in the transtension environment (Fig.4E), the structural arc between strike-slips had a smooth shape. Under these conditions, extension at the back of the arc was the highest.

Structural patterns of this type are found in areas of the highest constriction and crustal pile-up where a symmetry axis between diverging strike-slips of different signs is usually present.

A good illustration is the Late Cenozoic structure of the Central Asia region (Fig.5), which many authors believe might have been formed by the indentation of a rigid Hindustan block into Eurasia.

The northwest-trending dextral faults are in the west, whereas the sinistral ones are situated in the east. The axis of symmetry (axis of "streaming") that separates domains of contrasting strike-slips lies in the zone of greatest penetration of the Hindustan indenter. More specifically, one can detect here two minor axes, one running between the Talas-Fergana and Altyntag strike-slips and the other one in the south of Lake Baykal. The angle formed by symmetrical conjugated strike-slips that cut off wedge-shaped blocks is about 110^0. This is more than the experimental value ($65\text{-}70^0$) of the double shearing angle, suggesting the faults and were rotated toward the axis of maximum extension. The structural map shows antithetic rotations of structural lines that manifested themselves in the development of transverse to wrench faults and strike-slip-related compression strucrures. Finally, there is solid evidence of plastic deformation (in plan) of wedge-shaped blocks, especially in the arcuate north Tien Shan arc having the Tarim depression at the rear. Such bent tectonic wedges can be ascribed to moving apart and turning sideways of the neighbouring blocks.

Thus, the data suggest the presence of longitudinal, of sublatitudinal orientation with lengthening of the Central Asian region both to the west and east. This is also supported by Molnar and Tapponnier (1976).

Indeed, a similar structural pattern can be observed in the Caucasian syntaxis (Fig.6), positioned at the front of the maximum penetration by the Arabian Plate in the foldbelt. [Kopp, 1979, 1991, 1997, Nikitin, 1987 and Rastsvetaev, 1973, major strike-slip zones were many. These zones trend at a small angle (20-30^0) with reference to the general structural plan and locally assumed an almost longitudinal orientation; associated with multiple folds and thrusts. These facts suggest that either it was a transpressive strike-slip deformation, or else it was a rotation of faults with subsequent superposition by structures of compression. The Caucasus is radically different from the Central Asian region (north of the Himalaya and Tibet), where the strike-slip stress regime was either manifested directly or was in interaction with the normal fault stress regime. The Greater Caucasus strike-slip faults were

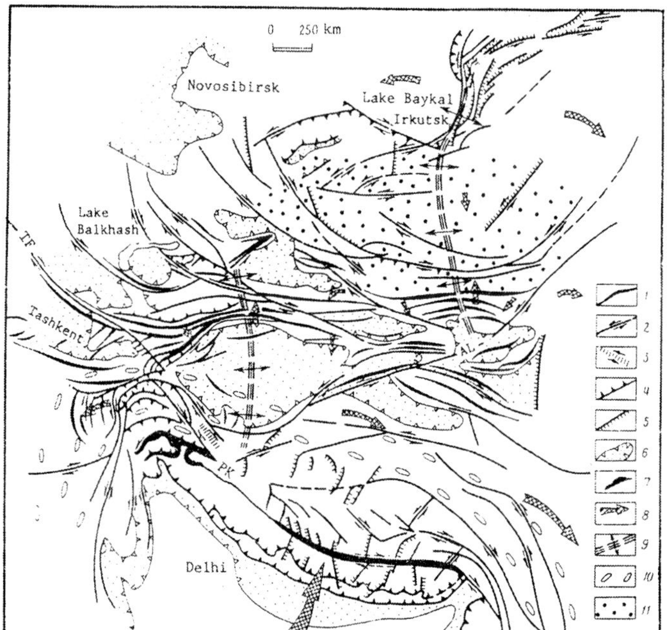

Figure 5: Central Asia: example of domains in the course of deformation experienced by longitudinal elongation resulting from two-directional (anti- and clockwise) rotation of two conjugated sets of blocks which were displaced along both dextral and sinistral strike-slips. Key: 1, lines of the fold strike; 2, strike-slips; 3, simple shear zones; 4, thrust and reverse faults; 5, normal faults; 6, molasse basins; 7, ophiolites; 8, direction of the lateral squeezing-out of the deformed material; 9, "axis of streaming", i.e. axis of symmetry of blocks laterally squeezed in opposite directions; areas under the setting : 10, transpression, 11, transtension. Strike-slip faults: TF, Talas-Fergana; PK, Pamir-Karakorum; AL, Altyntagh

formed in the Late Miocene-Quaternary times. The sinistral strike-slips at the convexity of the Dagestan arc are pre-Akchagylian in age, whereas the parallel ones of Kobystan and the Kura depressions are of Upper Pliocene or Quaternary.

At the center of the Greater Caucasus, there was an axis of streaming formed out due to replacement of dextral strike-slips by sinistral ones. Both left- and right-lateral faults trend in opposite directions away from the place of the tight constriction (deformation neck?) of the Greater Caucasus. Attenuations of major strike-slips were associated with transverse folds which were strongly pressed near the faults dipping gradually with distance: an evidence of antithetic rotation of the respective fault flanks. An arcuate Terek-Sunzha anticlinal zone lacked the necessary space to sag fully into the opening sphenochasm. The narrow neck of the Greater Caucasus, together with the signs of rotation and bending of blocks in plan, represents megaboudinage structure in the horizontal plane. Longitudinal lengthening of the meganticlinorium was compensated at periclines by the zones of transverse folds (Kopp, 1979, 1997).

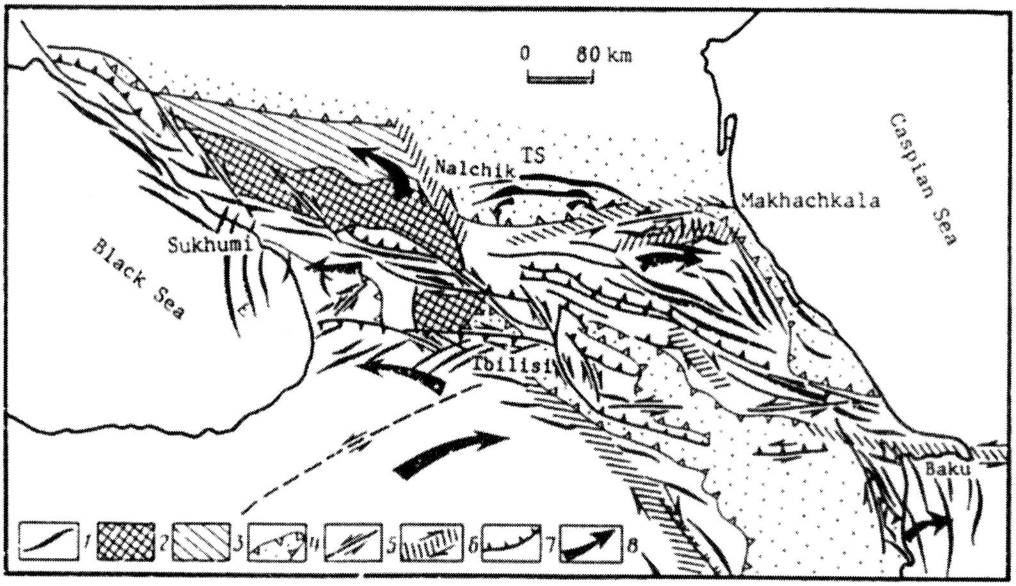

Figure 6: Greater Caucasus: another example of domains experienced by longitudinal elongation in the course of deformation resulting from two-directional (anti- and clockwise) rotation of two conjugated sets of blocks which were displaced along both dextral and sinistral strike-slips. Key: 1, lines of the fold strike; 2, Paleozoic crystalline core of the Greater Caucasus; 3, North Caucasian monocline; 4, molasse basins; 5, strike-slips; 6, simple shear zones; 7, thrusts; 8a, direction of lateral extrusion and regional block movement; 8, direction of local block movement; TS, Terek-Sunzha anticlinal chain

Problem of space for the rotating blocks - the effect of eccentricity. Antithetically rotating blocks tend to indent the space around them with their corners (Fig.2). Oriented across the axis of maximum compression, an "elevated" imaginary line is formed (in the plane of the given figure) (Fig.2B,C). Each rotating block acted eccentric that converted the rotating movement into translational motion. The mass excess accumulating at the blocks was partly "deposited" as thrusts or other compression structures whose amplitude gradually diminished with distance from the strike-slip. Intrablock gaps favour structures of extension. Such compensation of the horizontal shift (transition of a strike-slip in to a thrust or an extension structure) can be established throughout. However, "dissipation" of the conjugated gaps were due to movements solely along the horizontal plane. It could have happened as follows.

1. An arcuate, listric distortion of the fault line appeared concurrent with the rotation. It can be primarily be associated with the cycloidal configuration of the trajectories of maximum tangential stresses (Cummings, 1974; Tapponnier and Molnar, 1976). In addition, the curvature of the fault line was capable of growing with deformation and was graphically proved by Freund (1974) in comparison with the layers that slipped at the fold's limb (Fig.7). Such an approach may prove rather productive; if the geometry of various fold bends is known. It helps to identify the dependence between a strike-slip parameters (the fault line configuration in plan, the amplitude of displacement, rotation angle, etc.), and the deformation along the longer axis of the ellipsoid (Fig.7, II).

The geometric analysis carried out in this work leads to the following conclusions: (a) Completely conjugated (directed to both sides) curve can be formed only in the absence of any additional stresses along the longer ellipsoid axis (case of direct strike-slip deformation), as is symbolized by a concentric fold (Fig.7, IIC). When the deformation along the long axis of the ellipsoid is characterized by the presence of pronounced absolute extension, typical of transtension (a similar fold, Fig.7, IIB), or, compression (i.e., transpression) in the core of a concentric fold, fig. 7, IID, then line of strike-slip curves only in one direction. There will be then no smooth attenuation of strike-slips at both ends, and the horizontal movement can be to a great extent, compensated by deformations in the vertical plane; (b) strike-slips which evolve in different settings of transpression or transtension, can curve in opposite directions towards each other; (c) in transition to a transtension zone, amplitude grows abruptly due to the plastic flow along longitudinal ("sublatitudinal" as given in the figure) segments of listric strike-slips. In a transpression zone, amplitudes are diminished; (d) transverse shortening of the deformed volume can decrease towards the transtension zone.

Attempt has been made to summarize the above specific kinematic features of strike-slip deformation in various structural settings on special scheme (Fig.8). Comparing the deformation geometry of the curvature of strike-slips in plan with the known geometry of the trajectories of maximum tangential stresses [Cummings, 1976; Tapponnier, Molnar, 1976], one will inevitably see the similarity. This is an added proof of the validity of these conclusions. Above-mentioned features also have been registered in natural geologic settings. For example, in the region of Central Asia (Fig. 5) there are major strike-slips bending towards each other in Mongolia and Baykal and the Pamir-Tibet areas. Due to this the pattern of conjunction of strike-slips of different signs in the north and south of the regions are totally different: In the

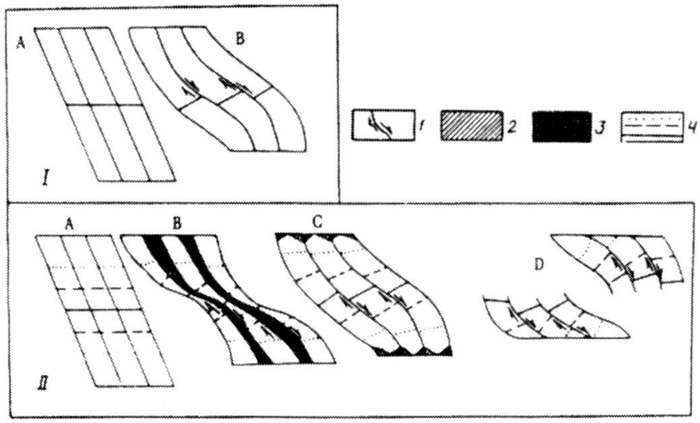

Figure 7: Deformation curvature of strike-slip fault lines (on the analogy with folds - explanations see in the text). I - gradual attenuation of strike-slips, after Freund (1974]) : A - before deformation B after deformation; II - dependence between the type of strike-slips attenuations and their configuration) and kinematics of bending with consideration for the sign of stresses along the long axis, axis c, of the deformation ellipsoid: A - before deformation, B - the analogy with a similar fold (extension along axis c), C - the analogy with a concentric fold (zero stress along axis c), D - plastic deformation at the core of the concentric or diapir fold (compression along the axis c). Key: 1, strike-slips; 2, areas with concentrated compressing stresses; 3, areas of concentratred extending stresses; 4, imaginary predeformation markers of initial longitudinal trend.

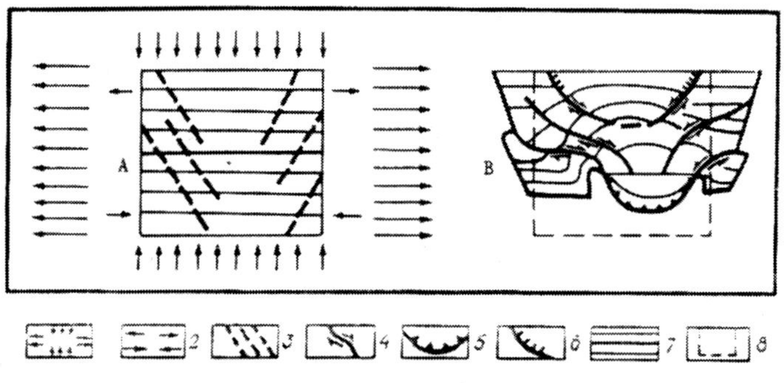

Figure 8: A general scheme showing the dependance between the plan configuration of strike-slips and the kinematic type of rotation of their walls with consideration of the sign of stresses along axis c. Key: 1, orientations of horizontal lengthening and shortening; 2, the sign of auxiliary stresses along axis c; 3, location of strike-slips at the start of deformation; 4, strike-slips; 5, thrust and reverse faults; 6, normal faults; 7, imaginary predeformation markers of initial longitudinal trend; 8, original contour of the deformed domain.

north, they form south convex arcs, whereas in the south they are conjugated at much more acute angle. In the transition zone, major strike-slips (Talas-Fergana and Altyntag) show typical double curves of fault lines. In the Pamirs and Tibet, the pattern of structural lines is rather intricate, with strike-slips surrounded by antithetic, and synthetic rotations (Pamirs, Kuang Lung). Therefore the general correspondence between the region's structural pattern (Fig. 5) and that of the hypothetical, and theoretical models seems apparent (Fig.8).

Thus, using solely the unique features of the plan configuration of strike-slips and structural lines, and drawing on no data of the associated compression or extension structures (in practice this must be done), one can easily conclude that in the southern part of Central Asia (Pamirs, southern Tibet, Himalaya), strike-slip faulting occurred under transpression; in Mongolia under transtension; and in the boundary zone (Tien Shan, Kuang Lung), it was strike-slip faulting again. The transpressive style of deformation over the margin of the Indian plate is indicative of lack of space, thus confirming its possible indentation into Eurasia.

2. Dissipation of the mass excess at the termination of the strike-slip can occur due to its splitting into several fanning branches (Fig. 9) (Chinnery, 1966; Freund, 1974). Such

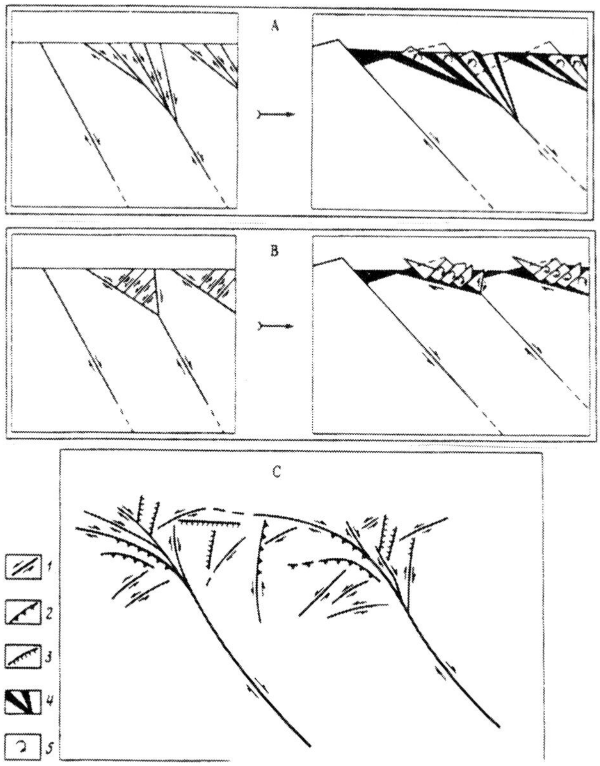

Figure 9: Dissipation of strike-slip displacement through fan-like splitting of the major fault into minor strike-slips. A-B - general diagrams (left - before deformation, right - after it): A - fan-shaped splitting of the fault line [after Chinnery (1966) and Freund (1974)]; B - emergence of minor strike-slips of the opposite sign ("antitethic" faulting); C - generalized structural pattern of the major strike-slip attenuation area. Key: 1, strike-slips; 2, thrust and reverse faults; 3, normal faults and other extension structures; 4, gaps between divergent blocks; 5, directions of block rotations.

splitting in plan is demonstrated by many major strike-slip faults, i.e. the North Anatolian, Levantine and Kopetdaghian.

3. It can be assumed that owing to the effect of eccentricity, elimination of mass excess can be contributed by smaller antithetic strike-slip of an inverse sign. As may be seen

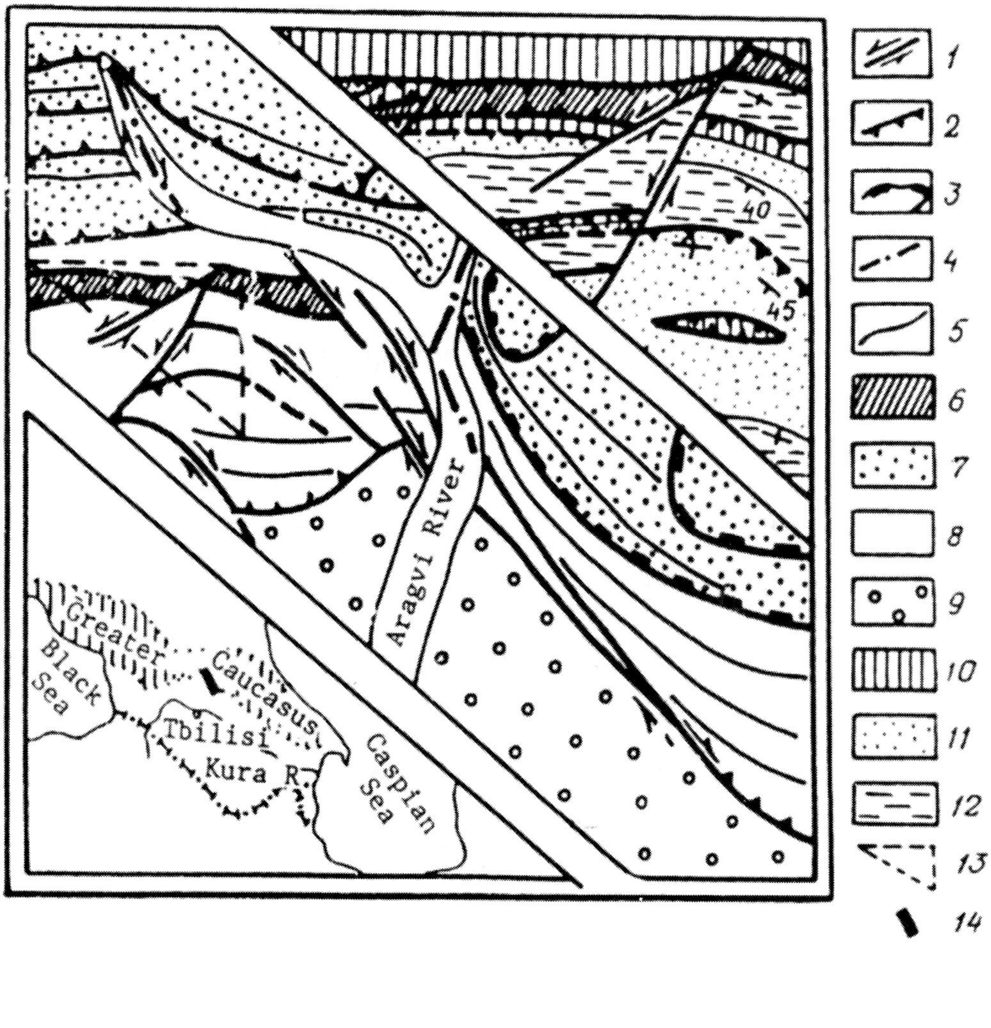

Figure 10: The structure of the north attenuation area of the East Kartly right-lateral strike-slip zone in the southern slope of the Greater Caucasus. Key: 1, strike-slips; 2, thrust and reverse faults; 3, nappes; 4, faults covered with Quaternary deposits; 5, fold axes and marker horizons; 6, Lower Jurassic strata of the cover of the Georgian rigid block; 7, Cretaceous flysch; 8, Paleogene flysch; 9, Neogene molasses; 10, Upper Middle Eocene shaly clays (Ildokani formation); 11, Upper Eocene carbonate flysch (above-Ildokani formation); 12, Upper Oligocene-Lower Miocene shaly clays (Maikop series); 13 - the boundary of the zone shown in the inset above; 14, location of the area (inset below)

(Fig.9B) such compensation of the strike-slip displacement goes together with a certain widening of its zone (as in the formation of the fan). Since small-amplitude inverse-signed strike-slips rotate in the opposite direction, such deformation can take place only at the synthetic rotation of the major strike-slip's walls.

All these complex features can affect the structure of the attenuation areas in specific strike-slips (Fig.9C). For an of example (Fig.11), we may refer to the northern tip of one of the right-lateral strike-slip zones of the East Caucasus, with the Kartly depression in the east. In the eastern flank of the strike-slip is the Alisisgorskiy flysch nappe. Its presence there is due to the shift of shear's flank southwards, concurrent with the general vergence. In the west of the nappe, there is a synthetic rotation of fold axes parallel to the strike-slip.

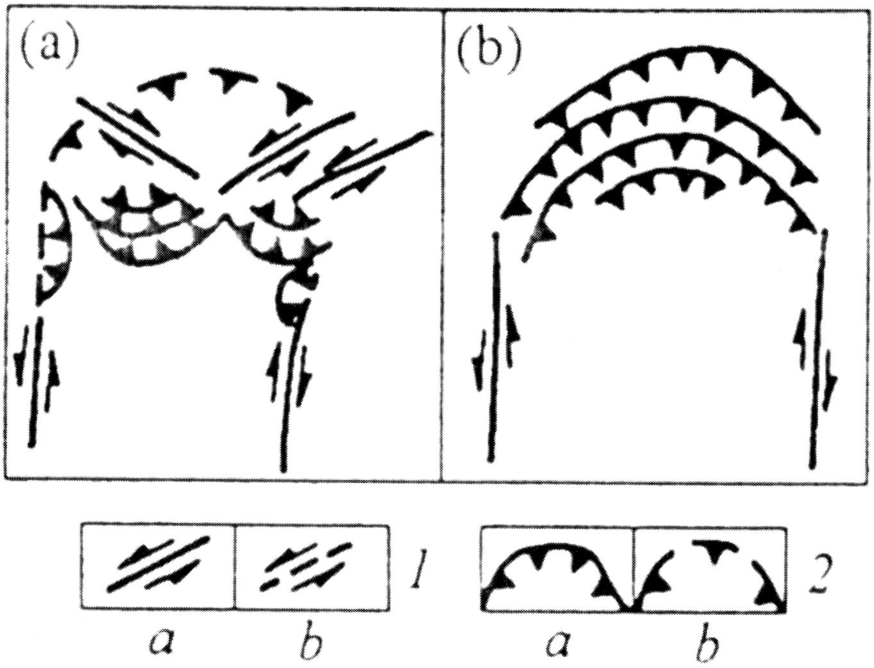

Figure 11: Typical structural patterns of (a) underthrust and (b) overthrust oroclines. Key: 1, strike-slips [necessarily present (a) and possibly present (b)]; 2, thrusts and other structures of the compression (necessarily present (a) and possibly present (b)).

The western wall of the latter was displaced to meet the regional vergence and apparently lacked space. This is proved by a surfacing tectonic wedge of Lower Jurassic rocks of the Georgian rigid block . Here, are also north-vergent thrusts anomalous to the southern slope of the Greater Caucasus. The western side of the strike-slip has been rotated in an antithetic mode: Instead of being of the regional west-northwest trend, the structures are oriented here latitudinally or even with some swing to the east-northeast. The registered deformation of the longitudinal lengthening may be attributed to the lack of space for the given side of the strike-slip. The principal strike-slip zone splits into a fan; in addition, the heavily constricted western side shows a set of minor left-strike-slips.

Compensation of the longitudinal elongation of fold-belt segments is recognised by their curving in plan as well as in formation of transverse compression structures [Kopp, 1979]. In the above-cited examples, similar structures are found at periclines of the Caucasus and Kopetdagh (Figs.3,6).

The structures of transverse advancement

The sections of thinning and longitudinal extension of folded zones are connected with their bulges, which arise during transverse displacements. Such flexures of folded zones are usually called structural arcs (oroclines). The arcs are limited on the flanks by synthetic strike-slips, oriented, as a whole, transversely to the general tectonic trend.

The morphology of oroclines substantially depends on the direction of tectonic transport (into or from the inside of the folded zone) and the kinematics of flexure, which, as for ordinary vertical folds, can be longitudinal or transverse (buckling or bending in the horizontal plane respectively) as well as it can be accompanied by rotation in the horizontal plane.

Underthrust and overthrust oroclines. In the case of movement of masses into the folded belt, the belt compresses and, correspondingly. converges, Such movement of masses, which usually take place with underthrusting, are met with significant resistance from the deformed rocks. In contrast, a movement, directed from inside of the folded belt, i.e., to the less dislocated area, is free and can be recogised as a widescale overthrust. The latter increases the width of the folded belt, but in doing so it decreases the effect of its common compression. Correspondingly, there are morphological differences between oroclines of the underthrust and overthrust types (Fig.11).

Underthrust oroclines have irregular angular outlines on their concave side (Fig.11A). To the convex side, the irregularities are smooth. The complex festoony pattern in the rear of the arc, due to buckling in the horizontal plane, is connected with the reactive roll back of the frontal part of the underthrusted block from the collision zone to the less crumpled area or deeper submerged foredeep. The general vergence has the same direction, and only on the convex side of the oroclines it changes on the opposite side. There are other signs of increased horizontal compression: significant amplitudes of orogenic uplift; high-pressure metamorphism and squeezing of the nappes of the remobilized basement. mass excess, accumulated ahead of the front of the underthrusting block, causes its lateral extrusion.

For the formation of overthrust oroclines (Fig.11B), the possibility of their free displacement along the Earth's surface determines the following specific features: smoothness of the flexure; in some places, strong stretching in the direction of the tectonic flow; absence of signs of the deficient space on the concave side, (and in contrast, the extension structures often develop here); lesser compression in the frontal part, but sharp increase of the degree of overthrusting; uniform vergence; the thin-skinned tectonics; and absence of big strike-slips and other structures, also testify to the lateral squeezing-out of the rocks, in the frontal zone.

Underthrust and overthrust oroclines are usually conjugate, and the difference of their structural patterns are noticed in comparing adjoining sides and arcs. The classical example of underthrust oroclines is enormous in Peri-Indian arc, which surrounds the Indian fragment of the continental crust on the Indo-Australian plate . This flexure can be named an arc only, conventionally, because it has an irregular shape due to its complication by a series of more proper overthrust arcs: the Himalaya, the Suleiman Mountains, and others which are convex towards the south. The presence of large diagonal wrench zones (Talass-Fergana, Dzhungarian, Altyntag and others), with longitudinal lengthening and lateral escape also conform it. The Makran (Fig.12, IIa) and Zond oroclines conjugated with the Peri-Indian one, have quite different morphology. The smooth flexure in the direction of the movement, the absence of the roll back features of deformed material and deficit of space in the back arc region testifiy that they were freely lengthened subparallel to the surface of the Earth. The degree of the orogenic uplift in the front of both oroclines is much smaller than that of the Himalayan-Tibet region. The presence of common decollement of the cover is inherent in the front of the oroclines. Essentially, the entire folded region of the Makran is the overthrust of the Phanerozoic cover removed to the south for some hundreds of kilometers.

The types of the flexure in the horizontal plane. The variant of the transverse flexure (bending) in a horizontal plane of oroclines is usually described in the literature. There are two synthetic strike-slips (or wide simple shear zones) with approximately equal amplitudes (Fig.12, Ia). A typical example of such structures is the above-mentioned Makran orocline (Fig.12, IIa). From the flanks it is limited by the two complexly built simple shear zones in (100-150 km) which the amplitude of the individual faults measures for some tens of kilometers, and their total effect is for about 300-400km. The shearing is associated with overthrusting subparallel to them. The shear lines along the flanks of the orocline tend to bend from the diagonal to transverse (to the folded belt) strike. So, the line of the Chaman sinistral strike-slip fault is arcuate to the west: the strike of fault varies from the northeast to the middle part. The shears on the west flank of the Makran structural arc have a similar configuration.

Thus, many features of the structures of transverse advancement (Table 1) are found in the structure of the Makran orocline. It is obvious that mass movement was directed to the south and had an overthrust nature. The overthrusting and convexity to the south of the structural arc was possibly accompanied by extension in its rear. This is confirmed by the presence of a deep molasse basins (Kharan and Khamun-i-Mashkel,) and also by the wide distribution of earthquakes associated with a focal extension mechanism (Chandra, 1984).

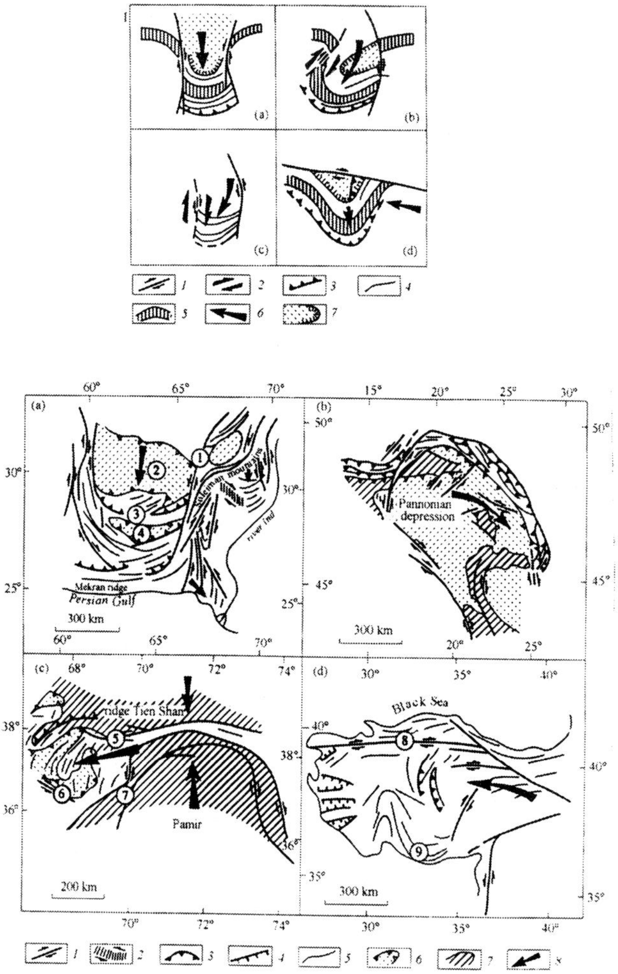

Figure12: Structural patterns characteristic of upthrust oroclines under the conditions of the different types of the flexuring in horizontal plane. I - the standard kinematic model: a - bending (transverse flexuring); b and c - bending with rotation in horizontal plane; and d - buckling (longitudinal flexuring). Key: 1, strike-slips; 2, simple shear zones; 3, thrusts; 4, strike-slip-related folds; 5, pre-strike-slip structural element, initially of longitudinal strike; 6, directions of block movements; 7, molasse basins, connected with shears.

II - regional examples: a - Makran arc, b - Carpathian loop. C - Tadzhik, d - Anatolian arcs. Key: 1, strike-slips; 2, simple shear zones; 3, thrusts; 4, normal faults; 5, fold axes and lines of strike; 6, molasse basins; 7, pre-Alpine basement; 8, directions of block movements. Figures in circles show tectonic elements. 1 - Chaman left strike-slip; 2 - Kharan intermontane basin; 3 - Chagay volcanic belt (anticlinorium); 4 - Hamun-i-Mashkel intermontane basin; 5 - Gissar-Iliak right strike-slip fault; 6 - Kuh-I-Alborz left strike-slip fault; 7 - Darvas left strike-slip fault; 8 - North Anatolian right strike-slip fault; 9 - structural arc of West Taurus.

The next variant in which the bending combines with rotation in the horizontal plane (Fig.12 IB and IC), is widespread and has been described to a lesser extent (except for the work by Caire [1979]). In this case one of the flank of strike-slips has a significanty bigger amplitude and, correspondingly, shows an antithetic appearrance, while the second offers to be synthetic or tends to disappear totally. The appearance of the orocline in this case depends on the proportion between the processes of the bending and formation of the new structures.

In the first case, when the flexure dominates (Fig. 12, IB) the structural loop is formed i.e. the Carpathian arc. To the north, in the region of the Vienna Basin, Sudeten, and West Carpathians, the flank of the loop is accompanied by the neotectonic sinistral strike-slips [Royden et al., 1983]. Their orientation from the northwest to the southeast of the Carpathian varies from the northeast to the sublatitudinal one, corresponding with a clockwise rotation. This point of view is confirmed by many palaeomagnetic results. The soutn flank of the arc has been subjected to clockwise rotation form 100^0 to 110^0, and the north flank has been rotated in the opposite direction angles varying form 40^0to50^0 [Bazhenov and Burtman, 1990; Burtman, 1984]. The region of anticlockwise movement coincides on the whole with the domain of sinistral shears. The idea that the transverse flexure of the Carpathian arc was accompanied by turning in a clockwise rotation was discussed by Caire [1979] and Tapponnier [1977].

In the case of intensive formation of new strike-slip-related compression structures, folds and overthrusts are formed near the antithetic shear, having biggest amplitude which gradually levels out in the opposite direction. As a result, confined (forced) migrations can occur. A typical example is the Tadjik (Fig.12, IIC), formed in the Neogene as a result of squeezing-out of Mesozoic-Cenozoic rocks out of the Pamir syntaxis [Luk'yanov, 1991]. The folds are limited by two shear zones (the dextral Gissar-Ilyak zone to the north and the sinistral Kuh-i-Alborz zone to the south), forming a series of subparallel arcs, convex to the west. These features display the fact that arcuate folds were bent due to transverse flexure. However, a number of facts (the better expression and the less width of the north dextral zone compared with the southern one, strongly diffused in space; the tightening of folds of the Tadjik depression on the north and its gradual submergence and fan-like divergence to the south; the plan asymmetry of the folded zones, the south flanks of the arcs much shortened and tucked-in are evidences for the progressive movement of the mass complicated by anticlockwise rotation. The palaeomagnetic data (Bazhenov and Burtman, 1990) show the presence of such a rotation (to 90°) only in the most eastern part of the Tadjik depression near the Pamirs.

Finally, the variant of the longitudinal flexure (buckling) in the horizontal plane (Fig.12, ID) is possible. On the flanks of such arcs concurrent shears can be absent, but the presence of large antithetic strike-slips in its rear, where they provide the detachment of the bending lithoplate from its basement, is evident. This can be due to longitudinal strike-slip (Fig.12, ID), or two longitudinal or diagonal strike-slip of different signs, which "dammed" the structural arc from both sides and turned its flanks on one another.

The Taurus arc curving (Fig.12, IID), took place due to the displacement of southern Anatolia along the North Anatolian dextral shear (Pavoni, 1961). It was found that the dextral amplitude of the North Anatolian fault did not exceed 100 km (Sengör et al., 1982). It is not enough to straighten the Central Taurus arc into the direct longitudinal structure. The palaeomagnetic data point to rotation in an anticlockwise direction near the North Anatolian fault (Burtman, 1984); however, the character of the distribution of the measured values can attest both longitudinal bending of the east flank of the structural arc and the rotation of all Anatolia as a rigid plate .

The mechanism of buckling is clearly determined for the Himalayan orocline. Both of its ends are sharply, and sometimes under the right angle, rest against the large shear zones of the framework of India: the Chaman-Darvas sinistral zone on the west (Fig.13) and the Semangku-Sagain dextral zone on the east. Nearer to the orogenic belt, these shear zones are split llike a fan and the sublatitudinal branches of the shear are developed. The strike-slips acquire the shape of an arc, directed by convexity to the north. Correspondingly sides of strike-slips adjacent to the India flanks experience antithetical rotation: clockwise for the flanks of the sinistral strike-slips of Punjab and anticlockwise for the flanks of the dextral strike-slips of Assam. This configuration of shear rotation can be directly correlated with the buckling of the flanks of the Himalayan structural arc and turning up of its ends to the north. Simultaneously, its apex experienced displacement to the south due to excess mass, located to the north from the Himalaya and the gravitational dragging into the pre-Himalayan foredeep. The absence of any large synthetic strike-slips on the flanks of the Himalayan arc is in accordance with the mechanism of buckling since the overthrusting occured along all its perimeter. The secondary character of the flexure was reliably found by palaeomagnetic data (Klootwijk et al., 1985).

Paragenetic combinations of the structures of transverse advancement and longitudinal lengthening

In real conditions the structures of longitudinal lengthening and transverse advancement can be combined to one on other. The strike-slips of the different kinematic types, bound together by paragenetical relations, can also coexist. So, the curving in the plan of block is limited from the one side by the antithetic strike-slips and from the other side by the synthetic ones. Correspondingly, paragenesis of rotation and its associated structural patterns can be derived. The corresponding examples have been already described above (Fig.12, IIB and IIC). Besides, this the paragenetic combinations of structures of longitudinal and transverse protrusions, can coexist which are related to the horizontal deformation of a simple shear or some heterogeneity caused by an obstacle to the movement in some directions. The flank, which moves to meet the obstacle, tends to turn along axis of foldbelt, i.e antithetically. The other flank, moving in a favorable direction (for example, forward a regional gravitational slope) moves rapidly and can turn back transversely to the foldbelt, i.e. antithetically. The latter process is accompanied by an increased overthrusting (but decreased compression) and possibly by the strike-slip lines. The paragenesis widely spreads in the folded zones with a destined vergence. It was specially analyzed for southern slope of Greater Caucasus (Fig.10).

These and other paragenetic combinations have been discussed in publications of the author (Kopp,1991,1995, 1997).

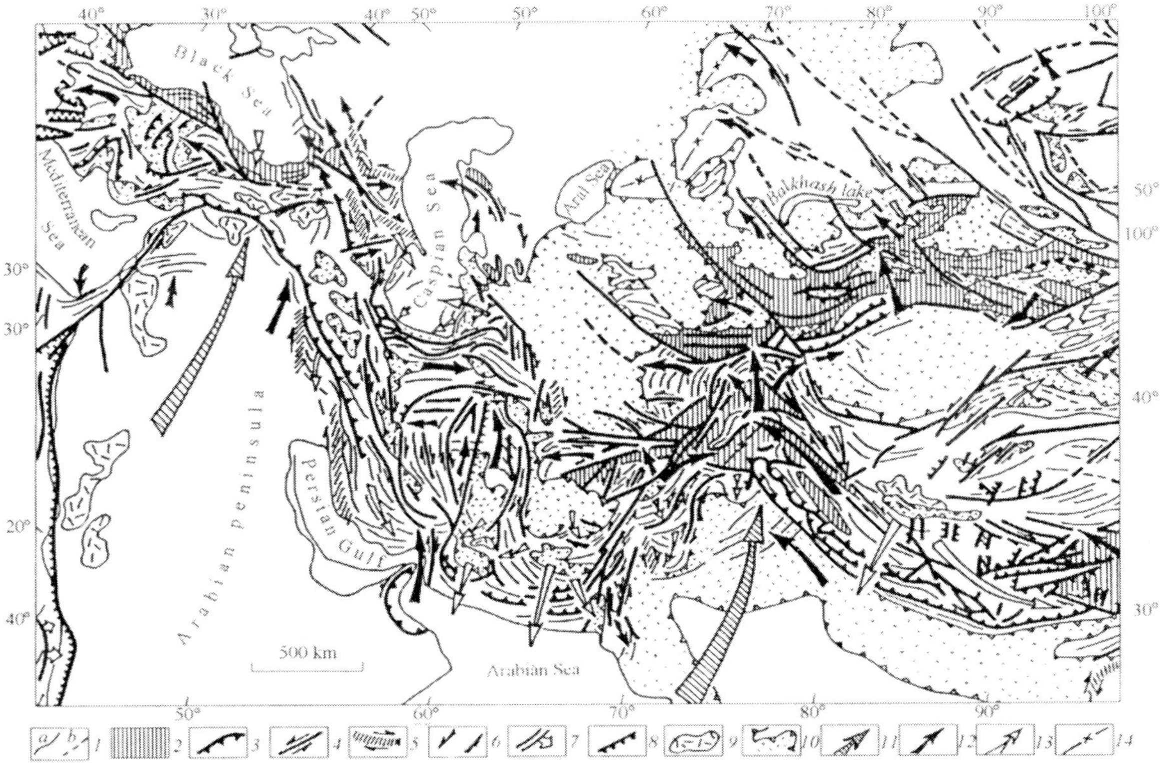

Figure 13: Geokinematic map of the Central part of the Alpine-Himalayan orogenic belt, compiled from the results of the structural patterns. Key: 1-3, compression structures: 1, lines of strike and fold axes (a - alpine; b - activated at alpine epoch); 2, neotectonic uplifts; 3, thrusts; 4-6, wrench structures: 4, strike-slip faults; 5, simple shear zones; 6, simple shear zones without clear boundaries; 7-9, extension structures: 7, axes of spreading; 8, normal faults; 9, regions of orogenic volcanism; 10, molasse basins; 11, directions of displacement and rotation of plate-indenters; 12-13, direction of lateral squeezing-out of deformed material: 12, refer to "stable" Eurasia; 13, refer to neighbouring blocks within orogenic belt; 14 - axes of gently sloping platform uplifts.

Geokinematic map of the Mediterranean-Central Asian collisional belt

The methodical principles described in the present work and in the preceding work of the author (Kopp, 1979, 1991, 1995, 1997) allow to compile the geokinematic map of the Mediterranean-Central Asian orogenic belt for the collision stage of the development (Oligocene/Quaternary), a part of which is shown in the Fig.13.

The work, confirms the idea that the collisional deformations of the Mediterranian-Central Asian belt were connected with the indentation of the three Gondwana blocks. Adriatic, Arabian and Indian, in Eurasia (Tapponnier, Molnar,1976). These blocks with strong crumpling in the immediate zone of the collision and in the cores of the diapiric folds, overturned in the horizontal plane (the "festoons" of the Indian frame). It is clear that the problem of the rigidity of the blocks, deforming the Eurasia, still needs special study. Be it as it may, each such indenter produced its own field of the deformation in the boundaries of the Eurasia, towards the interior of where of the typical zones are present: (1) the two symmetrical zones of the synthetic simple shears on the flanks, (2) the zone of the horizontal pumping and vertical squeezing-out of the rocks separating the two subzones: the subzone of the tectonic denudation, represented by the roots of the overthrusts, and the subzone of the supercharged rock masses, formed by overthrusts and the festoonlike overthrust arcs; (3) the zone of the lateral escape of the rocks where the region of the tectonic denudation, represented by antithetically turned axis of the extension and longitudinal strike-slips, and two symmetrical regions of the tectonic accumulation of the squeezed rocks and (4) the zone of the compensational extension of the external part of the region of the dynamic influence. Opposite of each of the Gondwana blocks not only the major kinematic zones but even the individual elements of structure can be found. The regular recurrence of similar structures in the same places of the zones of the emplacement of the different indenters is sometimes wonderful. It shows not only the workability of the assumed procedure, but also an important arguement for the orogenic origin of these structures. A number of typical examples of such recurrence are given in the work by the author (1997).

The applied procedure can reveal additional factors of the squeezing-out of the rocks (the Caucasus, the Pamirs and others): the role of the lateral escape of the crust and its bending along the Gondwana blocks, and their rotation and coordinated movements.

Further, the comparative analysis allows to reveal the sufficient asymmetry of the zones of the dynamic influence on all blocks. In the internal parts the dextral strike-slips, tangentially bent around the indenter are better pronounced, whereas in the external ones the sinistral strike-slips are present. Correspondingly, the place of the most lateral pumping of the rocks can displace the western flanks of the Gondwana blocks (the Alps, the Caucasus, the Pamirs), whereas the place of its unloading, can be accompanied by the extension, displacing its eastern flanks (the Pannonian depression, the Central Iran, the Tibet). Due to this the eastern part of the zone of the lateral escape is always strongly enlarged. The conclusion that the asymmetry was caused by the rotation of internal region anticlockwise, and the external one clockwise, is significant. In other words, the movement to the north and the simultaneous anticlockwise rotation of the Gondwana blocks caused the reactive flow of the rocks in the opposite (southeastern) direction. So, present data independently and in more detail confirm the early stated idea of Carey, (1955) about the anticlockwise rotation of the Apulian, Arabian and Indian microcontinents.

CONCLUSIONS

In analyzing the wrench fault kinematics there emerges a problem of space for volume of rock masses lengthening in the horizontal plane. Two alternative variants are possible: (a) axial lengthening of the foldbelt and (b) its curving in the horizontal plane (transverse advancement). Characteristic features of each kind of the deformations are considered . The method of structural pattern analysis allows to recognize the kinematic type of structures (wrench fault, thrust, etc.) and also the tectonic movements of the whole set of blocks. The technique of structural analysis was developed to find out the movements and deformations which were typical for the foldbelts, (e.g. longitudinal, transverse and rotational block displacements relative to the general structural trend, curving in horizontal plane (bending, buckling), and lateral squeezing-out and compensational supercharge of deformed material. The above structural patterns in various regions of the Mediterranean-Central Asian belt can be considered typically.

On the basis of the structural pattern analysis with the use of the palaeomagnetic data the geokinematic map (scale 1:5 000 000 in original) of the Mediterranean-Central Asian orogenic belt was compiled by the author. Geokinematic analyses of the Mediterranean-Central Asian orogenic structures according to structural patterns make it possible to confirm independently about the result of emplacement of the Gondwana microcontinents (Apulian, Arabian and Indian) into Eurasia. Moreover, additional emplacement centres (Caucasus, Kopetdagh, Pamirs, etc.) were established and block rotational movements were discovered.

ACKNOWLEDGEMENTS

I am thankful to the Russian Foundation of Basic Research (projects number 96-05-65870, 97-07-90074, 99-05-65366, 00-05-64513 : 00-15-98531) as well as Peri-Tethys Programme (grant number 95-96-41) for financial support. I particularly thank Valentin Burtman, Mikhail Goncharov, Yuri Leonov, Aleksei Lukjanov and Leonid Rastsvetajev for their suggestions.

REFERENCES

Atlas Tethys,1993. Paleoenvironmental maps. Eds. J. Dercourt, L.E. Ricou and B. Vrielynck. Paris, 1993

Bazhenov M.L. and Burtman V.S., 1990. Structural arcs of the Alpine belt: Carpathians - Caucasus - Pamirs. Moskow: Nauka, 168 pp. (in Russian).

Bazhenov M.L.,1987. Paleomagnetism of Cretaceous and Paleogene sedimentary rocks from the Kopetdagh and its tectonic implications. Tectonophysics, 136: 223-235.

Burke K., Sengör A.M.C.1986. Tectonic escape in the evolution of the continental crust. In Reflection seismology: The Continental Crust, Geodin. Ser., v.14, 41-53.

Burtman, V.S. 1984. Kinematics of the Carpathian structural loop, Geotectonics, Engl. Transl., no.3, 17-31.

Burtman, V.S., 1989. Kinematics of the Arabian syntaxis. Geotectonics, Engl.Transl., 23: 139-145.

Caire, A.,1979. Geotectonique giratoire, Geol. Mijnbouw,., 58(2), 241-252.

Carey, S.W.,1955.The orocline concept in geotectonics/The paper and proceeding of the Royal Society of Tasmania.,vol.89,255-288.

Carter, J., B. Luyendyk and R. Terres,1987. Neogene clockwise tectonic rotation of the Eastern Transverse Aandes, California, suggested by paleomagnetic vectors, Geol. Soc. Am. Bull, 98, 2, 199-206.

Chinnery, M.A.,1966.Secondary faulting.1.Theoretical aspects.Can.J.Earth Sci.,3:167-174.

Cloos, E.,1957. Experimental analysis of fracture patterns. Bull. Geol. Soc. Amer., v. 66, P.241-256.

Cloos, H.,1928 Uber antitetische Bewegungen. Geol. Rundschau, v.19, no 3.

Cobbold, P.R., and Davy, P.H.,1988 Indentation tectonics in nature and experiments, 2, Central Asia. Bull. Geol. Univ. Uppsala, New. Ser., 14: 143-162.

Cummings, D.,1976. Theory of plasticity applied to faulting. Mojave Desert, Southern California. Bull. Geol. Soc.Amer., v. 87, no 5:720-724.

Chandra, U. Focal mechanism solutions for earthquakes in Iran // Phys. Earth and Planet. Inter. 1984. Vol.34, 1/2. P.9-16.

Freund, R.,1974. Kinematics of transform and transcurrent faults. Tectonophysics, v.21, no 1/2:93-134.

Höeppener, R., E., Kalthnoff, and P. Schrader.1969. Zur physikalischen Tektonik: Bruchbildung bei vershidenen Deformationen im Experiment. Geol. Rundschau, v. 59:179-193

Klootwijk, C.T., P.J. Conaghan and C. Mc. A. Powell, 1985. The Himalayan arc: Large-scale continental subduction, oroclinal bending and back-arc spreading. Earth Planet. Sci. Lett., 75: 167-183.

Kopp, M.L.,1979. On the origin of transverse folded zones of epigeosynclinal fold belts (illustrated by the eastern Eurasian Alpine belt). Geotectoniks,. no 2, pp.94-107 (in Russian).

Kopp, M.L.,1991. The structural patterns, connected with the longitudibal movements within the folded belts (the Mediterranean-Himalayan belt as an example), Geotectonics, Engl. Transl.,vol.25, no. 1, 15-27.

Kopp, M.L.,1995. Transverse displacements in the folded belts and structural patterns associated with Hum (Alpine-Himalayan belt). - Geotectonics, Engl. Transl. vol.28,No 4: 35-51.

Kopp, M.L.,1997. Lateral escape structures in the Alpine-Himalayan collisional belt. Moscow.: Scientific World. 314 p. (Transactions of GIN RAS; Vol. 506) (in Russian; res. Engl.)

Krymus, V.N.,1966. Fault tectonics of Kopetdagh. In: Tectonika Turkmenii. M., Nauka, p.186-191 (In Russian).

Lukjanov, A.B.,1991. Ductile deformation and tectonic flow in the lithosphere. Moscow,Nauka, 144p. (Transactions of GIN RAS;.Vol.460) (in Russian).

Nikitin, M.Yu.,1987. Neotectonics of Dagestan Mountains. In: Geology and Useful Resources of the Great Caucasus. Moscow, Nauka: 221-238.(in Russian).

Ognev, V.N.,1937. Tectonics of Turkmenskaya SSR. In : Trudy XVII sessii MGK. M., t.2, (In Russian).

Pavoni, N.,1961. Die nordanatolische Horizontal-verschiebung, Geol. Rundsch., 51(1): 120-140.

Rastsvetaev, L.M.,1966. Faults of Kopetdagh and their relationship with folds. Geotectonics, no. 3: 93-108 (In Russian).

Rastsvetayev, L.M.,1973. Some features of Late Alpine structure of orogenic zones of the southern USSR and tectonic stresses of the Recent time. In: Recent Tectonics, Recent Deposits and Man. MGU Press, issue 5: 57-108.(in Russian)

Royden, L.,F.Horvath, and I.Rumpler,1983. Evolution of the Pannonian basin system, Tectonophysics, 2 (11): 63-90.

Sengör , A.M.C., Burke K., Dewey J.1982. Tectonics of the North Anatolian transform fault. - Multidisciplinary approach to earthquake prediction. Wiesbaden: 3-22.

Tapponnier, P., and P. Molnar, 1976. Slip-line field theory and largescale continental tectonics. Nature, 264 (5584): 319-324.

Tapponnier, P., 1977. Evolution tectonique du sysleme Alpien Mediterranea: Poinconnement et ecrasement rigide-Plastique, Bull.Soc. Geol.,Fr., 19: 437-460.

Tapponnier, P., Peltzer G., Armijo R., 1986. On the mechanics of the collision between India and Asia. - Collis. tecton.: 115-157.

Wilcox, R.E., T.R. Harding, and D.R.Scely, 1973. Basic wrench tectonics, Ann. Accos. Petr. Geol. Bull, 57: 74-96

Woodcock, H.,Fischer, M.,1986. Strike-slip duplexes.J.Struct.Geol.vol.87: 725-735.